Oxford Lecture Series in
Mathematics and its Applications 31

Series Editors
John Ball   Dominic Welsh

OXFORD LECTURE SERIES
IN MATHEMATICS AND ITS APPLICATIONS

1. J. C. Baez (ed.): *Knots and quantum gravity*
2. I. Fonseca and W. Gangbo: *Degree theory in analysis and applications*
3. P. L. Lions: *Mathematical topics in fluid mechanics, Vol. 1: Incompressible models*
4. J. E. Beasley (ed.): *Advances in linear and integer programming*
5. L. W. Beineke and R. J. Wilson (eds): *Graph connections: Relationships between graph theory and other areas of mathematics*
6. I. Anderson: *Combinatorial designs and tournaments*
7. G. David and S. W. Semmes: *Fractured fractals and broken dreams*
8. Oliver Pretzel: *Codes and algebraic curves*
9. M. Karpinski and W. Rytter: *Fast parallel algorithms for graph matching problems*
10. P. L. Lions: *Mathematical topics in fluid mechanics, Vol. 2: Compressible models*
11. W. T. Tutte: *Graph theory as I have known it*
12. Andrea Braides and Anneliese Defranceschi: *Homogenization of multiple integrals*
13. Thierry Cazenave and Alain Haraux: *An introduction to semilinear evolution equations*
14. J. Y. Chemin: *Perfect incompressible fluids*
15. Giuseppe Buttazzo, Mariano Giaquinta and Stefan Hildebrandt: *One-dimensional variational problems: an introduction*
16. Alexander I. Bobenko and Ruedi Seiler: *Discrete integrable geometry and physics*
17. Doina Cioranescu and Patrizia Donato: *An introduction to homogenization*
18. E. J. Janse van Rensburg: *The statistical mechanics of interacting walks, polygons, animals and vesicles*
19. S. Kuksin: *Hamiltonian partial differential equations*
20. Alberto Bressan: *Hyperbolic systems of conservation laws: the one-dimensional Cauchy problem*
21. B. Perthame: *Kinetic formulation of conservation laws*
22. A. Braides: *Gamma-convergence for beginners*
23. Robert Leese and Stephen Hurley: *Methods and algorithms for radio channel assignment*
24. Charles Semple and Mike Steel: *Phylogenetics*
25. Luigi Ambrosio and Paolo Tilli: *Topics on Analysis in Metric Spaces*
26. Eduard Feireisl: *Dynamics of Viscous Compressible Fluids*
27. Antonín Novotný and Ivan Straškraba: *Introduction to the Mathematical Theory of Compressible Flow*
28. Pavol Hell and Jarik Nesetril: *Graphs and Homomorphisms*
29. Pavel Etingof and Frederic Latour: *The dynamical Yang-Baxter equation, representation theory, and quantum integrable systems*
30. Jorge Ramirez Alfonsin: *The Diophantine Frobenius Problem*
31. Rolf Niedermeier: *Invitation to Fixed-Parameter Algorithms*

Invitation to Fixed-Parameter Algorithms

Rolf Niedermeier

Institut für Informatik, Friedrich-Schiller-Universität Jena

OXFORD
UNIVERSITY PRESS

This book has been printed digitally and produced in a standard specification in order to ensure its continuing availability

OXFORD
UNIVERSITY PRESS

Great Clarendon Street, Oxford OX2 6DP

Oxford University Press is a department of the University of Oxford.
It furthers the University's objective of excellence in research, scholarship, and education by publishing worldwide in

Oxford New York

Auckland Cape Town Dar es Salaam Hong Kong Karachi
Kuala Lumpur Madrid Melbourne Mexico City Nairobi
New Delhi Shanghai Taipei Toronto

With offices in

Argentina Austria Brazil Chile Czech Republic France Greece
Guatemala Hungary Italy Japan South Korea Poland Portugal
Singapore Switzerland Thailand Turkey Ukraine Vietnam

Oxford is a registered trade mark of Oxford University Press in the UK and in certain other countries

Published in the United States by Oxford University Press Inc., New York

Reprinted 2008

ISBN 978-0-19-856607-6

PREFACE

This book grew out of my *habilitationsschrift* at the University of Tübingen in 2002. Currently, there is only one monograph dealing with the issue of fixed-parameter algorithms: Rod G. Downey and Michael R. Fellows' groundbreaking monograph *Parameterized Complexity* (1999). Since then there have been numerous new results in this field of exactly solving combinatorially hard problems. Moreover, Downey and Fellows' monograph focuses more on structural complexity theory issues than on concrete algorithm design and analysis. By way of contrast, the objective of this book is to focus on the algorithmic side of parameterized complexity, giving a fresh view of this highly innovative field of algorithmic research.

The book is divided into three parts:

1. a broad introduction that provides the general philosophy and motivation;
2. a part on algorithmic methods developed over the years in fixed-parameter algorithmics, forming the core of the book; and
3. a final section discussing the essentials of parameterized hardness theory, focusing first on $W[1]$-hardness, which parallels NP-hardness, then stating some relations to polynomial-time approximation algorithms, and finishing up with a list of selected case studies to show the wide range of applicability of the methodology presented.

The book is intended for advanced students in computer science and related fields as well as people generally working with algorithms for discrete problems. It has particular relevance when studying ways to cope with computational intractability as expressed by NP-hardness theory.

The reader is recommended to start with Part I, but Parts II and III do not need to be read in the given order. Thus, from Chapter 7 on (with a few exceptions) there are almost no restrictions concerning the chosen order. The material presented can be used to form a course exclusively dedicated to the topic of fixed-parameter algorithms as well as to provide supplementary material for an advanced algorithms class.

We believe that the concept of fixed-parameter tractability is fundamental for the algorithmics of computationally hard discrete problems. Due to the ubiquity of the proposed problem parameterization approach discussed here, fixed-parameter algorithms should be seen as basic knowledge for every algorithm designer. May this book help to spread this news.

ACKNOWLEDGEMENTS

To all who helped, particularly the unnamed ones!

The first hint that I should study parameterized complexity came from Klaus-Jörn Lange. He pointed me to strange things like "towers of complexity" or "parameterization of languages by the slice". Shortly afterwards, during my 1998 stay at Charles University, Prague, Jaroslav Nešetřil showed me a bunch of papers by Rod G. Downey and Michael R. Fellows. This prompted my first steps in researching fixed-parameter algorithms in enjoyable cooperation with Peter Rossmanith.

However, I am most grateful to my (partially former) Ph.D. students who share(d) their ideas and time with me. I list them in alphabetical order: Jochen Alber, Michael Dom, Jiong Guo, Jens Gramm, Falk Hüffner, and Sebastian Wernicke. They helped me a lot in countless ways and without them this book would not exist. In addition, I greatly profited from working with my graduate students Nadja Betzler, Britta Dorn, Frederic Dorn, Erhan Kenar, Hannes Moser, Amalinda Oertel, David Pricking, Daniel Raible, Marion Renner, Christian Rödelsperger, Ramona Schmid, Anke Truss, and Johannes Uhlmann. All of them have been infected with fixed-parameter algorithmics in the broadest sense.

Finally, special thanks go to Hans Bodlaender, Mike Fellows, Henning Fernau, Edward Hirsch, and Ton Kloks for collaborations from which I have profited greatly.

I thank Rod Downey for making the connection with Oxford University Press and the staff at Oxford University Press for a smooth and enjoyable cooperation.

I apologize for omitting further names here—there are too many to name them all and it would be too dangerous to forget one of them.

CONTENTS

Part I

Foundations

A fixed-parameter algorithm is one that provides an *optimal* solution to a discrete combinatorial problem. As a rule, such a problem is *NP*-hard and that is why one must accept exponential running times for fixed-parameter algorithms. The fundamental idea is to restrict the corresponding, seemingly unavoidable, "combinatorial explosion" that causes the exponential growth in the running time of certain problem-specific parameters. It is hoped then that these parameters (in the concrete application behind the problem under consideration) might take only relatively "small" values, so that the exponential growth becomes affordable; that is, the fixed-parameter algorithm *efficiently* solves the given "parameterized problem".

As an example of "parameterization", consider the problem of placing as few queens as possible to attack all the squares on a chessboard. There is a way to place only five (which is optimal) queens on an 8×8 chessboard to do this. Here, a natural parameter is the size k of the solution set we search for, that is, the set of queens to be placed. Hence for 8×8 chessboards $k = 5$. What about general $n \times n$ chessboards? Can we find a minimum solution efficiently?

A "more serious" example is the following. Assume that one wants to establish transmission towers; the towers will be located on inhabited buildings, and each such building must be reachable by at least one transmission tower. In addition, assume that if a tower in location u can reach location v, then also one at v can reach u. Then, given all pairs that can reach each other, how many transmitters are needed to cover all the buildings? Again, a natural parameter to consider is the number of transmitters needed. Thus the task is to find a small number of transmission tower locations such that all buildings can be reached.

Both examples are instantiations of an *NP*-hard graph problem called DOMINATING SET:

> **Input:** An undirected graph $G = (V, E)$ and a nonnegative integer k.
> **Task:** Find a subset of vertices $S \subseteq V$ with k or fewer vertices such that each vertex in V is contained in S or has at least one neighbor in S.

An optimal solution to DOMINATING SET can be found in $O(n^{k+1})$ steps by simply trying all size-k subsets of the vertex set V of size n. According to parameterized complexity theory, there is little hope of doing significantly better than this. Fortunately, however, for restricted classes of graphs we can do better. For instance, for *planar* graphs (that is, graphs that can be drawn in the plane without edge crossings) DOMINATING SET can be solved in $O(8^k \cdot n)$ time. Note that DOMINATING SET remains *NP*-hard when restricted to planar graphs. Another algorithm even finds a solution in $O(c^{\sqrt{k}} \cdot n)$ time for some (larger) constant c. This is what we understand by fixed-parameter algorithms—the superpolynomial factor in the running time depends exclusively on the parameter k. Finally, again in case of planar graphs, there are simple data reduction rules that—in polynomial time—can shrink an original input graph with n vertices into a new one with only $O(k)$ vertices such that the search for an optimal solution can be done within the size $O(k)$ instance. All these results lead to the fundamental conclusion that the combinatorial explosion can be confined to the parameter k only, the central goal to be achieved by fixed-parameter algorithms. Generally speaking, a fixed-parameter algorithm solves a problem with an input instance of size n and a parameter k in

$$f(k) \cdot n^{O(1)}$$

time for some computable function f depending solely on k. That is, for every fixed parameter value it yields a solution in polynomial time and the degree of the polynomial is independent from k.

Fixed-parameter algorithms have been scattered around the literature for decades. As a method of algorithm design and analysis, parameterized complexity was systematized by Rod G. Downey and Michael R. Fellows and some of their co-authors during the 1990s. In particular, they developed a theory of parameterized computational complexity, which is a strong mathematical tool for guiding fixed-parameter algorithm design. In this book, we make use of parameterized computational complexity theory to the extent that is necessary to learn about the design and analysis of algorithms. More structural complexity-theoretic aspects are neglected in this work. Fixed-parameter algorithms are introduced as a valuable alternative to complement other algorithmic approaches for attacking hard combinatorial problems, such as approximation or heuristic algorithms.

Fixed-parameter algorithms adhere to a very natural concept when trying to solve hard combinatorial problems. In the following we give a concise description of the very basic ideas and objectives behind this work and parameterized complexity analysis. The focus of Part I is on encouraging the reader to adopt a parameterized view of the study of computationally hard problems. Besides simple motivating examples and the presentation of the elementary concepts needed throughout the book, the breadth of the parameterized complexity approach is illustrated by means of an extensive discussion of the *NP*-complete graph problem VERTEX COVER. Having dealt with this perhaps most popular parameterized problem, we finally move on and finish with a general discussion on the "art" of parameterizing problems.

1

INTRODUCTION TO FIXED-PARAMETER ALGORITHMS

Computationally hard problems are ubiquitous. The systematic study of computational (in)tractability lies at the heart of computer science. Michael R. Garey and David S. Johnson's monograph *Computers and Intractability* from the late 1970s was a landmark achievement in this direction, providing an in-depth treatment of the corresponding theory of *NP*-completeness. With the theory of *NP*-completeness at hand, we can prove meaningful statements about the computational complexity of problems. But what happens after we have succeeded in proving that a problem is *NP*-hard (that is, "intractable"), but which nevertheless must be solved in practice? In other words, how do we cope with computational intractability?

The approach followed in this book is based on worst-case analysis of deterministic, exact algorithms to solve hard problems. In the course of dealing with intractable problems, we will also refer to several other related algorithmic methodologies, such as approximation algorithms, average-case analysis, randomized algorithms, and purely heuristic methods. Each approach has advantages and disadvantages. With regard to exact *fixed-parameter algorithms*, the advantages are

- guaranteed optimality of the solution; and
- provable upper bounds on the computational complexity.

The disadvantage is that we have to take into account

- exponential running time factors.

Clearly, exponential growth quickly becomes prohibitive when running algorithms in practice. Fixed-parameter algorithmics provides guidance on the feasibility of the "exact algorithm approach" for hard problems by means of a refined, two-dimensional complexity analysis. The fundamental idea is to strive for better insight into a problem's complexity by exploring (various) problem-specific parameters to find out how they influence the problem's computational complexity. Ideally, we aim for statements such as "if some parameter k is small in problem X, then X can be solved efficiently". For instance, in the case of the *NP*-complete graph problem VERTEX COVER we know that if the solution set we are searching for is "small", then this set can be found efficiently whatever the graph looks like and however big it is—the exponential factor in the running time amounts to less than 1.28^k, where the parameter k is the size of the solution set sought.

We begin with three brief case studies of computationally hard problems. In doing so, we give a concise overview of how to cope with their computational intractability, providing the first examples of the fixed-parameter approach.

1.1 The satisfiability problem

The CNF-SATISFIABILITY problem for Boolean formulae in conjunctive normal form may be considered the "drosophila[1] of computational complexity theory". This fundamental *NP*-complete problem has been the subject of research on exact algorithms for decades and it continues to play a central role in algorithmic research—the annual "SAT" conference is devoted to theory and applications of satisfiability testing. The problem is defined as follows.

> **Input:** A Boolean formula F in conjunctive normal form.
> **Task:** Determine whether or not there exists a truth assignment for the Boolean variables in F such that F evaluates to true.

Example 1.1 The formula

$$(x_1 \vee x_2) \wedge (\neg x_1 \vee x_2 \vee x_3) \wedge (\neg x_2 \vee \neg x_3)$$

is satisfiable, satisfied by the truth assignment $T(x_1) = \mathit{true}$, $T(x_2) = \mathit{false}$, $T(x_3) = \mathit{true}$. An alternative satisfying assignment is $T(x_1) = \mathit{false}$, $T(x_2) = \mathit{true}$, $T(x_3) = \mathit{false}$. By way of contrast, there is no satisfying assignment for the formula

$$(x_1 \vee \neg x_2) \wedge (x_1 \vee x_2) \wedge (\neg x_1 \vee x_2 \vee x_3) \wedge (\neg x_2 \vee x_3) \wedge (\neg x_1 \vee \neg x_3).$$

Numerous applications such as VLSI design and model checking make CNF-SATISFIABILITY of real practical interest. Often, however, in concrete applications the corresponding problem instances turn out to be much easier than one might expect given the fact of CNF-SATISFIABILITY's *NP*-completeness. For instance, large CNF-SATISFIABILITY instances arising in the validation of automotive product configuration data are efficiently solvable. Hence the question arises of whether we can learn more about the complexity of CNF-SATISFIABILITY by means of studying various problem parameters. In particular, we are less interested in empirical results and more in provable performance bounds related to various parameterizations of CNF-SATISFIABILITY. In this way, we hope to obtain a better understanding of the problem properties that might be of algorithmic use.

Formally, an input of CNF-SATISFIABILITY is a conjunction of m *clauses* where each clause consists of a disjunction of *literals*, that is, negated or non-negated Boolean variables. Let there be n different variables occurring in the formula. We list a set of various "parameterizations" of CNF-SATISFIABILITY.

[1] *Drosophila melanogaster* is a fruit fly and is one of the most valuable organisms in biological research. It has been used as a model organism for research for almost a century.

The point subsequently is to see how the running time bounds of solving algorithms depend on the given parameters. The central question is whether and how we can confine the seemingly unavoidable combinatorial explosion to the respective parameter. In particular, how small can the corresponding exponential term be kept? We list some results from the literature given in the end of this chapter.

Parameter "clause size". The maximum number k of literals in any of the given clauses is a very natural formula parameter. For $k = 3$ (that is, 3-CNF-SATISFIABILITY), however, the problem remains *NP*-complete, whereas it is polynomial-time solvable for $k = 2$ (that is, 2-CNF-SATISFIABILITY). Thus, for upperbounding the combinatorial explosion, this parameterization seems of little help.

Parameter "number of variables". The number n of different variables occurring in a formula significantly influences the computational complexity. Since there are 2^n different truth assignments, in essentially[2] this number of steps CNF-SATISFIABILITY can be solved. When restricting the maximum clause size by some k, for instance, $k = 3$, better bounds are known. The currently best upper bound for a deterministic algorithm solving 3-CNF-SATISFIABILITY is below 1.49^n. The currently best randomized algorithm for 3-CNF-SATISFIABILITY has expected running time below 1.33^n.

Parameter "number of clauses". If the number of clauses in a formula can be bounded from above by m, then CNF-SATISFIABILITY can be solved in 1.24^m steps.

Parameter "formula length". If the total length (that is, counting the number of literal occurrences in the formula) of the formula F is bounded from above by $\ell := |F|$, then CNF-SATISFIABILITY can be solved in 1.08^ℓ steps.

The above parameterizations do not suffice to explain all the cases of good behaviour of CNF-SATISFIABILITY in many practical situations. There are application scenarios where none of them would lead to an efficient algorithm. Hence the following way of parameterizing CNF-SATISFIABILITY seems to offer good prospects.

Parameters exploiting "formula structure". There are several recent exact algorithms with exponential bounds depending on certain *structural* formula parameters. These parameters are based on structural graph decompositions where a graph is associated with the formula structure—for instance, one obtains so-called variable interaction graphs. The basic idea is that the way in which different variables (or literals) occur in common clauses—and thus how they interact—strongly influences the complexity of finding a satisfying assignment if any exists. Then structure and width

[2]We neglect polynomial-time factors in the running time analysis throughout the book whenever they play a minor role in our considerations.

concepts for graphs (as considered later in this book) are exploited to derive tractability results for formulae with particular structural properties on the graph side.

Finally, from the parameterized complexity *theory* point of view the following parameterization is of particular relevance.

Parameter "weight of assignment". Here it is asked whether a formula has a satisfying assignment with *exactly* k variables being set to true. Although the parameterization with this "weight parameter" seems rather artificial, it plays a major role in characterizing parameterized intractability and the corresponding structural complexity theory. It serves as an anchor point in the parameterized hardness program similar to the role that CNF-SATISFIABILITY plays in classical *NP*-completeness theory. Thus, in particular, it is considered very unlikely that this version of CNF-SATISFIABILITY can be solved in $f(k) \cdot |F|^{O(1)}$ steps. So far it is only known how to achieve the trivial running time $n^{O(k)}$ by testing all $\binom{n}{k}$ candidate truth assignments—it seems that the combinatorial explosion cannot be confined to a function exclusively depending on k. Note that this hardness already holds true when considering the weighted version of 2-CNF-SATISFIABILITY.

In contrast, if one asks whether there is a satisfying assignment with *at most* k variables set true, then this weighted 2-CNF-SATISFIABILITY can be easily solved using a size-2^k search tree: as long as there are clauses with only positive literals, take an arbitrary one of these clauses and branch into the two cases, setting either of the variables true (at least one of the two variables must be set true). In each branch, simplify the formula according to the variable chosen to be set true. After that, a formula remains where every clause has at least one negative literal and this instance is trivial to solve. Clearly, this method generalizes to arbitrary constant clause sizes.

To summarize, different ways of parameterizing CNF-SATISFIABILITY lead to a better understanding of the problem's inherent complexity facets. In particular, it is hoped that structural parameterizations in the future will yield new insights into tractable cases of CNF-SATISFIABILITY; but, in any case, there surely is *no "best parameterization"*. As a rule, the nature of the complexity of hard computational problems will probably almost always require such a multi-perspective view in order to gain better insight into practically relevant, efficiently solvable *special cases*. To cope with intractability in a mathematically sound way, it seems that we have to pay the price of shifting our view from considering the input through just one pair of glasses (mostly these glasses are called "complexity measurement relative to input size") to a multitude of pairs of glasses, each of them with a different focus (that is, parameter).

CNF-SATISFIABILITY, as defined above, is a *decision problem* with answer either "yes" or "no", usually also providing a satisfying truth assignment if it exists. In most application scenarios considered in this book, however, we will have to deal with *optimization problems* where the goal is to minimize or maximize a certain solution value. An optimization version of CNF-SATISFIABILITY is MAXIMUM (CNF-)SATISFIABILITY, in which one is asked to find a truth assignment that simultaneously satisfies as many clauses as possible. Obviously, SATISFIABILITY is a special case of MAXIMUM SATISFIABILITY in the sense that here one asks whether all clauses can be satisfied. Analogous parameterizations to those for SATISFIABILITY can also be investigated for MAXIMUM SATISFIABILITY.

Parameter "clause size". Even for maximum clause size $k = 2$ (MAXIMUM 2-SATISFIABILITY) the problem remains *NP*-complete.

Parameter "number of variables". Analogously to SATISFIABILITY, MAXIMUM SATISFIABILITY can be solved in 2^n steps, where n denotes the number of variables appearing in the given formula. It is an open problem to obtain better bounds even if the maximum clause size is 3 (MAXIMUM 3-SATISFIABILITY). A recent breakthrough shows, however, that MAXIMUM 2-SATISFIABILITY can be solved in 1.74^n steps.

Parameter "number of clauses". If the number of clauses in a formula can be bounded from above by m, then MAXIMUM SATISFIABILITY can be solved in 1.33^m steps. MAXIMUM 2-SATISFIABILITY can be solved in 1.15^m steps.

Parameter "formula length". If the total length of the formula is bounded from above by ℓ, then MAXIMUM SATISFIABILITY can be solved in 1.11^ℓ steps. MAXIMUM 2-SATISFIABILITY can be solved in 1.08^ℓ steps.

We are not aware of investigations of MAXIMUM SATISFIABILITY concerning parameterizations according to "formula structure" as discussed for SATISFIABILITY.

1.2 An example from railway optimization

In our second example, we deal with a graph problem that is directly motivated by an application in railway optimization. Although the fixed-parameter approach in this case so far is not able to prove tractability according to a reasonable parameterization, the preprocessing technique presented is of great interest as a core algorithmic tool in parameterized complexity studies. The problem is defined as follows.

> **Input:** A set of trains, a set of train stations, and for each train the stations where it stops.
> **Task:** Find a minimum size set of train stations such that each train stops in at least one of the selected stations.

The motivation is that at the selected stations S one wants to build some supply stations for trains. To save costs, the set S shall be as small as possible. The formalization as a graph problem leads to an *NP*-complete version of the DOMINATING SET problem in bipartite graphs. To do so, one associates trains with one vertex set and stations with the other vertex set; one draws an edge between a train vertex and a station vertex iff the train stops in this station:

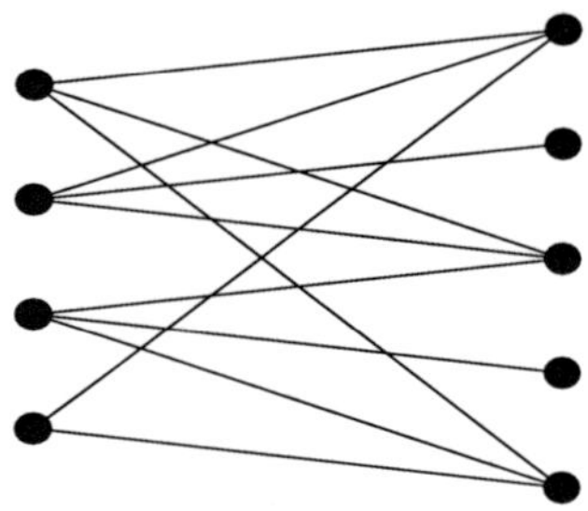

FIG. 1.1. An input instance with four trains (left-hand side) connected to their respective stopping stations (right-hand side).

Input: An undirected, bipartite graph $G = (V_1 \cup V_2, E)$ with disjoint vertex sets V_1 and V_2 and edge set E.
Task: Find a minimum size set $S \subseteq V_2$ such that each vertex in V_1 has an adjacent edge connecting it to some vertex in S.

Example 1.2 Figure 1.1 shows an example with four trains (drawn on the left hand-side and forming the set V_1) and five stations (drawn on the right-hand side and forming the set V_2). In this simple example, one may easily verify that an optimal solution S can be built by (arbitrarily) choosing exactly two of the first, the third and the fifth vertex on the right-hand side.

Here, a natural parameter to look at is the size k of set S, that is, $k := |S|$. According to parameterized complexity theory, it is unlikely that we can show a running time that is exponential exclusively with respect to k—the complexity behaviour is similar to that described in Section 1.1 for SATISFIABILITY with respect to the parameter "weight of assignment". Furthermore, it is open to find and study other reasonable parameterizations of the problem. Nevertheless, simple data reduction by polynomial-time preprocessing rules seems to suffice to solve the problem for many practical input instances. For the two subsequent data reduction rules there is no proven performance guarantee in terms of fixed-parameter tractability—later we will explore data reduction rules that lead to so-called problem kernels that have such a desirable performance guarantee. These two rules are based on neighborhood considerations. To this end, define for a vertex $v \in V_1 \cup V_2$ the set of its neighbors $N(v) := \{ u \mid \{u, v\} \in E \}$.

Train Rule. For $t, t' \in V_1$: if $N(t') \subseteq N(t)$, then remove t (together with its adjacent edges) from the input instance.

Station Rule. For $s, s' \in V_2$: if $N(s') \subseteq N(s)$, then remove s' (together with its adjacent edges) from the input instance.

The interpretation of the Train Rule is that train t' stops only in stations where train t also stops. Thus, as soon as we "cover" train t' by one station we automatically cover train t as well. Hence we can remove t from further considerations when searching for an optimal solution S. The Station Rule works analogously

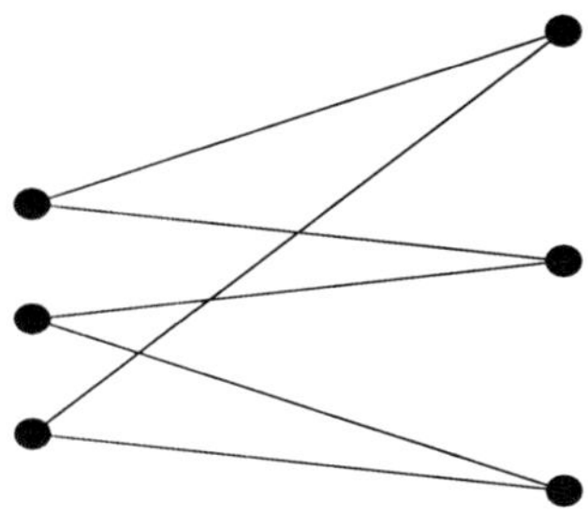

FIG. 1.2. Reduced instance.

to the Train Rule, interchanging stations with trains. Note, however, that after applying the Train Rule or Station Rule not all optimal solutions can necessarily be found any more.

Example 1.3 (Continued.) Refer to Figure 1.2 to see the resulting *reduced instance* after having applied both rules as often as possible to the input instance in Figure 1.1. Using this reduced instance, it is significantly easier to detect what an optimal solution S is. Clearly, two out of three vertices on the right-hand side suffice.

An important point in applying these data reduction rules is that they always guarantee to find *at least one* optimal solution (with the help of the simpler reduced instance). Note, however, that there is no longer a guarantee of finding *all* optimal solutions.

The correctness and polynomial-time realizability of the above two rules is clear. The nice thing about these data reduction rules is that, as a matter of practical experience, for input instances such as those occurring in railway optimization they usually seem to suffice to transform the overall input instance into a collection of small connected components. Each of these connected components then can be solved quickly by simple exhaustive search, thus providing an optimal solution to the original problem. The drawback from a theoretical point of view is that we cannot guarantee (that is, prove) the effectiveness of the above two rules based on a rigid mathematical analysis. For other NP-complete problems such an analysis is possible: fixed-parameter algorithmics offers the concept of "reduction to a problem kernel" to give a theoretically sound framework for analyzing the virtue of data reduction. An in-depth treatment of this issue is given in Chapter 7; reduction to a problem kernel together with data reduction rules also plays an important role in the next (and final) case study in the following section.

Finally, it is worth emphasizing that data reduction is usually considered and used as a form of preprocessing before the "main algorithm" starts. It has been shown, however—both theoretically and empirically—that it is often beneficial to apply data reduction rules over and over again during the course of the whole algorithm. The point is that after a few steps of the main algorithm the input

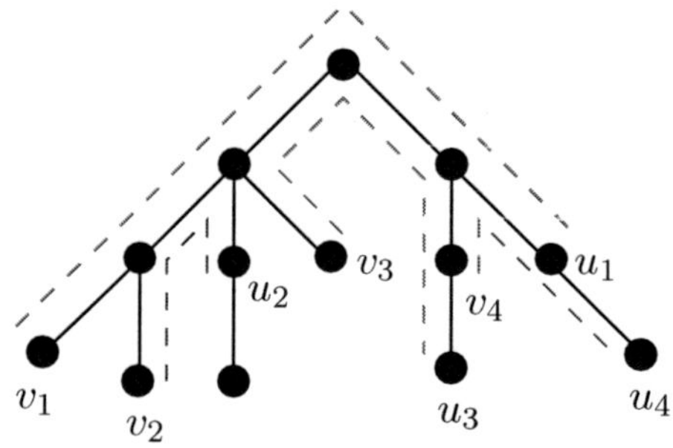

FIG. 1.3. An instance of MULTICUT IN TREES with four demand paths.

instance may have changed such that data reduction again applies. As a rule, data reduction is frequently so effective that there is rarely a case where one would not want to exploit it.

1.3 A communication problem in tree networks

Our third and last introductory example gives a concrete fixed-parameter algorithm together with its simple analysis. More specifically, we consider the problem MULTICUT restricted to trees. It is motivated by applications in communication networks. The problem is studied on general graphs as well; it is one example of a relatively small number of graph problems that remain *NP*-complete even when restricted to trees. MULTICUT IN TREES is defined as follows.

> **Input:** An undirected tree $T = (V, E)$, $n := |V|$, and a collection H of m pairs of nodes in V, $H = \{(u_i, v_i) \mid u_i, v_i \in V, u_i \neq v_i, 1 \leq i \leq m\}$.
> **Task:** Find a minimum size subset E' of E such that the removal of the edges in E' separates each pair of nodes in H.

Each pair of nodes in a tree uniquely determines a corresponding path which we subsequently refer to as a *demand path.*

Example 1.4 Figure 1.3 gives an example instance for MULTICUT IN TREES with four pairs of nodes, that is, four demand paths. There is an optimal solution which removes two edges as marked in Figure 1.4.

By trying all possibilities (and using the fact that for trees $|E| = n-1$) we can solve the problem in 2^{n-1} steps. Can we do better if there exists a small solution set E'? Consider the parameter $k := |E'|$ and study how k influences the problem complexity. Note that clearly $k < n$ but in some application scenarios $k \ll n$ seems to be a reasonable assumption. Here, the following simple algorithm works. Assume that the given tree is (arbitrarily) rooted—that is, choose an arbitrary node $r \in V$ and direct all edges from r to its neighbors (that is, children), from r's children to their children, and so on. In Figure 1.3 the chosen node r is drawn at the top of the picture. Consider a pair of nodes $(u, v) \in H$ such that the uniquely determined path p between u and v has maximum distance from

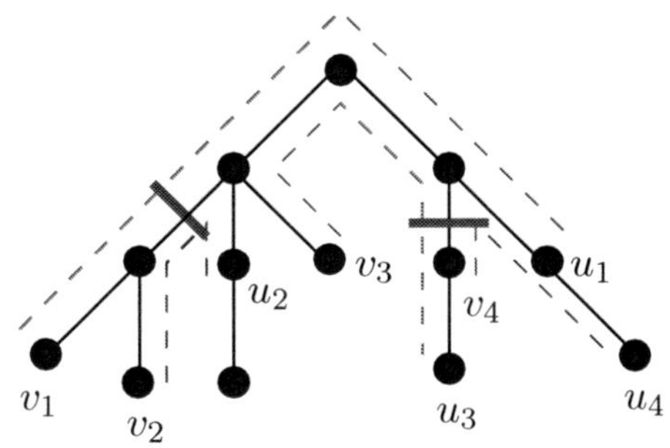

FIG. 1.4. An optimal solution that removes the two marked edges.

the root r. In Figure 1.3 (u_2, v_2) or (u_4, v_4) is such a pair. Call w the node on p that is closest to r; that is, w is the least common ancestor of u and v.

Then one may easily see (using the tree structure) that there is an optimal solution that contains at least one of the two path edges connected to w. This implies that we can build a search tree of depth bounded by k: simply search for w as specified above and then branch into the two cases (at least one of them *has to* yield an optimal solution) of taking one of the two neighboring path edges of w into the solution edge set to be constructed. Each of these two cases represents the deletion of one of the two edges. After deleting an edge, we remove all no longer connected node pairs from H. Iterating this process at most k times, we obtain a search tree of size bounded by 2^k, and thus we have a fixed-parameter algorithm for MULTICUT IN TREES with respect to parameter k. Note that in a straightforward way, by taking both edges, one also obtains a polynomial-time factor-2 approximation algorithm. Nothing better is known. An in-depth treatment of depth- and thus size-bounded search trees is given in Chapter 8.

Additionally, there is a set of simple data reduction rules for MULTICUT IN TREES. In the description of the subsequent rules, we often *contract an edge e*. Let $e = \{v, w\}$ and let $N(v)$ and $N(w)$ denote the sets of neighbors of v and w, respectively. Then, contracting e means that we replace v and w by one new vertex x and we set $N(x) := N(v) \cup N(w) \setminus \{v, w\}$. We occasionally consider paths P_1 and P_2 in the tree and we write $P_1 \subseteq P_2$ when the node set (and edge set) of P_2 comprises that of P_1. Now, the correctness of the following four data reduction rules and their polynomial-time realizability is easy to see.

Idle Edge. If there is a tree edge with no path passing through it, then contract this edge.

Unit Path. If a demand path has length one (in other words, it consists of exactly one edge), then the corresponding edge e has to be in the solution set E'. Contract e and remove all demand paths passing e from H and decrease the parameter k by one.

Dominated Edge. If all demand paths that pass edge e_1 of T also pass edge e_2 of T, then contract e_1.

Dominated Path. If $P_1 \subseteq P_2$ for two demand paths, then delete P_2.

These rules transform a MULTICUT IN TREES instance (T, H) with parameter k into a "reduced" instance (T', H') with parameter k' where $k' \leq k$. The rules guarantee that (T, H) has a solution of size at most k iff (T', H') has a solution of size at most k'. Notably, none of the above four data reduction rules makes explicit use of the parameter value k. That is, to apply them, parameter k does not need to be known in advance. Since this turns out to be a crucial advantage in practical applications, this sort of rule is termed *parameter-independent data reduction*. By way of contrast, consider the following two parameter-dependent data reduction rules for MULTICUT IN TREES.

Disjoint Paths. If an instance of MULTICUT IN TREES has more than k edge-disjoint demand paths, then there is no solution with parameter k.

Overloaded Edge. If more than k distinct length-two demand paths pass through an edge e, then contract e, remove all demand paths passing e, and decrease parameter k by one.

Again, the correctness of these rules is easy to verify. By using the above six data reduction rules together with two more parameter-dependent, slightly more complicated, rules it is possible to show—using a fairly demanding mathematical analysis—that a reduced instance of MULTICUT IN TREES has a size that can be upper-bounded by a function solely depending on k. In words of parameterized complexity analysis, these data reduction rules lead to a *problem kernel* for MULTICUT IN TREES with respect to parameter k. Recall that, unfortunately, such a problem kernel could not be proven for the railway optimization problem studied in Section 1.2; there is strong theoretical evidence that this will not be possible. Problem kernelization is a key issue in fixed-parameter algorithm design and it will be explored in depth in Chapter 7.

In summary, polynomial-time preprocessing by data reduction rules is not only a tool of central importance for designing fixed-parameter algorithms but will almost always be beneficial in *any* approach attacking computationally hard problems. The railway optimization problem from the previous section is only one important example in this respect.

1.4 Summary

The leitmotif of parameterized algorithm design and analysis is the search for a better explanation of what causes the hardness of *NP*-complete and related problems. The method of attack is to do a two-dimensional complexity analysis. Besides a problem's size itself, a parameter is also singled out and we investigate whether and how the seemingly unavoidable combinatorial explosion can be solely confined to this parameter. Fixed-parameter algorithmics is based on worst-case analysis and it leads to exponential-time algorithms providing optimal solutions for the given problems. The running times are analyzed in a mathematically rigorous way and several fundamental techniques, such as bounded search trees and data reduction by preprocessing, play a key role. As a rule, "small" parameter values are required for there to be hope for efficient practical implementations of fixed-parameter algorithms. Compared with polynomial-time approximation algorithms, the advantage of fixed-parameter algorithms lies in the guaranteed optimality of the solution; compared with purely heuristic methods, the advantage of fixed-parameter algorithms lies in the rigorous mathematical analysis with provable (worst-case) performance bounds.

It is important to notice that there are usually many ways of parameterizing a problem. We saw several parameterizations of the SATISFIABILITY problem, but the MULTICUT IN TREES problem also has more than one reasonable parameterization. For instance, an alternative parameter here is the maximum number of demand paths passing through an edge or node of the tree. Hence parameterized complexity analysis allows for a multitude of views of one and the same problem. Each of these views might require different techniques and algorithms. Moreover, it seems clear that various parameterizations lead to the feasibility of different, practically relevant, special cases of the considered problem. There appears to be no way to avoid the study of more than one parameterization to get a more complete picture of the landscape of (in)tractability—finding "good" parameterizations of a problem is a task in its own. This is a rich playground for creative and innovative ideas.

The aim of this work is not to list as many parameterized problems as possible together with (if they exist) their fixed-parameter algorithms, but to give, in a sense, an application-oriented introduction to the prosperous field of developing and analyzing efficient fixed-parameter algorithms. To achieve this, we start with presenting fundamental concepts, ideas, and observations in Part I. In Part II we then exhibit key techniques in the development of fixed-parameter algorithms. We reinforce the insights from Part II in Part III by several case studies from various application fields, ranging from graph to logic problems. In addition, Part III contains more theoretical considerations concerning parameterized complexity theory and connections to the related field of polynomial-time approximation algorithms.

At the necessary risk of being incomplete and biased, the whole presentation is kept within clear, relatively small dimensions. If you follow the references given to the already extensive literature, this book might serve (and it is meant as) a springboard into the fast developing field of research on fixed-parameter algorithms.

1.5 Exercises

1. Determine how many queens are necessary to attack all squares on chessboards of sizes 5×5, 6×6, and 7×7.
2. Discuss the various parameterizations of SATISFIABILITY and the potential practical usefulness of the corresponding exact algorithms mentioned in Section 1.1. Are there parameters that can be related to each other (in at least one way)?

3. Give an easy argument why 3-SATISFIABILITY can be solved in less than 2^n steps, n being the number of Boolean variables.
4. Give an easy argument why MAXIMUM 2-SATISFIABILITY can be solved in less than 2^m steps, m being the number of clauses.
5. Find simple data reduction rules for SATISFIABILITY, analyze their running time, and prove their correctness.
6. Find simple data reduction rules for MAXIMUM SATISFIABILITY. Give at least one rule that applies to SATISFIABILITY but not to MAXIMUM SATISFIABILITY.
7. Give running time estimates for the two data reduction rules in Section 1.2.
8. Consider a graph with vertices arbitrarily colored either black or white. We look for a minimum size set S of vertices such that all black vertices have at least one neighbor in S. Describe several simple data reduction rules for such an input instance.
9. Show that the first four data reduction rules for MULTICUT IN TREES also lead directly to a size 2^k search tree.
10. Give running time estimates for the four data reduction rules in Section 1.3.
11. For MULTICUT IN TREES a minimum size set of edges to be removed is sought. Consider two "node variants" of MULTICUT IN TREES. For the first, a minimum size of nodes to be removed (to disconnect all given node pairs) is sought—so-called UNRESTRICTED NODE MULTICUT IN TREES. For the second, we additionally require that the nodes to be removed are not part of any of the given node pairs—so-called RESTRICTED NODE MULTICUT IN TREES.
    (a) Show that UNRESTRICTED NODE MULTICUT IN TREES is polynomial-time solvable.
    (b) RESTRICTED NODE MULTICUT IN TREES is *NP*-complete. Give a fixed-parameter algorithm with search tree size 2^k, where k denotes the number of removed nodes.

1.6 Bibliographical remarks

Michael R. Garey and David S. Johnson's opus (Garey and Johnson, 1979) is a continuing source of hard problems for this work. Most *NP*-completeness results we encounter here can be found there. Parameterized complexity has been chiefly developed by Rod G. Downey and Michael R. Fellows in their monograph (Downey and Fellows, 1999). Their book summarizes work till the end of the 1990s. It is broader in scope than this book; in particular, it contains much more material concerning structural complexity and related theoretical issues than considered here.

Results for DOMINATING SET IN PLANAR GRAPHS as discussed in the introduction to Part I can be found in Alber *et al.* (2005*b*), Alber *et al.* (2002), and Alber *et al.* (2004). An early example of a fixed-parameter algorithm is given in Dreyfus and Wagner (1972) for the so-called STEINER TREE problem in graphs. A very recent improvement of this algorithm can be found in Mölle *et al.* (2005).

We assume familiarity with basic notions of algorithms and complexity as can be found in standard textbooks such as Cormen *et al.* (2001), Kleinberg and Tardos (2005), and Skiena (1998). In-depth treatments of approximations algorithms can be found in Ausiello *et al.* (1999), Hochbaum (1997), and Vazirani (2001), average-case analysis (as opposed to worst-case analysis, which is done in this book) is treated in Hofri (1995), randomized algorithms are the subject of Mitzenmacher and Upfal (2005) and Motwani and Raghavan (1995), and expositions on heuristic methods are to be found in Michalewicz and Fogel (2004) and Pearl (1984). Finally, the book by Juraj Hromkovič (Hromkovič, 2002) spans the whole field of algorithmics for hard problems and also covers a little parameterized complexity analysis, relating it to the concept of pseudo-polynomial-time algorithms.

The mentioned upper bound 1.28^k for VERTEX COVER derives from Chen *et al.* (2001) and Chandran and Grandoni (2005). The *NP*-completeness results for SATISFIABILITY and MAXIMUM SATISFIABILITY are contained in Garey and Johnson (1979). The proceedings of the annual "SAT" conference appear in Springer's Lecture Notes in Computer Science series. Example applications of SATISFIABILITY in automotive product configuration and the "easy" solvability of the corresponding formulae can be found in Küchlin and Sinz (2000) and Sinz *et al.* (2003). A survey on research results for SATISFIABILITY is Dantsin *et al.* (2001). The upper bound 1.49^n for 3-SATISFIABILITY with respect to the number of variables is from Dantsin *et al.* (2002). The randomized algorithm with upper bound 1.33^n is described in Iwama and Tamaki (2003). The upper bounds with respect to the number of clauses and formula length are due to Hirsch (2000). The article by Szeider (2004*b*) surveys several SATISFIABILITY algorithms exploiting formula structure. The parameter "weight of assignment" and its relevance in parameterized intractability results are deeply explored in Downey and Fellows (1999). The breakthrough showing that MAXIMUM 2-SATISFIABILITY can be solved in 1.74^n steps is due to Williams (2004). Further bounds for MAXIMUM SATISFIABILITY and MAXIMUM 2-SATISFIABILITY are contained in Bansal and Raman (1999), Chen and Kanj (2004), and Gramm *et al.* (2003*a*).

The example of railway optimization and the corresponding data reduction rules is taken from Weihe (1998) and Weihe (2000). The parameterized hardness of the problem follows from the theory developed in Downey and Fellows (1999) (refer to RED-BLUE DOMINATING SET). MULTICUT IN TREES was introduced and studied in Garg *et al.* (1997), where *NP*-completeness, *MaxSNP*-hardness, and polynomial-time factor-2 approximability (even for the more general weighted version) was shown. Our presentation concerning fixed-parameter issues follows Guo and Niedermeier (2005*b*).

Many survey papers on parameterized complexity and algorithmics are available: see Alber *et al.* (2001), Downey *et al.* (1999), and Raman (1997) for older and Downey (2003), Fellows (2002), Fellows (2003*a*), Fellows (2003*b*), Grohe

(2002), and Niedermeier (2004) for more recent ones. A survey with the slightly different focus of "exact algorithms" is Woeginger (2003). The surveys Downey (2003), Fellows (2002), and Fellows (2003*b*) put special emphasis on the connection to approximation algorithms and, in particular, polynomial-time approximation schemes (PTASs). The survey by Grohe (2002) is directed towards a database audience and the surveys Fellows (2003*a*) and Niedermeier (2004) deal with the diverse possibilities of problem parameterization.

2

PRELIMINARIES AND AGREEMENTS

In this chapter we briefly sketch some of the notation used throughout the book. To avoid lengthy and tiring material dealing mostly with definitions, advanced, more problem-specific concepts and notions will be introduced in the course of the exposition in the various chapters where needed. Also, we assume a sound basic knowledge of (discrete) mathematics and familiarity with the very fundamentals of algorithms and complexity. Refer to the bibliographical remarks in Section 2.5 for pointers to more complete presentations of the subsequently described definitions and concepts.

2.1 Basic sets and problems

Finite alphabets are usually denoted by Σ. We often deal with the set of non-negative integer numbers, also denoted by $\mathbb{N}$. By way of contrast, $\mathbb{R}$ refers to the real numbers, where $\mathbb{R}^+$ is the subset of all positive reals.

This work is about exact algorithms that solve computationally hard problems. We present problems in an "input-task-style". This first of all refers to *decision problems* where it is asked whether, for a given input instance of a problem, the answer is "yes" or "no". The task is to find the correct answer. A standard example is given with the basic SATISFIABILITY problem as described in Section 1.1. As a rule, however, there exists a naturally corresponding *optimization problem* which tries to minimize or maximize a certain cost value. For instance, a natural optimization version of SATISFIABILITY is the MAXIMUM SATISFIABILITY problem as described in Section 1.1. All algorithms in this work can be used not only to output "yes" or "no" to answer the decision question, but can also be easily adapted to constructively output a desired solution object—most of the time they already do. Moreover, in most cases the algorithms can also be modified to deliver *optimal* solutions (that is, not only fulfilling the given constraints, but being optimal among all these solutions) to the corresponding optimization problem. We do not make a sharp distinction between decision or optimization for the parameterized problems introduced below (Section 3.1) because it will always be made clear from the context what is meant. As a matter of fact, in a certain sense one may identify decision and parameterized problems in many cases.

2.2 Model of computation and running times

The *Random Access Machine* (*RAM*) model of computation will be used to give a machine-independent description of algorithms, sometimes using pseudocode in the same style as is common in standard textbooks on algorithms and complexity. Particular features of RAM computation are that each "simple" operation (basic arithmetic, assignments, if-statements, etc.) takes one time unit, as does every access to one memory cell (with some reasonable word size). In particular, the word size is big enough to hold all numbers occurring in the algorithms presented. None of the described RAM features will be misused in the sense that the corresponding algorithms could not be implemented on existing computers in an efficient way. The RAM model employed, in order to simplify the analysis of algorithms, will not distinguish between different levels of the memory hierarchy (cache versus disk etc.).

We use the familiar "*big Oh notation*" to upper-bound the running times of our algorithms. Hence we ignore constant factors, but, if appropriate, we point to cases where the constant factors involved in the algorithms are large and thus might threaten or destroy the practical usefulness of the algorithms. Recall that the use of "Θ" instead of "O" means that there is in addition a corresponding lower bound. The use of "o" instead of "O" means that we refer to an asymptotically strict less than constant factor growing function following the "o". For example, we have that $n = o(n \log n)$. All running time analysis in this work is *worst-case* analysis, that is, the presented bounds hold over *all* input instances of a given problem. At a few points, partially based on experimental experience, we will indicate that the given bounds or the worst-case analysis as such appear to be too pessimistic when the algorithm is applied in practice—to analyze mathematically any kind of average-case complexity, however, is beyond the scope of this work. Indeed, due to the inherent difficulties of average-case analysis starting with the "simple" problem to define what a practically relevant average case is and continuing with difficult mathematical problems, relatively little is generally known about average-case complexity.

2.3 Strings and graphs

A *word* or, equivalently, a *string* over a finite alphabet Σ, is a sequence of elements from Σ. Strings play a particularly important role in fixed-parameter algorithms related to computational biology. Here, we also make use of the *Hamming distance* measure $d_H(s, t)$ between two strings s and t of equal length L. This is defined as

$$|\{\, p \mid s[p] \neq t[p], 1 \leq p \leq L \,\}|,$$

where $s[p]$ denotes the character at position p in string s.

Many computational problems that we study are graph problems. An *undirected graph* G is a pair (V, E), where V is a finite set of *vertices* and E is a finite set of *edges* which are *unordered* pairs of vertices. Almost all graphs considered in this work are undirected—ordered pairs of vertices yield directed edges and thus *directed* graphs. Furthermore, all our graphs are *simple* (that is, there is at most one edge between each pair of vertices) and do not contain *self-loops* (that is, edges from a vertex to itself are forbidden). The *degree* of a vertex is the number of edges incident to it. A graph is *d-regular* if every vertex has degree exactly d. The (open) *neighborhood* of a vertex v in graph $G = (V, E)$ is defined as $N(v) := \{ u \mid \{u, v\} \in E \}$. In a d-regular graph each neighborhood has size exactly d. The *closed neighborhood* $N[v]$ of v is then $N(v) \cup \{v\}$. Finally, for a vertex set $U \subseteq V$ we define $N(U) := \bigcup_{v \in U} N(v)$ and $N[U] := \bigcup_{v \in U} N[v]$. A *path* in a graph is a sequence of pairwise distinct vertices so that subsequent vertices are adjacent in the graph. If the last vertex is adjacent to the first vertex in the sequence, we have a *cycle*. A graph is *connected* if every pair of vertices is connected by a path. If a graph is not connected then it naturally decomposes into its *connected components*. *Trees* are undirected connected graphs that contain no cycles. Sometimes we deal with *rooted* trees. These arise when choosing one distinct vertex (the *root*) and directing all edges from the root to its neighbors—which are then called its *children*—analogously directing all edges between the root's children and their neighbors different from the root, and so on.

For a graph $G = (V, E)$ and a set $V' \subseteq V$, the *subgraph* of G *induced* by V' is denoted by $G[V'] = (V', E')$, where $E' := \{ \{u, v\} \in E \mid (u \in V') \wedge (v \in V') \}$. By way of contrast, a *subgraph* $H = (V'', E'')$ of $G = (V, E)$ simply fulfills the conditions that $V'' \subseteq V$, $E'' \subseteq E$, and the endpoints of edges in E'' must be contained in V''. A *subdivision* of a graph is obtained by replacing some of the edges with pairwise internally disjoint paths. A *contraction* of an edge $e = \{u, v\}$ is the replacement of u and v with a single new vertex whose incident edges are the edges other than e that were incident to u or v. As a rule, after having contracted an edge, we will delete any double edges that might possibly exist. *Minors* are a core concept of modern graph theory. A graph H is a minor of a graph G if a copy of H can be obtained from G by deleting and/or contracting edges of G. Two graphs $G = (V, E)$ and $G' = (V', E')$ are *isomorphic* if there exists a one-to-one mapping $g : V \to V'$ such that $\{u, v\} \in E$ iff $\{g(u), g(v)\} \in E'$.

In this work we often deal with a special class of graphs called *planar graphs.* A graph G is called planar if it can be drawn in the plane such that no two edges cross. A particular crossing-free drawing of a graph is called a *plane embedding* of that graph and a *plane graph* is a planar graph together with its embedding in the plane. Planar graphs are sparse in the sense that the well-known *Euler formula* says that a planar graph with n vertices has at most $3n - 6$ edges—a general graph may contain up to $n(n-1)/2$ edges. The *faces* of a plane graph are the maximal regions of the plane that contain no point used in the embedding. A *triangular face* is a face enclosed by only three edges (which is the smallest number possible). A plane graph where every such face boundary is a cycle of three edges is called a *triangulation.*

Planar graphs have a famous characterization due to Kasimir Kuratowski. A graph is planar iff it does not contain a subdivision of K_5 or $K_{3,3}$. Herein, K_i, $i \geq 1$, denotes the *complete graph* with i vertices and all vertices adjacent to each other. A graph is *bipartite* if each vertex is in exactly one of two subsets of the vertex set of the graph and within each subset no two vertices are adjacent to each other. Then $K_{i,j}$, $i, j \geq 1$, denotes the *complete bipartite graph* with i vertices in one subset and j vertices in the other subset and each pair of vertices from different subsets adjacent.

2.4 Complexity and approximation

Efficiency of algorithms from a theoretical point of view means algorithms running in time polynomial in the input size. By way of contrast, a vast number of important computational problems have so far resisted efforts to find efficient solutions. These problems are mostly known as *NP-hard* problems and they form a central topic of basic research in theoretical computer science. The must-read reference here is Michael R. Garey and David S. Johnson's monograph *Computers and Intractability*. Formally, *NP*-hardness refers to decision problems and not to optimization problems. Nevertheless, adopting a somewhat sloppy but simplifying point of view, we often speak of the *NP*-hardness of optimization problems. This is justified by the fact that optimization problems can be turned into decision problems by introducing a threshold value as an additional part of the input and asking whether there exists a solution above (in the case of maximization problems) or below (in the case of minimization problems) the threshold value. The mother of all *NP*-hard problems is the SATISFIABILITY problem as already considered in Section 1.1. An *NP*-hard problem is *NP-complete* if it can be solved in polynomial time by a nondeterministic Turing machine—the (complexity) class of all problems solvable by this means is denoted by *NP*. By way of contrast, the computational complexity class P contains all problems that can be solved in polynomial time by a deterministic Turing machine. Note that polynomial time on a deterministic Turing machine is equivalent to polynomial time on a RAM. All *NP*-complete problems are closely related to each other by the concept of *polynomial-time reducibility*. That is, an instance I of any *NP*-complete problem A can be transformed into an instance I' of any *NP*-complete problem B in polynomial time such that I has a yes-answer concerning A iff I' has a yes-answer concerning B. All computational problems considered in this work are *decidable* (or *computable*); that is, there is an algorithm for solving the given problem in finite time.

NP-hard decision problems generally need exponential (or worse) running time to be solved exactly—which appears to be impractical in many cases. A less ambitious goal in attacking the corresponding optimization versions of the *NP*-hard problems, as pursued by *approximation algorithms,* is to strive for approximate instead of optimal solutions. The hope is that the solution obtained is "not too far from the optimum". The quality of the approximation (with respect to the given optimization criterion) is measured by the *approximation ratio*. An approximation algorithm for a problem has approximation ratio $\rho(n)$ if for any input size n, the cost C of the solution produced by the approximation algorithm is within a factor of $\rho(n)$ of the cost C^* of an optimal solution:

$$\max\{C/C^*, C^*/C\} \leq \rho(n).$$

The goal is to make $\rho(n)$ as small as possible. For instance, the best known approximation ratio for MULTICUT IN TREES (see Section 1.3) is $\rho(n) = 2$, meaning that in polynomial time a solution can be constructed that is at most twice as large as the optimal one. By way of contrast, a formalization of the "domination problem" in the railway optimization scenario discussed in Section 1.2 leads to the best known approximation ratio $O(\ln n)$. A *polynomial-time approximation scheme* (*PTAS*) for an optimization problem is an algorithm which, for each $\epsilon > 0$ and each problem instance, returns a solution with approximation ratio $1 + \epsilon$. The polynomial running time of such an algorithm, as a rule, depends crucially on $1/\epsilon$. If the time bound depends only *polynomially* on $1/\epsilon$, then the approximation scheme is called *fully polynomial* (*FPTAS*) which is considered to be the best degree of approximability one can hope for. We mention in passing that a class *MaxSNP* (also known as *APX*) of optimization problems can be "syntactically" defined together with a reducibility concept. The point is that *MaxSNP-hard* problems are unlikely to have polynomial-time approximation schemes. For example, the optimization version of MULTICUT IN TREES (see Section 1.3) is *MaxSNP*-hard.

2.5 Bibliographical remarks

Many important concepts have been introduced in this chapter in a very informal, sometimes sloppy, way. Hence this chapter must be seen as a brief reminder of the things the reader should mostly have seen (and learned) already; if not, refer to more detailed textbooks on algorithms, complexity, and discrete mathematics. There are many good sources that provide the prerequisites needed for working through this book. A very limited selection is as follows. Concerning fundamentals and advanced material on design and analysis of algorithms, refer to Aho *et al.* (1974), Cormen *et al.* (2001), and Kleinberg and Tardos (2005). The "algorithm design manual" (Skiena, 1998) provides a quick overview of algorithmic results and methods for many important problems. Fundamentals of theoretical computer science and models of computation can be found in Hopcroft *et al.* (2001). Valuable introductions to discrete mathematics and closely related issues are Biggs (1989) and Matoušek and Nešetřil (1998). Particular sources for graph theory with relations to algorithmic issues are available in Brandstädt *et al.* (1999), Diestel (2005), Golumbic (2004), and West (2001). *NP*-completeness and computational complexity theory are explored in Garey and Johnson (1979) and Papadimitriou (1994). Approximation algorithms and corresponding theoretical issues are treated in Ausiello *et al.* (1999), Hochbaum (1997), and Vazirani (2001). Ausiello *et al.* (1999) contains an extensive list of problems together with their corresponding approximability properties and results—an up-to-date on-line version is maintained through a web site. The recent monograph by Hromkovič (2002) provides an extensive survey on various ways (including approximation and exact algorithms) to cope with computational intractability. Related texts can also be found in Atallah (1999). The approximation ratio 2 and the *MaxSNP*-hardness results for MULTICUT IN TREES are due to Garg *et al.* (1997). The approximation ratio for domination and related problems appears in Feige (1998).

3

PARAMETERIZED COMPLEXITY THEORY—A PRIMER

We briefly sketch general aspects of the theoretical basis of the study of parameterized complexity. The focus of this chapter, however, lies on the algorithmic side of fixed-parameter tractability, and the consideration of complexity-theoretic issues remains very limited here. A more formal treatment follows in Part III.

3.1 Basic theory

The *NP*-complete minimization problem VERTEX COVER is the best studied problem in the field of fixed-parameter algorithms:

> **Input:** A graph $G = (V, E)$ and a nonnegative integer k.
> **Task:** Find a subset of vertices $C \subseteq V$ with k or fewer vertices such that each edge in E has at least one of its endpoints in C.

Figure 3.1 illustrates some basic graph problems occurring in the course of this section.

In what follows, let $n := |V|$ denote the number of graph vertices. VERTEX COVER is *fixed-parameter tractable*: there are algorithms solving it in $O(1.28^k + kn)$ time. By way of contrast, consider the also *NP*-complete maximization problem CLIQUE:

> **Input:** A graph $G = (V, E)$ and a nonnegative integer k.
> **Task:** Find a subset of vertices $C \subseteq V$ with k or more vertices such that C forms a clique, that is, C induces a complete subgraph of G.

CLIQUE appears to be *fixed-parameter intractable*: it is *not* known whether it can be solved in time $f(k) \cdot n^{O(1)}$, where f may be a computable but arbitrarily fast growing function only depending on k. The well-founded conjecture is that no such fixed-parameter algorithm for CLIQUE exists. The best known algorithm solving CLIQUE runs in time $O(n^{ck/3})$, where c is the exponent on the time bound for multiplying two integer $n \times n$ matrices (currently, $c = 2.38$). Note that $O(n^{k+1})$ is the running time for the straightforward algorithm that just checks all size-k subsets. The decisive point is that k appears in the exponent of n, and there seems to be no way "to shift the combinatorial explosion only into k", independent from n.

The observation that *NP*-complete problems like VERTEX COVER and CLIQUE behave completely differently in a "parameterized sense" lies at the very heart of parameterized complexity, a theory pioneered by Rod G. Downey and Michael R. Fellows. In the remainder of the book, we will mainly concentrate

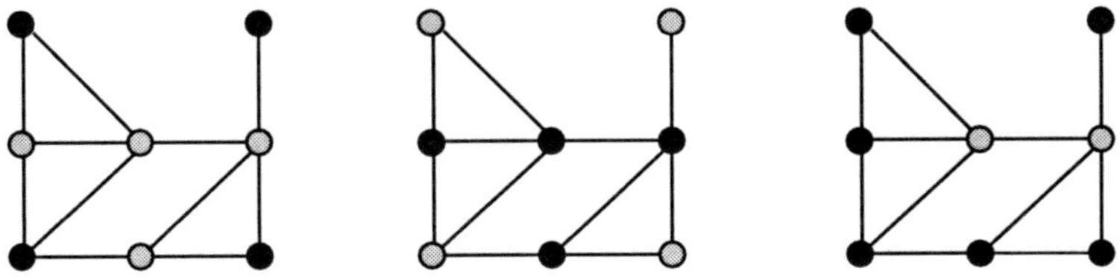

FIG. 3.1. From left to right, the three central parameterized problems VERTEX COVER, INDEPENDENT SET, and DOMINATING SET with optimal solution sets marked. Their parameterized complexity ranges from *FPT* over *W[1]*-complete to *W[2]*-complete. The parameter is always defined as the size of the solution set and the respective solution sets are marked by light shading (opposite to the dark shading of the remaining vertices).

on the world of fixed-parameter tractable problems such as those exhibited by VERTEX COVER. Here, we only briefly sketch some very basic ideas from the theory of parameterized intractability in order to provide some background on parameterized complexity theory and the ideas behind it.

We begin with some basic definitions of parameterized complexity theory.

Definition 3.1 *A* parameterized problem *is a language* $L \subseteq \Sigma^* \times \Sigma^*$*, where* Σ *is a finite alphabet. The second component is called the parameter of the problem.*

In practically all the examples in this work the parameter is a nonnegative integer or a set of nonnegative integers. Hence we will usually write $L \subseteq \Sigma^* \times \mathbb{N}$ instead of $L \subseteq \Sigma^* \times \Sigma^*$. In principle, however, the above definition leaves open the possibility of also defining more complicated parameters; for instance the subgraphs one is searching for in a graph. For $(x, k) \in L$, the two dimensions of parameterized complexity analysis are constituted by the input size n, that is, $n := |(x, k)|$ and the parameter value k (usually a nonnegative integer) or its size.

Definition 3.2 *A parameterized problem* L *is* fixed-parameter tractable *if it can be determined in* $f(k) \cdot n^{O(1)}$ *time whether or not* $(x, k) \in L$*, where* f *is a computable function only depending on* k*. The corresponding complexity class is called FPT.*

In the most general form of fixed-parameter tractability we have a multiplicative connection between $f(k)$ and the polynomial running time part. The running time for VERTEX COVER is of the form $f(k) + n^{O(1)}$. Using "asymptotic considerations" or, more importantly, by data reduction through preprocessing, an algorithm with $f(k) \cdot n^{O(1)}$ running time can be transferred into an algorithm with $g(k) + n^{O(1)}$ running time, g again being a function depending only on parameter k. See Chapter 7 in Part II for more on this.

Fixed-parameter tractability is the key notion in this book. In Definition 3.2, one may assume any standard model of sequential deterministic computation such as deterministic Turing machines or RAMs. For the sake of convenience, if not stated otherwise, we will always take the parameter, denoted by k, as a nonnegative integer encoded with a unary alphabet. Unary encoding avoids "dirty tricks" in the analysis of the computational complexity of the algorithms that might be possible using binary encoding.

A core tool in the development of fixed-parameter algorithms is polynomial-time preprocessing by *data reduction rules*, often yielding a *reduction to a problem kernel*. Here, the goal is, given any problem instance x with parameter k, to transform it into a new instance x' with parameter k' such that the size of x' is bounded from above by some function depending only on k, $k' \leq k$, and (x, k) has a solution iff (x', k') has a solution. Observe that data reduction (as we have already discussed for MULTICUT IN TREES in Section 1.3) as defined here is not to be confused with the subsequently defined (Definition 3.3) concept of parameterized reduction. The first is an algorithmic tool important for fixed-parameter tractability and the second is a complexity-theoretic tool important for fixed-parameter intractability.

Proving nontrivial, absolute lower bounds on the computational complexity of problems appears to be very difficult and has made relatively little progress so far. Hence it is probably not surprising that up to now there is no proof that *no* $f(k) \cdot n^{O(1)}$ time algorithm for CLIQUE exists. In a more complexity-theoretical language, this can be rephrased by saying that it is unknown whether CLIQUE $\in$ *FPT*. Analogously to classical complexity theory, Downey and Fellows developed some way out of this quandary by providing a reducibility and completeness program. The completeness theory of parameterized intractability involves significantly more technical effort than the classical one. We very briefly sketch some integral parts of this theory in the following.

To prove the "relative hardness" of parameterized problems, we first need a reducibility concept:

Definition 3.3 *Let $L_1, L_2 \subseteq \Sigma^* \times \mathbb{N}$ be two parameterized problems. We say that L_1* reduces to L_2 by a (standard) parameterized (many-one-)-reduction *if there are functions $k \mapsto k'$ and $k \mapsto k''$ from $\mathbb{N}$ to $\mathbb{N}$ and a function $(x, k) \mapsto x'$ from $\Sigma^* \times \mathbb{N}$ to Σ^* such that*

1. *$(x, k) \mapsto x'$ is computable in $k'' \cdot |(x, k)|^c$ time for some constant c and*
2. *$(x, k) \in L_1$ iff $(x', k') \in L_2$.*

Notably, most reductions from classical complexity turn out *not* to be parameterized ones. Consider the *NP*-complete maximization problem INDEPENDENT SET (also known as STABLE SET).

> **Input:** A graph $G = (V, E)$ and a nonnegative integer k.
> **Task:** Find a subset of vertices $I \subseteq V$ with k or more vertices that form an independent set, that is, I induces an edgeless subgraph of G.

Refer to Figure 3.1 for an example. The well-known reduction from INDEPENDENT SET to VERTEX COVER, which is given by letting $G' = G$ and $k' = |V| - k$

for a graph instance $G = (V, E)$, is not a parameterized one in the sense of Definition 3.3.

This is due to the fact that the reduction function determining the new parameter k' depends strongly on the instance G itself, hence contradicting the definition of a parameterized reduction that requires that k' solely depends on k. The reductions from INDEPENDENT SET to CLIQUE and vice versa, however, which are obtained by simply passing the original graph over to the *complementary* one[3] for $k' = k$, are indeed parameterized ones. Therefore, these problems are of comparable degree of difficulty in terms of parameterized complexity.

Now, the "lowest class of parameterized intractability" can be defined as the class of parameterized languages that are equivalent to the so-called SHORT TURING MACHINE ACCEPTANCE problem (also known as the k-STEP HALTING problem).

> **Input:** A nondeterministic Turing machine M with its transition table, an input word x, and a nonnegative integer k.
> **Task:** Find out whether M accepts x in a computation of at most k steps.

This is the parameterized analogue to the TURING MACHINE ACCEPTANCE problem (where the bound on the number of steps is polynomial in the input size $|x|$ instead of being k)—the basic generic *NP*-complete problem in classical complexity theory. SHORT TURING MACHINE ACCEPTANCE can trivially be solved in $O(n^{k+1})$ time, where n denotes a bound on the total input size—this is done by exhaustively exploring all k-step computation paths—and it would be surprising if this can be significantly improved. Herein, observe that the alphabet size is not bounded from above by a constant. Together with the above introduced reducibility concept, SHORT TURING MACHINE ACCEPTANCE can now be used to define the lowest class of parameterized intractability, that is, $W[1]$.

Definition 3.4 *The class of all parameterized languages that reduce by a standard parameterized reduction to* SHORT TURING MACHINE ACCEPTANCE *is called* $W[1]$. *A problem which lets* SHORT TURING MACHINE ACCEPTANCE *reduce to it by a parameterized reduction is called* $W[1]$-hard*; if, additionally, it is contained in* $W[1]$*, then it is called* $W[1]$-complete.

The fundamental conjecture that $FPT \neq W[1]$ is very much analogous (but clearly "weaker") to the conjecture that $P \neq NP$. Note that $W[1]$ is originally defined in a "circuit-based", more technical way. This is also where the "W" stems from—it refers to the *weft* of Boolean circuits, that is, the maximum number of unbounded fan-in gates on any path from the input variables to the output gate of the (decision) circuit. The weft has also been interpreted as the "logical depth" of a problem. Clearly, by definition, SHORT TURING MACHINE ACCEPTANCE is $W[1]$-complete. Other problems that are $W[1]$-complete

[3]That is, the *complement graph* has the same set of vertices and there is an edge between a pair of vertices iff there is no edge between them in the original graph.

include CLIQUE and INDEPENDENT SET, where the parameter is the size of the relevant vertex set. Also, for constant maximum clause size, the aforementioned parameterization of CNF-SATISFIABILITY by the weight of an assignment (see Section 1.1) gives a $W[1]$-complete problem. $W[1]$-hardness gives a concrete indication that a parameterized problem with parameter k is unlikely to allow for a solving algorithm with $f(k) \cdot n^{O(1)}$ running time, that is, restricting the combinatorial explosion to the parameter seems illusory. A parameterized reduction from any of these $W[1]$-complete problems to VERTEX COVER would lead to a collapse of the classes $W[1]$ and FPT.

As a matter of fact, a whole hierarchy of parameterized intractability can be defined, $W[1]$ only being the lowest level. In general, the classes $W[t]$ are defined based on the number of alternations between unbounded fan-in AND- and OR-gates in Boolean circuits. For instance, consider the NP-complete minimization problem DOMINATING SET.

> **Input:** A graph $G = (V, E)$ and a nonnegative integer k.
> **Task:** Find a subset of vertices $S \subseteq V$ with k or fewer vertices such that each vertex in V is contained in S or has at least one neighbor in S.

Refer to Figure 3.1 for an example. DOMINATING SET is known to be $W[2]$-complete with respect to parameter k. There exists a very rich structural theory of parameterized complexity somewhat similar to classical complexity. Observe, however, that in some respects parameterized complexity appears to be, in a sense, "orthogonal" to classical complexity: for instance, consider the maximization problem VAPNIK-CHERVONENKIS DIMENSION (VC DIMENSION):

> **Input:** A family $\mathcal{C}$ of subsets of some universe U and a nonnegative integer k.
> **Task:** Find out whether or not there is a subset $S \subseteq U$ with $|S| \geq k$ such that for each $T \subseteq S$ there exists a subset $C \in \mathcal{C}$ such that $S \cap C = T$.

VC DIMENSION has applications in learning theory and is not known (and not believed) to be NP-hard but to lie somewhere above P but below the class of NP-complete problems. It is $W[1]$-complete with respect to the parameter k, though. Thus, although VC DIMENSION in the classical sense appears to be easier than VERTEX COVER (which is NP-complete), the opposite appears to be true in the parameterized sense, because VERTEX COVER is in FPT.

From a more algorithmic point of view (as pursued in this book), it is usually sufficient to distinguish between $W[1]$-hardness and membership in FPT. Thus, for an algorithm designer who is unable to show fixed-parameter tractability of a problem, it may be "sufficient" to give a reduction from, for example, CLIQUE to the given problem using a parameterized reduction. This then proves that, unless $FPT = W[1]$, the problem does not allow for an $f(k) \cdot n^{O(1)}$ time algorithm. One piece of circumstantial evidence for this unlikeliness is the result showing that $FPT = W[1]$ would imply a $2^{o(n)}$ time algorithm for the NP-complete 3-SATISFIABILITY problem (where n denotes the number of variables of the given

Boolean formula), which would mean a major (and so far considered unlikely) breakthrough in computational complexity theory. All known upper bounds for 3-SATISFIABILITY are of the form $2^{O(n)}$; see also Section 1.1.

3.2 Interpreting fixed-parameter tractability

Most of this book concentrates on the world inside the class *FPT* of fixed-parameter tractable problems and the potential it offers. We therefore finish this chapter with a discussion of *FPT* under some application-relevant aspects. Note that in the definition of *FPT* the function $f(k)$ may take unreasonably large values such as

$$2^{2^{2^{2^{2^{2^{2^k}}}}}}.$$

For instance, there are concrete examples from model checking showing that unless $P = NP$, there is no algorithm for evaluating so-called monadic second-order queries on trees in $f(k) \cdot n^{O(1)}$ time for any elementary function f (that is, function $f(k)$ cannot be bounded from above by

$$2^{k^{\cdot^{\cdot^{\cdot^k}}}}$$

where the "tower of ks" shall have a height upper-bounded by any constant), where k denotes the query size and n is the tree size. This limits or even excludes the practical usefulness of the corresponding known fixed-parameter tractability results. Thus, showing that a problem is fixed-parameter tractable does not necessarily lead to an efficient algorithm, and this may not even be true for tiny values of parameter k. In fact, many problems that are classified fixed-parameter tractable still await more efficient, practical algorithms. In this sense, we must distinguish clearly two different aspects of fixed-parameter tractability: the theoretical part, which consists in classifying parameterized problems along the W-hierarchy (i.e. proving membership in *FPT* or hardness for $W[1]$), and the algorithmic part of actually finding efficient algorithms for problems inside the class *FPT*.

The famous *Graph Minor Theorem* by Robertson and Seymour, for instance, provides a great tool for the classification of graph problems. It states that, for a given family of graphs $\mathcal{F}$ which is closed under taking minors, membership of a graph G in $\mathcal{F}$ can be checked by analyzing whether a certain finite "obstruction set" appears as a minor in G. Moreover, the GRAPH MINOR ORDER TESTING problem is in *FPT*; more precisely, for a fixed graph G of n vertices there is an algorithm with running time $f(|H|) \cdot n^3$ that decides whether a graph H is a minor of G or not. As a matter of fact, the set $\mathcal{F}_k$ of graphs with vertex covers of size at most k is closed under taking minors. Hence there exists a finite obstruction set $\mathcal{O}_k$ for $\mathcal{F}_k$. The above method then yields the existence of a fixed-parameter algorithm for VERTEX COVER. The problematic thing is that the function f appearing in the GRAPH MINOR ORDER TESTING algorithm grows very fast and

k	$f(k) = 2^k$	$f(k) = 1.32^k$	$f(k) = 1.28^k$
10	$\approx 10^3$	≈ 16	≈ 12
20	$\approx 10^6$	≈ 258	≈ 140
30	$\approx 10^9$	≈ 4140	≈ 1650
40	$\approx 10^{12}$	$\approx 6.7 \cdot 10^4$	$\approx 2.0 \cdot 10^4$
50	$\approx 10^{15}$	$\approx 1.1 \cdot 10^6$	$\approx 2.3 \cdot 10^5$
75	$\approx 10^{22}$	$\approx 1.1 \cdot 10^9$	$\approx 1.1 \cdot 10^8$
100	$\approx 10^{30}$	$\approx 1.1 \cdot 10^{12}$	$\approx 5.3 \cdot 10^{10}$
500	$\approx 10^{150}$	$\approx 1.9 \cdot 10^{60}$	$\approx 4.1 \cdot 10^{53}$

Table 3.1 *Comparing the combinatorial explosions of various* VERTEX COVER *algorithms with respect to parameter k (size of vertex cover set) for functions $f(k)$ found in the literature.*

the constants hidden in the O-notation are huge. Moreover, finding the obstruction set in the Graph Minor Theorem, in general, is highly non-constructive. Thus the above-mentioned method may serve only as a classification tool. Other tools for deciding containment in FPT are discussed in greater detail in Part II.

Adopting the viewpoint that fixed-parameter tractability should be understood as an approach to cope with inherently hard problems, it is necessary to aim for practical, efficient fixed-parameter algorithms. In the case of VERTEX COVER, for example, it is easy to come up with an algorithm of $O(2^k \cdot n)$ running time using a simple search tree strategy. The base of the exponential term in k has been further improved and is now below 1.28. Table 3.1 gives concrete values for these. From this table we can conclude that improvements in the exponential growth of f can lead to significant changes in the running time, and, therefore, deserve investigation. Observe, however, that when comparing theoretical upper bounds the worst case is assumed. The behaviour of various algorithms for practical problem instances may be much better. In addition, Table 3.1 neglects the potentially different administrative overheads that may come along in the implementations of the various algorithms. Still, it indicates that improving the growth of function $f(k)$ is a worthwhile effort to undertake.

Finally, to demonstrate the problematic nature of the comparison "fixed-parameter tractable" versus "fixed-parameter intractable", let us compare the growth of the two (arbitrarily chosen) functions 2^{2^k} and $n^k = 2^{(k \log n)}$. The first refers to fixed-parameter tractability and the second to fixed-parameter intractability. It is easy to verify that assuming input sizes n in the range from 10^3 up to 10^{15}, the value of k where 2^{2^k} starts to exceed n^k is in the small range $\{6, 7, 8, 9\}$, meaning that in practical applications a fixed-parameter algorithm needs not be superior to a straightforward "check-all-possibilities" algorithm.

A striking concrete example in this direction is that of computing the *treewidth* of graphs (as further discussed in Part II). For constant k, there is a famous result giving a linear-time algorithm to compute whether a graph has treewidth at most k. More precisely, the algorithm has running time $2^{\Theta(k^3)} \cdot k^{O(1)} \cdot n$. As a consequence, a known simpler $O(n^{k+1})$ time algorithm appears to be more practical. This shows how careful one must be with the term "fixed-parameter tractable" in an applied sense, since in practical cases with reasonable input sizes a fixed-parameter intractable problem might turn out to have a more effective solving algorithm than a fixed-parameter tractable one. It is the function f that counts.

3.3 Exercises

1. Describe parameterized reductions
   (a) from VERTEX COVER to DOMINATING SET;
   (b) from INDEPENDENT SET to SHORT TURING MACHINE ACCEPTANCE;
   (c) from MAXIMUM CUT to MAXIMUM SATISFIABILITY;
   (d) from INDEPENDENT SET to WEIGHTED SATISFIABILITY, the latter as defined in Section 1.1 in the point "parameter weight of assignment".

   Herein, the parameter is always k as given in the definitions of the various problems. Moreover, MAXIMUM CUT is defined as follows.

   Input: A graph $G = (V, E)$ and a nonnegative integer k.
   Task: Find a subset of vertices $S \subseteq V$ such that at least k edges have one endpoint in S and one endpoint in $V \setminus S$.
2. Briefly argue whether or not the following problems are fixed-parameter tractable.
   (a) To color the vertices of a graph with k colors, where k is the parameter.
   (b) 3-HITTING SET:
   Input: A collection $\mathcal{C}$ of subsets of size three of a finite set S and a nonnegative integer k.
   Task: Find a subset $S' \subseteq S$ with $|S'| \leq k$ such that S' contains at least one element from each subset in $\mathcal{C}$?
   (c) To find a length-ℓ common subsequence of k given strings over a constant size alphabet. The parameter is ℓ. (What about parameterization with k?)
3. Show that if there is a reduction to a problem kernel for a (decidable) problem then it is fixed-parameter tractable. Does the reverse direction also hold?

3.4 Bibliographical remarks

Downey and Fellows (1999) is clearly the standard reference for parameterized complexity. Almost everything described in this chapter also appears there in much greater detail. The so-far best search tree size for VERTEX COVER appears in Chen *et al.* (2005). Nešetřil and Poljak (1985) describe the $O(n^{ck/3})$ time algorithm for CLIQUE. Concerning the complexity of the multiplication of two integer matrices, refer to Coppersmith and Winograd (1990). Boppana and Sipser (1990) give an overview of the little progress so far made in computational complexity theory concerning the derivation of absolute lower bounds. For general information on computational complexity refer to Papadimitriou (1994). Lists of parameterized complexity classification results following the Garey–Johnson style can be found in Cesati (2004) and Downey and Fellows (1999). You can learn more about the problem of computing the VAPNIK–CHERVONENKIS DIMENSION by looking at Blummer *et al.* (1989) and Papadimitriou and Yannakakis (1996). Its $W[1]$-completeness is shown in Downey and Fellows (1995). The Graph Minor Theory is due to Neil Robertson and Paul D. Seymour, obtained in a series of seminal papers. Downey and Fellows (1999) gives a more thorough treatment of this topic. The linear-time algorithm to determine the treewidth of a graph (for constant treewidth) is due to Bodlaender (1996). The alternative $O(n^{k+1})$ algorithm with smaller involved constant factors appears in Arnborg *et al.* (1987). Recent new developments concerning parameterized intractability classes can be found in Downey *et al.* (2003) and Flum *et al.* (2004).

4

VERTEX COVER—AN ILLUSTRATIVE EXAMPLE

VERTEX COVER has so far been the most popular fixed-parameter tractable problem. It is an important problem in combinatorial optimization and graph theory and it is easy to grasp. Many aspects of fixed-parameter algorithms can naturally be developed and illustrated using VERTEX COVER as a running example. Hence VERTEX COVER also plays a prominent role in this book. We start with recalling the definition.

> **Input:** A graph $G = (V, E)$ and a nonnegative integer k.
> **Task:** Find a subset of vertices $C \subseteq V$ with k or fewer vertices such that each edge in E has at least one of its endpoints in C.

VERTEX COVER has seen quite a long history in the development of fixed-parameter algorithms. Surprisingly, many papers have published fixed-parameter results on VERTEX COVER that are worse than the $O(2^k n)$ time algorithm that follows directly from the elementary search tree method already described in Kurt Mehlhorn's 1984 textbook on graph algorithms. The corresponding simple observation, which is also used in the ratio-2 approximation algorithm for VERTEX COVER, is as follows: each edge $\{u, v\}$ must be covered. Hence u or v must be in the cover (perhaps both). Thus building a search tree where we branch so as to bring either u or v into the cover, deleting the respective vertex together with its incident edges and continuing recursively with the remaining graph by choosing the next edge (which can be any of the still existing edges), solves the problem. The search tree with depth at most k has less than 2^{k+1} nodes, each of which can be processed in linear time using suitable data structures. In recent times research has been very active into reducing the size of the search tree, the best known result now being better than 1.28^k. The basic idea behind all of these papers is to use two fundamental techniques for developing efficient fixed-parameter algorithms, namely *depth-bounded search trees* and *data reduction rules.* Both will be explained in greater detail in Part II.

To improve the search tree size, intricate case distinctions with respect to the degree of graph vertices have been designed. As for reducing the problem kernel, a very elementary idea is as follows: assume that there is a graph vertex v of degree $k+1$, that is, $k+1$ edges have endpoint v. Then, to cover all these edges, one must bring either v or all its neighbors into the vertex cover. In the latter case, however, the vertex cover would then have size at least $k+1$—too much if we are only interested in covers of size at most k. Hence, without branching we *have* to bring v and all other vertices of degree greater than k into the cover. One can easily conclude that after doing this preprocessing, the remaining graph may consist of at most k^2+k vertices and at most k^2 edges. Observe, however, that this data reduction rule is *parameter-dependent* in the sense that it makes decisive use of the parameter k. A well-known theorem of George L. Nemhauser and William T. Trotter can be used to construct an improved reduction to a problem kernel resulting in a "kernel graph" of only $2k$ vertices, as will be discussed in Section 7.4. In particular, this data reduction works in a *parameter-independent* fashion; that is, the knowledge of the value of the parameter k is not necessary.

In the following sections, we shed some light on the diverse opportunities offered by fixed-parameter complexity studies. VERTEX COVER is the example used to illustrate this.

4.1 Parameterizing

In our discussions of VERTEX COVER so far, the parameter k has always denoted the size of the vertex cover set to be found. This is not the only way in which the parameter can be chosen—it is probably the most immediate way, though. Let us discuss some other possible choices for what we term the *parameter*.

As a first choice, consider the *dual parameterization* $n-k$, that is, the question is now whether or not there is a vertex cover of size $n-k$, where n denotes the total number of graph vertices and k again denotes a nonnegative integer given as part of the input. It is a well-known fact that a graph has a vertex cover of size k iff it has an independent set of size $n-k$. Hence the question of whether or not a graph has a vertex cover of size $n-k$ is equivalent to the question whether or not a graph has an independent set of size k. The latter problem, however, is known to be $W[1]$-complete (see Section 3.1), and thus is unlikely to be fixed-parameter tractable. In this sense, therefore, one can say that the parameterization with $n-k$ is "hopeless" concerning desirable fixed-parameter tractability results (the explicit parameter value still being k, which then denotes the number of vertices *not* being part of the vertex cover).

There is a further subtle point brought up by this discussion, namely the distinction between *minimization* problems (such as VERTEX COVER) and *maximization* problems (such as INDEPENDENT SET) which might also influence the choice of the parameterization to be applied to the given problem. In several settings, in order to guarantee a small parameter value (and, in particular, often for maximization problems) it might make particular sense to choose a dual parameterization $n-k$ because then k might be small whereas $n-k$ might be not.

Another reasonable parameterization for VERTEX COVER is the following, where we restrict, for the time being, our attention to planar graphs. VERTEX COVER is *NP*-complete also when restricted to planar graphs. For planar graphs we have the famous four-color theorem—every planar graph has a coloring (that is, neighboring vertices have to have different colors) using only four colors. Moreover, there exists a polynomial-time algorithm for constructing a four-coloring. From this fact, however, it easily follows that every planar graph has a vertex cover of size at most $\lfloor 3n/4 \rfloor$—at least the vertices from the largest color class (of size at least $\lceil n/4 \rceil$) do not have to be part of a vertex cover set. Hence the question of whether or not a given planar graph possesses a vertex cover of size $\lfloor 3n/4 \rfloor - k$ comes up naturally (again, k is a given nonnegative integer and n is the total number of vertices). To the best of our knowledge, however, it is open whether VERTEX COVER in planar graphs with this parameterization is fixed-parameter tractable or whether it is $W[1]$-hard. This way of parameterizing problems is known as *parameterizing above* (for this case more correctly, respectively, *below*) *guaranteed values*.

Finally, a completely different way of parameterizing VERTEX COVER is to consider the structure of the input graph. If the given graph allows for a tree decomposition of width w, then it is well-known (see Part II) that VERTEX COVER can be solved in $O(2^w \cdot n)$ time independent of the size k of the cover set we are searching for. Informally speaking, the treewidth of a graph measures how tree-like a graph is. So, trees have width 1. Restricted to trees as input graphs, VERTEX COVER can be solved trivially in linear time by simple dynamic programming starting at the leaves of the tree. In conclusion, VERTEX COVER is also fixed-parameter tractable when parameterized by *treewidth* w.

To summarize, VERTEX COVER has natural and obvious practically relevant parameterizations. In Chapter 5 we will further explore on the issue of choosing parameters for a given problem.

4.2 Specializing

In the preceding section we encountered VERTEX COVER on *planar* graphs, a still *NP*-complete special case of the general problem on unrestricted graphs. Due to results concerning approximation algorithms, where VERTEX COVER on planar graphs can be better approximated than in the general case,[4] one might expect that it is easier to cope with it on planar graphs. Indeed, this is the case. There are $O(c^{\sqrt{k}} + kn)$ time algorithms for VERTEX COVER on planar graphs, where c can be upper-bounded by $2^{4\sqrt{3}}$. This is an (asymptotic) exponential speedup in comparison with the best known bound $O(1.28^k + kn)$ for general graphs, as mentioned before. Here, it is important to notice the huge difference between the exponential base values $c = 2^{4\sqrt{3}} \approx 121.8$ and 1.28, which, at first sight, makes this result a purely theoretical one. Recent experimental studies, however, reveal that for the mathematically proven constant c there is probably quite some room for improvement and/or the average-case behaviour of the corresponding algorithm seems to be much better. These encouraging improvements for planar graphs raise the general issue of investigating VERTEX COVER on various special graph classes, which has been the subject of recent and ongoing research.

[4]There are linear-time algorithms giving approximation ratio-2 for the unweighted case (as considered here) as well as for the weighted case of VERTEX COVER on general graphs. Both results can be improved to an approximation ratio that is asymptotically better: $2 - ((\log\log|V|)/(2\log|V|))$. By way of contrast, Brenda S. Baker gave a polynomial-time approximation scheme (PTAS) for VERTEX COVER and related problems in planar graphs.

4.3 Generalizing

There are many, more general "relatives" of VERTEX COVER. The most immediate ones are WEIGHTED VERTEX COVER problems, where we put, for example, nonnegative integer or real weights on the graph vertices. Then, the task is to search for a minimum weight vertex cover set, where we sum up the weights of the vertices in the set. Recent studies also give efficient fixed-parameter search tree algorithms for some WEIGHTED VERTEX COVER problems, but, in comparison with unweighted VERTEX COVER, other branching strategies for the corresponding search trees have been designed. Very recently, a still more general version, so-called CAPACITATED VERTEX COVER, with applications in drug design, has been considered. Here, the point is that each vertex can only cover a pre-specified number of edges incident on it; this is called the capacity of a vertex. Hence choosing a vertex to be in the cover set may not suffice to cover all its incident edges. This problem has recently been shown to be fixed-parameter tractable with respect to the parameter solution size; the corresponding combinatorial explosion (so far) is much worse than for VERTEX COVER.

Another method of generalization is to consider a broader graph concept, that is, *hypergraphs*. By way of contrast with standard graphs, here edges may connect more than two vertices. For instance, if we allow three vertices per edge, then VERTEX COVER becomes the so-called 3-HITTING SET problem. Developing efficient fixed-parameter algorithms for 3-HITTING SET is possible but requires new ideas (see Part II).

Another possibility for generalizing VERTEX COVER is as follows. Consider classes of graphs that are defined through a base graph class. Then a graph is in such a graph class if a base graph can be generated from it by deleting at most k vertices. In particular, if the base graph class consists of all edgeless graphs, then this exactly yields the class of graphs with a vertex cover of size at most k. Thus, a recognition algorithm for this graph class yields an algorithm for the parameterized VERTEX COVER problem. The corresponding fixed-parameter tractability results seem to be of more theoretical interest because the exponential functions involved so far (for instance, k^k and worse) in the running times are too big for most practical purposes.

4.4 Counting or enumerating

So far, parameterized complexity has been focusing on the consideration of decision or search problems. Counting problems, being a standard theme in classical computational complexity, have largely been neglected. Consider the graph consisting of n vertices and $n/2$ vertex-disjoint edges. Clearly, an optimal vertex cover of this graph has size $k = n/2$ and there are exactly 2^k different optimal vertex covers. Generally, it may become difficult to exactly count the number of optimal vertex covers of a given graph. The problem arises that the common search tree approaches to solving VERTEX COVER in a fixed-parameter way must be extended such that it is ensured that the same vertex cover set is not generated by two different paths through the search tree, thus avoiding double counting. There is a fixed-parameter search tree strategy for exactly counting vertex covers with exponential running time factor $O(1.47^k)$.

More general considerations of parameterized counting, in particular including hardness results, have recently been undertaken. Specifically, a whole complexity theory of parameterized counting has been initiated.

In some applications it might be of interest not only to determine one vertex cover of size at most k but to enumerate all of them, and then perhaps to choose the one most suitable for the application at hand. Hence, enumeration of all solutions is an important aspect worth consideration. Note that the aforementioned (see introduction to Chapter 4) data reduction technique due to Nemhauser and Trotter can also be used for enumeration. If one needs to enumerate all optimal vertex covers of a given graph (let an optimal vertex cover have size k), then clearly the trivial 2^k search tree is optimal because, again considering the example graph from above, there are 2^k optimal vertex covers of this graph and 2^k is an upper bound for the number of optimal vertex covers of every graph. It is a different question, however, to investigate whether a "compact description" of all vertex covers of size at most k can be computed with combinatorial explosion $o(2^k)$. By using *minimal* vertex covers (in the set inclusion sense) only this seems possible, as value $O(1.8^k)$ has been very recently reported.

4.5 Lower bounds

Giving lower bounds for the computational complexity of problems is a core challenge in theoretical computer science. Unfortunately, it is also a very hard problem with relatively little success so far. Nevertheless, it is worth pursuing this issue also in the context of fixed-parameter algorithms. Besides the "relative lower bounds" given by the $W[1]$-hardness program as such the following questions also merit attention:

- Can VERTEX COVER on general graphs be solved in $(1+\epsilon)^k \cdot n^{O(1)}$ time for arbitrary $\epsilon > 0$ or is there a minimum ϵ-value?
- Can VERTEX COVER on general graphs be solved in $2^{o(k)} \cdot n^{O(1)}$ time?

There is most likely a negative answer to the second question: VERTEX COVER cannot be solved in $2^{o(k)} \cdot n^{O(1)}$ time unless 3-SATISFIABILITY can be solved in $2^{o(n)}$ time, where n is the number of variables of the Boolean 3-CNF formula. Note that the currently best deterministic algorithm for 3-SATISFIABILITY runs in basically $O(1.49^n)$ time. The first question appears to be open.

4.6 Implementing and applying

All our discussions up to now have had a strong theoretical flavor. The proven worst-case performance bounds on the time complexity of algorithms using the O-notation indicate their practical usefulness to only some limited extent. Empirical confirmation of theoretical findings is needed. Fixed-parameter algorithms for VERTEX COVER (enriched and combined with heuristic strategies) and for VERTEX COVER on planar graphs have actually been implemented and tested.

The results are encouraging, but this field as a whole is still in its infancy, VERTEX COVER being one of the rare examples with some practical experiences. Usually a lot of algorithm engineering is necessary to turn a theoretically efficient algorithm into a practically useful tool. In particular, it is foreseeable that the theoretically best search tree algorithms for VERTEX COVER, which are based on highly complicated case distinctions, need to be simplified for practical applications. The administrative (and intellectual) overhead caused by these fine-grained case distinctions does not seem to pay off in practice. Hence the question of "re-engineering" these case distinctions as much as possible arises where the challenge is to simplify the case distinctions as much as possible and, at the same time, to deteriorate the (worst-case) bounds on the search tree size as little as possible. An in a sense "orthogonal" effort is to investigate how far the development of complicated case distinctions as employed in the case of VERTEX COVER search trees can be mechanized. Initial results have shown that several case distinctions can be led back to relatively few and simple core concepts in the sense that, together with "limited exhaustive search", they can be automatically generated by feeding these core concepts into a program computing favorable branching strategies.

We close with a subtle point. It is still comparatively easy in the academic setting to get fixed-parameter algorithms implemented and to test them on some "toy examples" such as given by random graphs. It may become a significantly more time-consuming and challenging task to experiment with "real data", that is, graphs derived from real-world applications together with solving practically relevant problems. Observe that, especially for exponential-time algorithms such as those we deal with, final "fine tuning" is a must in order to make them run as fast as possible for the particular application at hand. Also, there should be a serious comparison with different approaches that may also exist (especially with those of "practical people"). The task of actually finding, for at first sight a more academic problem such as VERTEX COVER (although there are nice textbook examples of applications), a real-life, useful, industrial-level application is a challenge of great importance. This topic deserves still more attention than it already has, although first steps in this direction have been taken. Another issue of increasing importance is to study the potential benefits of making *FPT* techniques and algorithms work on parallel machines. For instance, search-tree based algorithms should be relatively easy to parallelize—distributing the load among several processors by assigning them different parts of the search tree is an obvious and promising thing to do. Research into new parallel *FPT* methods is in its infancy, though.

4.7 Using vertex cover structure for other problems

As VERTEX COVER is a well-investigated parameterized problem with efficient fixed-parameter algorithms, the question arises whether this can help in solving other hard (parameterized) graph problems. For instance, by definition it trivially follows that every vertex cover set is also a dominating set, that is, a set S of vertices such that every other graph vertex has at least one neighbor in S. Hence, "vertex cover size" bounds "dominating set size" from above. A set being an optimal vertex cover may be far from being an optimal dominating set, though: for example, a complete graph with n vertices has a minimum vertex cover of size $n-1$ but a minimum dominating set of size 1.

Using "vertex cover structure", however, seems more useful for the "dual" problem of Dominating Set, the so-called Nonblocker:

Input: A graph $G = (V, E)$ and a nonnegative integer k.
Task: Find a subset of vertices $S \subseteq V$ with k or more vertices such that each vertex in S has at least one neighbor in $V \setminus S$.

The set S is called *nonblocking.* Clearly, Nonblocker is NP-complete. The basic idea to solve Nonblocker using vertex cover structure is as follows. Herein, note that in a graph with no isolated vertices every minimal (not necessarily minimum) vertex cover is a nonblocking set as well.

1. Compute a minimal vertex cover S of G.
2. Distinguish two cases.
   (a) If $|S| \geq k$, then S is a desired nonblocking set.
   (b) If $|S| < k$, then the underlying graph G has a particular structure that is easily accessible to dynamic programming: it is "path-like"; more specifically, its so-called "pathwidth" is bounded above by $k-1$.

The pathwidth of a graph is directly related to the concept of path decomposition and, more generally, tree decomposition. Here, we just note that a path decomposition of pathwidth at most $k-1$ can be constructed using the vertex cover structure by forming the sets

$$S \cup \{v_1\}, S \cup \{v_2\}, \ldots, S \cup \{v_n\},$$

where $V = \{v_1, v_2, \ldots, v_n\}$. This path decomposition allows a dynamic programming technique to solve Nonblocker in $O(4^k \cdot n)$ time. We refer to Chapter 9 for the definitions and methods related to path and tree decompositions of graphs.

In summary, with the help of vertex cover structure we either directly obtain a solution for Nonblocker or are helped to build an efficient fixed-parameter algorithm with respect to parameter k solving Nonblocker.

Further examples of using vertex cover structure are known, including the problems

- to determine whether an n-vertex graph can be colored using at most $n-k$ colors, the parameter being k; and
- to determine whether a graph has a spanning tree with at least k internal vertices.

We refer to the the bibliographical remarks for more details.

We close with two remarks. First, note that all the above problems in which vertex cover structure was applied are derived from maximization problems with respect to parameter k. It would be interesting to see applications for minimization problems, where small parameter values seem more common. Second, the above scenario is not limited to vertex cover structure; something like "dominating set structure" etc. might be interesting as well. Such investigations are only at the very beginning and should be extended in future research.

4.8 Exercises

1. Find a search tree smaller than 2^k for VERTEX COVER. Bound its size from above.
2. Give a simple parameterized reduction from VERTEX COVER to MULTICUT IN TREES. Note that it suffices to use MULTICUT IN TREES restricted to trees that form a star, that is, trees with maximal path length two.
3. Consider a graph that can be transformed into a tree by deleting k edges. Assume that this set of edges is given. Give a fixed-parameter algorithm for VERTEX COVER restricted to this class of graphs, parameterized by k.
4. What is the maximum size that an optimal vertex cover set may have in an n-vertex planar graph?
5. Show that INDEPENDENT SET restricted to planar graphs is fixed-parameter tractable with respect to the parameter "size of the independent set sought".
6. Consider VERTEX COVER with positive integer weights on the vertices. Show how this general version can be solved using an algorithm for (unweighted) VERTEX COVER.
7. Given a bipartite graph $G = (V_1, V_2, E)$. Design a fixed-parameter algorithm that finds out whether G has a vertex cover that consists of k_1 vertices from V_1 and k_2 vertices from V_2. The running time of your algorithm should not exceed $1.5^{k_1+k_2} \cdot n^{O(1)}$.

4.9 Bibliographical remarks

For decades, the VERTEX COVER problem has played a central role in computational complexity theory and combinatorial optimization (Crescenzi and Kann, 1998). It was among the first *NP*-complete problems (Garey and Johnson, 1979). Downey and Fellows (1999) describe the history in the development of fixed-parameter algorithms for VERTEX COVER. Till the present day, VERTEX COVER has remained the problem which has drawn the highest level of attention in the development of efficient fixed-parameter algorithms. The now trivial search tree of size 2^k for VERTEX COVER derives directly from Mehlhorn (1984, p. 216); for some time, authors have been unaware of this fact, for instance, see Papadimitriou and Yannakakis (1996). Bounds for search tree size currently below 1.28^k appear in Chandran and Grandoni (2005), Chen *et al.* (2001), Chen *et al.* (2005), and Niedermeier and Rossmanith (2003*b*).

The fundamental theorem giving the size-$2k$ problem kernel for VERTEX COVER first appeared in Nemhauser and Trotter (1975); alternative proofs and descriptions are given in Bar-Yehuda and Even (1985) and Khuller (2002). The usefulness of the Nemhauser–Trotter theorem for parameterized complexity—originally used in the approximation algorithms field—was first pointed out in Chen *et al.* (2001). A new problem kernelization technique for VERTEX COVER called "crown rules" is described in Abu-Khzam *et al.* (2004), Fellows (2003*a*), and Langston and Suters (2005).

For the *W*[*1*]-completeness of INDEPENDENT SET, the dual problem of VERTEX COVER, refer to Downey and Fellows (1999). The four-color theorem for planar graphs is due to Appel and Haken (1977*a*) and Appel and Haken (1977*b*); refer to Robertson *et al.* (1997) for later improvements. The polynomial-time algorithm to four-color a graph is given in Robertson *et al.* (1996). The idea of parameterizing above guaranteed values was put forward in Mahajan and Raman (1999), there restricted to the problems MAXIMUM SATISFIABILITY and MAXIMUM CUT.

How to approximate WEIGHTED VERTEX COVER with a ratio 2 is shown in Bar-Yehuda and Even (1981). This can be (asymptotically) improved to an approximation ratio $2 - \log\log|V|/(2\log|V|)$ (Bar-Yehuda and Even, 1985; Monien and Speckenmeyer, 1985). Baker (1994) contains often cited PTAS results for VERTEX COVER and several other problems restricted to planar graphs.

Theoretical work on $c^{\sqrt{k}}$-algorithms for VERTEX COVER on planar graphs appears in Alber *et al.* (2003) and Alber *et al.* (2004)—accompanying empirical studies appear in Alber *et al.* (2005*a*).

The fixed-parameter complexity of WEIGHTED VERTEX COVER is studied in Niedermeier and Rossmanith (2003*b*). The approximability of CAPACITATED VERTEX COVER is dealt with in Guha *et al.* (2003) and its fixed-parameter tractability is shown in Guo *et al.* (2005*b*). The 3-HITTING SET problem is the subject of Niedermeier and Rossmanith (2003*a*). Other nontrivial generalizations of VERTEX COVER are explored in Nishimura *et al.* (2001).

The algorithm counting VERTEX COVER solutions is due to Peter Rossmanith [personal communication]. Parameterized counting algorithms are also studied in Arvind and Raman (2002). More general considerations concerning counting appear in Flum and Grohe (2004*b*) and McCartin (2002). The applicability of the Nemhauser–Trotter theorem for enumerating VERTEX COVERS is shown in Chlebík and Chlebíková (2004). Damaschke (2004) proposes a way to obtain a more compact description of all vertex covers of size at most k with combinatorial explosion $O(1.8^k)$.

The cited lower bound result for VERTEX COVER can be found in Cai and Juedes (2003); it is closely related to earlier work in Impagliazzo *et al.* (2001). The deterministic worst-case bound for 3-SATISFIABILITY stems from Dantsin *et al.* (2002).

Cheetham *et al.* (2003) investigates the parallelization of fixed parameter algorithms for VERTEX COVER. Prospective work in this direction is also done by Abu-Khzam *et al.* (2005).

Empirical tests and findings for VERTEX COVER (partially on planar graphs) appear in Abu-Khzam *et al.* (2004), Alber *et al.* (2005*a*), and Cheetham *et al.*

(2003). First approaches to automate the generation of search tree algorithms (so far not including VERTEX COVER, but the basic concept clearly applies to VERTEX COVER as well) based on case distinctions appear in Fedin and Kulikov (2004) and Gramm *et al.* (2004).

The example of using vertex cover structure to solve NONBLOCKER is attributed to Faisal Abu-Khzam in Fellows (2003*a*) and Prieto and Sloper (2003). Other applications for using vertex cover structure are also discussed there. The graph coloring application is explored in detail in Chor *et al.* (2004). The dynamic programming algorithm to solve DOMINATING SET, and thus NONBLOCKER, is developed in Alber *et al.* (2002) and Alber and Niedermeier (2002). The corresponding issues will be further explored in Chapter 9.

5

THE ART OF PROBLEM PARAMETERIZATION

For VERTEX COVER we have already seen in Section 4.1 that there is usually more than one natural way to parameterize the considered problem. As we will now see, it cannot be taken for granted that we will find the "right" parameterization for a problem. As a matter of fact, several reasonable and "equally valid" parameters to choose may exist, and which parameterization is to be preferred often depends on the application or on additional knowledge about the problem—if any is to be preferred at all. In what follows, we try to collect a sort of check list of questions to ask when confronted with a new problem where the goal is to solve it exactly by means of a fixed-parameter algorithm using a suitable parameterization.

5.1 Parameter really small?

Recalling VERTEX COVER in Section 4.1, the parameter "size of the vertex cover set" appeared as a natural choice, and, for general graphs, it seems plausible in many cases to assume that the size of the vertex cover set is significantly smaller than the total number of graph vertices. Hence this can be considered as a useful parameterization of VERTEX COVER. The situation definitely changes when we turn our attention to VERTEX COVER restricted to planar graphs. Observing the fact that planar graphs with n vertices have at most $3n - 6$ edges and an average vertex degree of less than six, it is no longer clear that the vertex cover sizes for planar graphs are really "small" compared with the total number of graph vertices. And, indeed, in experiments with "combinatorial random" planar graphs one may observe that minimum size vertex covers often contain about half of all graph vertices. Thus the consideration of other parameterizations, such as that by treewidth (see Section 4.1 and Part II) which refer to structural properties of the input graph, deserves attention.

For a second example, recall MULTICUT IN TREES from Section 1.3.

> **Input:** An undirected tree $T = (V, E)$, $n := |V|$, and a collection H of m pairs of nodes in V, $H = \{(u_i, v_i) \mid u_i, v_i \in V, u_i \neq v_i, 1 \leq i \leq m\}$.
> **Task:** Find a minimum size subset E' of E such that the removal of the edges in E' separates each pair of nodes in H.

There, we considered $k := |E'|$ as the parameter. One may expect k to be particularly small when there exist edges in the tree which are passed by many demand paths. Then a simple fixed-parameter algorithm with combinatorial explosion $O(2^k)$ as described in Section 1.3 can be perfectly all right. If the maximum number of demand paths passing through a tree node is small, however, this leads to another parameterization that might be superior to the one by parameter k as above.

So, let us consider the maximum number of paths passing through a node of the given tree. This measures an input property which is basically independent of the solution size $|E'|$. By means of dynamic programming, it can be shown that MULTICUT IN TREES is fixed-parameter tractable with respect to this parameter $d :=$ "maximum path number"—the combinatorial explosion amounts to 3^d. It is worth noting here that with respect to parameter d *edge-weighted* MULTICUT IN TREES remains fixed-parameter tractable, whereas this is open when considering parameter k.

Finally, perhaps particularly concerning maximization problems, it often appears reasonable to study *dual parameterizations*—for instance, if k is the optimization parameter (such as the size of an independent set sought), then the dual parameter k', where then $n - k'$ denotes the size of the solution set and n is an upper bound on its maximum size, might lead to a more fruitful parameterization.

5.2 Guaranteed parameter value?

This issue is closely related to the previous point. Once more reconsider VERTEX COVER restricted to planar graphs. In Section 4.1 we already noted that no minimum vertex cover can contain more than $\lfloor 3n/4 \rfloor$ of all n graph vertices due to the four-color theorem. For the dual problem of VERTEX COVER, INDEPENDENT SET, on planar graphs this means that every maximum independent set contains at least $\lceil n/4 \rceil$ vertices. We have the *guaranteed value* $\lceil n/4 \rceil$. Hence, the at first sight natural parameter "size of the independent set" is not to be considered as really small. There is an alternative parameterization, which makes sense: find an independent set of size $\lceil n/4 \rceil + k$. Unfortunately, the fixed-parameter complexity of INDEPENDENT SET in planar graphs is open with respect to parameter k chosen in this way. This alternative problem formulation for INDEPENDENT SET on planar graphs adheres to the concept of *parameterizing above guaranteed values.*

Two further problems with guaranteed values appear in the context of computational biology. The *NP*-complete MINIMUM QUARTET INCONSISTENCY problem appears in the construction of evolutionary (that is, leaf-labeled binary) trees:

> **Input:** A set S of n taxa and a set Q_S of $\binom{n}{4}$ quartet topologies such that there is exactly one topology for *every* quartet corresponding to S and a nonnegative integer k.
> **Task:** Determine whether there is an evolutionary tree T where the leaves are bijectively labelled by the elements from S such that the set of quartet topologies induced by T differs from Q_S in at most k quartet topologies.

Here, a *quartet* is a size-four subset $\{a, b, c, d\}$ of the set of taxa and the *quartet topology* for $\{a, b, c, d\}$ induced by T is simply the four leaves subtree of T induced by $\{a, b, c, d\}$. So, each quartet topology is simply an (unrooted) tree with four leaves. There is a polynomial-time algorithm that finds an optimal solution for all instances which fulfill $k < (n - 3)/2$. Thus we may define the guaranteed value $(n - 3)/2$.

Finally, the BETWEENNESS problem deals with a finite set of n elements $S = \{x_1, \ldots, x_n\}$ and a finite set of m constraints. Each constraint consists of a triplet $(x_i, x_j, x_k) \in S \times S \times S$. A candidate solution of BETWEENNESS is a total order "$<$" on the elements of S. A total order $x_{i_1} < x_{i_2} < \ldots < x_{i_n}$ *satisfies the constraint* (x_i, x_j, x_k) if either $x_i < x_j < x_k$ or $x_k < x_j < x_i$. That is, each constraint forces the second element x_j to be between the two other elements x_i and x_k, but does not specify the relative order of x_i and x_k. Formally, we arrive at the following definition of BETWEENNESS:

> **Input:** A finite set $S = \{x_1, \ldots, x_n\}$ of n distinct elements and a set of m constraints.
> **Task:** Determine whether there exists a total order of all elements from S such that all constraints are simultaneously satisfied.

The optimization version of BETWEENNESS is to find a total ordering satisfying the maximum number of constraints. The BETWEENNESS problem arises when analyzing certain mapping problems in molecular biology—for example, it occurs when trying to order markers on a chromosome, given the results of a radiation hybrid experiment. It is easy to find an assignment satisfying $\lceil m/3 \rceil$ out of the m constraints. What is the situation with respect to satisfying $\lceil m/3 \rceil + k$ constraints? Is BETWEENNESS with this parameter k fixed-parameter tractable? This question appears to be open.

In summary, guaranteed bounds for parameter values should always—when available—be taken into account when designing fixed-parameter algorithms. The particular difficulty herein is that to prove these bounds can be very hard; and, additionally, it is not always clear whether these bounds are optimal or whether even better guaranteed bounds might exist. An example of such a "problematic case" is MAXIMUM SATISFIABILITY, where one can easily check that *always* at least half of all clauses are satisfiable. In fact, however, even somewhat better guaranteed bounds exist.

5.3 More than one obvious parameterization?

Many problems naturally offer a whole selection of possible parameterizations and some parameterizations may make the design of a fixed-parameter algorithm "easy" and some may make it "hard". Different application settings and different side conditions arising in practice may require different parameterizations.

Consider the *NP*-complete CLOSEST STRING (or, equivalently, CONSENSUS STRING or CENTER STRING) problem. From a "linguistic" point of view, a "closest" string would only mean a string with minimum Hamming distance to the given strings. Beyond that, we use the term "closest" here for a string which has Hamming distance at most d to all given strings:

> **Input:** A set of k strings $s_1, s_2, \ldots, s_k$ over alphabet Σ of length L each, and a nonnegative integer d.
> **Task:** Find a string s such that $d_H(s, s_i) \leq d$ for all $i = 1, \ldots, k$.

Here, $d_H(s, s_i)$ denotes the Hamming distance between strings s and s_i. Consider the *two* natural parameters of CLOSEST STRING: the maximum Hamming distance d allowed and the number k of given input strings. Under the natural assumption that either d or k or both are (very) small (in particular, in biological applications it is appropriate to assume small d, for example, $d < 10$), it is important to ask whether efficient polynomial-time or, even better, linear-time algorithms are possible when d or k are constants. Put in slightly more general terms, this is the question of the fixed-parameter tractability of these problems. CLOSEST STRING can be solved in $O(d^d \cdot kd + kL)$ time, yielding a linear time search tree algorithm for constant d. Using an *integer linear program* formulation, one can observe that the problem is also fixed-parameter tractable with respect to k—the exponential term in k is huge, though. Thus there are two parameters with completely different fixed-parameter algorithms—the application must determine which one is to be preferred. In this particular case, so far, the parameterization with d seems to be the first choice in most cases.

The situation changes when moving on to generalized (and for applications in computational biology more relevant) versions of CLOSEST STRING, namely CLOSEST SUBSTRING and, further on, DISTINGUISHING SUBSTRING SELECTION. In CLOSEST SUBSTRING, the goal string now only needs to be a *substring* in each of the given strings, thus increasing the combinatorial complexity of the problem. And, indeed, with respect to the parameter k, CLOSEST SUBSTRING turns out to be $W[1]$-hard. For the still more general DISTINGUISHING SUBSTRING SELECTION problem, $W[1]$-hardness has been shown for parameterizations with respect to k and with respect to d. For both CLOSEST SUBSTRING and DISTINGUISHING SUBSTRING SELECTION, the only parameterization known to work in the sense of fixed-parameter tractability (for constant-size alphabet) is with respect to solution string length (simple enumeration of all possibilities); but this has limited applicability in many cases of practical interest.

If one has to design and analyze a fixed-parameter algorithm involving more than one parameter at the same time, things might become more difficult as readily as they might become easier. Two examples, one for each case, follow.

A close relative of VERTEX COVER is CONSTRAINT BIPARTITE VERTEX COVER:

> **Input:** A bipartite graph $G = (V_1, V_2, E)$ and two nonnegative integers k_1 and k_2.
> **Task:** Find two subsets $C_1 \subseteq V_1$ and $C_2 \subseteq V_2$ of sizes $|C_1| \leq k_1$ and $|C_2| \leq k_2$ such that each edge in E has at least one of its endpoints in $C_1 \cup C_2$.

The presence of *two* parameters and two vertex sets makes this problem quite different from the original VERTEX COVER problem. Thus, whereas the classical VERTEX COVER (with only one parameter!) restricted to bipartite graphs is solvable in polynomial time (because it is equivalent to a polynomial-time solvable maximal matching problem), by a reduction from CLIQUE it has been shown that CONSTRAINT BIPARTITE VERTEX COVER is *NP*-complete. The fact that the problem in a sense requires the "minimization with respect to two parameters" seemingly makes the problem significantly harder. It can be solved in time $O(1.40^{k_1+k_2} + (k_1 + k_2)n)$, but it is supposed that, due to its different combinatorial structure in comparison with VERTEX COVER, it should be very hard to get an exponential base close to the one there (which is 1.28).

On the contrary, looking back to CLOSEST STRING, it is clear that, in principle, it is easier to design an algorithm with running time $f_1(k,d) \cdot n^{O(1)}$ than to give one with running time only $f_2(k) \cdot n^{O(1)}$ or $f_3(d) \cdot n^{O(1)}$ with arbitrarily growing functions f_1, f_2, and f_3. So, to the best of the author's knowledge, for CLOSEST STRING only a size $k \cdot d$ *problem kernel* but no problem kernel with size only depending on d or k alone is known.

5.4 Close to "trivial" problem instances?

The *NP*-complete GRAPH COLORING problem is our first example of "parameterizing away from triviality":

> **Input:** A graph $G = (V, E)$ and a nonnegative integer k.
> **Task:** Determine whether there is a way to assign each graph vertex one of k possible colors such that no pair of adjacent vertices has the same color.

Note that in its decision version GRAPH COLORING is already *NP*-complete for $k = 3$ colors, so the parameterization by k is of no help. We mention in passing that the dual parameterization by asking whether an n-vertex graph can be colored with at most $n - k$ colors (the parameter is k) was shown to be fixed-parameter tractable. Coming back to the original version of GRAPH COLORING, for instance, restricted to the class of split graphs[5] (where GRAPH COLORING is known to be polynomial-time solvable) GRAPH COLORING is fixed-parameter tractable with respect to parameter ℓ on graphs that are formed from split graphs by adding or deleting at most ℓ *edges*. Thus, split graphs form the trivial case. By way of contrast, GRAPH COLORING is $W[1]$-hard with respect to parameter ℓ when deletion of at most ℓ *vertices* leads to a split graph. Interestingly, the problem is much harder in case of (the trivially 2-colorable) bipartite graphs instead of split graphs: GRAPH COLORING becomes *NP*-complete for graphs that originate from bipartite graphs by adding three edges or if two vertex deletions are needed to make a graph bipartite. In summary, the prospective idea

[5] A graph is called a *split graph* if its vertex set can be partitioned into two sets, one inducing a clique and the other inducing an independent set.

brought forward here is that the distance parameter ℓ measures a distance from a "triviality": $\ell = 0$ represents a special case that is solvable in polynomial time.

A second example of such a "distance from triviality" parameterization is given by studying exact solutions for the *NP*-complete TRAVELING SALESPERSON problem in the two-dimensional Euclidean plane:

> **Input:** A set of n points in the two-dimensional Euclidean plane with pairwise Euclidean distances.
> **Task:** Determine a shortest round-trip through all points, visiting each point once.

A trivial problem instance can be determined as follows. Consider a set of n points in the Euclidean plane. Determine their convex hull, which can be done in $O(n \log n)$ time by standard methods. If all points lie on the hull, then this gives the shortest tour. That is, this describes a trivial problem instance. It can be shown that the problem is solvable in $O(2^k \cdot k^2 \cdot n)$ time, where k denotes the number of points inside the convex hull. Thus the distance from triviality here is the number k of inner points.

Finally, we give an example related to SATISFIABILITY. Assume that a formula in conjunctive normal form has a *matching* between variables and clauses that matches all clauses. More precisely, this means that each clause is injectively mapped to one variable also appearing in this clause. Then, it is easy to observe that such a formula is satisfiable. Now, for a formula F, being a set of m clauses over n variables, define its *deficiency* as $\delta(F) := m - n$. The *maximum deficiency* then is

$$\delta^*(F) := \max_{F' \subseteq F} \delta(F').$$

Here, it can be proven that the satisfiability of a formula F can be decided in $O(2^{\delta^*(F)} \cdot n^3)$ time. Note that a formula F with $\delta^*(F) = 0$ has a matching as described above. Again, $\delta^*(F)$ is a structural parameter measuring the distance from trivial problem instances which are easily solvable.

We close with a remark concerning the already several times mentioned treewidth of graphs. A graph of treewidth 1 is a tree. Many otherwise hard graph problems such as VERTEX COVER or DOMINATING SET are linear-time solvable in trees. Thus, parameter "treewidth" measures the closeness to the trivial problem instance trees and, in this sense, is the most prominent example of a "distance from triviality" parameterization.

In conclusion, the examples given, taken together, show the versatility of the proposed parameterization methodology to choose as a parameter a certain distance measure that reflects the distance between simple (polynomial-time solvable) and hard problem instances and how it may open new views on well-known hard problems.

5.5 Exercises

1. Show that in any Boolean formula in conjunctive normal form at least half of the clauses can always be satisfied.
2. Design simple linear-time algorithms that solve VERTEX COVER and DOMINATING SET in trees.
3. MAXIMUM SATISFIABILITY is fixed-parameter tractable with respect to the number of clauses k to be satisfied.
   (a) Is this a reasonable parameterization; that is, may we expect small parameter values?
   (b) Is MAXIMUM SATISFIABILITY also fixed-parameter tractable with respect to parameter k' when we ask whether at least $\lceil m/2 \rceil + k'$ clauses can be satisfied, where m denotes the total number of clauses?
4. Consider the CLOSEST STRING problem.
   (a) Why is the modification where not the maximum distance but the sum of distances shall be "minimized" trivial?
   (b) Show that the following instance with $k = 4$, $d = 2$, and $L = 15$ has no solution: $AGCATAGCATTAATA$, $ATCATAGAATCAATA$, $AGCTTAGCATTAAAT$, $AGCATAACGTTTATA$.
   (c) Show that the following input instance with $k = 5$, $d = 3$, and $L = 10$ has a solution: $AACTTAGGTT$, $GGCCTCAGAT$, $TGCATCGGAC$, $AGCTTGAAAT$, $AGATTCGGGT$.
5. Consider the GRAPH COLORING problem.
   (a) Show that GRAPH COLORING is trivial for bipartite graphs. Describe a coloring algorithm.
   (b) Give a simple recursive algorithm that colors a planar graph with six colors. Hint: in a planar graph there always exists a vertex of degree at most five.
6. Detect "trivial instances" for VERTEX COVER, DOMINATING SET, CLOSEST STRING, MULTICUT IN TREES, etc. Find natural "distance from triviality" parameterizations for each of them.

5.6 Bibliographical remarks

The alternative parameterization for MULTICUT IN TREES has been studied in Guo and Niedermeier (2005*b*).

Mahajan and Raman (1999) first introduced the parameterization above guaranteed values, applying it to the problems MAXIMUM SATISFIABILITY and MAXIMUM CUT. The parameterized complexity of MINIMUM QUARTET INCONSISTENCY is investigated in Gramm and Niedermeier (2003). Its *NP*-completeness follows from Berry *et al.* (1999) and Jiang *et al.* (2000). The quartet cleaning algorithm from Berry *et al.* (1999) finds an optimal solution for input instances with $k < (n-3)/2$. The BETWEENNESS problem derives from Opatrny (1979) and Chor and Sudan (1998). Parameterization above a guaranteed value was proposed in personal communication by Benny Chor from Tel Aviv.

The *NP*-completeness of CLOSEST STRING is shown in Frances and Litman (1997). Its parameterized complexity has been investigated in Gramm *et al.* (2003*b*), presenting both discussed fixed-parameter algorithms. Some more background on biological applications of CLOSEST STRING is given in Lanctot *et al.* (2003). The generalizations CLOSEST SUBSTRING and DISTINGUISHING SUBSTRING SELECTION are investigated in Fellows *et al.* (2002), Marx (2005), and Gramm *et al.* (2003). The *NP*-completeness of CONSTRAINT BIPARTITE VERTEX COVER is shown in Kuo and Fuchs (1987). The corresponding fixed-parameter search-tree algorithm is developed in Fernau and Niedermeier (2001).

Cai (2003) initiated the study of parameterized GRAPH COLORING in the sense of "distance from triviality" parameterization (although it was not called this). The parameter "number of inner points" for the *NP*-complete TRAVELING SALESMAN problem in the Euclidean plane appears in Deĭneko *et al.* (2004). Szeider (2004*a*) studied the "maximum deficiency" parameterization of SATISFIABILITY. The general topic of "parameterizing by distance from triviality" has been proposed in Guo *et al.* (2004). Kloks (1994) is a monograph on the treewidth of graphs.

6

SUMMARY AND CONCLUDING REMARKS

The systematic study of parameterized complexity analysis is still a young and fast-growing field. Nevertheless, it has already contributed significant new concepts and methods to algorithms and complexity both in theory and practice. As is to be expected, it is not a miracle cure for all forms of computational intractability, but it clearly has its pros and cons. When compared to the more established fields of polynomial-time approximation and heuristic algorithms, fixed-parameter complexity analysis carries enough potential and challenges to deserve a still larger share of the attention in the research community than it receives today.

At this point, let us once more emphasize the numerous possibilities, but also decisions to be made, when attacking a problem the fixed-parameter way. The wealth of opportunities concerning parameterization, in some sense, also makes things more complicated. That is, generally one probably may not have *the* fixed-parameter algorithm for a problem because, at the same time, several alternative parameterizations of the same problem may make sense, the various options not being comparable with each others. This stands in contrast to approximation algorithms where the approximation ratio is tied to the optimization goal, whereas the value to optimize usually is only *one* possible parameterization of the given problem—consider CLOSEST STRING in Section 5.3 where the distance value d is the optimization goal but k, the number of input strings, yields an equally reasonable parameterization. Perhaps this also makes it more difficult to obtain a clean, unified (complexity) theory covering everything.

Coping with computational intractability is not an easy thing to do. A flexible response, as offered by parameterized complexity, is a worthwhile opportunity to take into consideration. Understanding the multi-dimensionality and technical difficulties of parameterized complexity as a valuable challenge for ongoing and future research is highly recommended.

Part II

Algorithmic Methods

This is a book about algorithms. Hence algorithm design and analysis techniques form its core part. So, this part of the book, which is by far the longest, tries to survey currently known fundamental methods and techniques used in the development of fixed-parameter algorithms. It provides a toolbox when trying to prove new or improved fixed-parameter tractability results. Let us illustrate this with a concrete example.

Recall our favorite problem VERTEX COVER.

> **Input:** A graph $G = (V, E)$ and a nonnegative integer k.
> **Task:** Find a subset $C \subseteq V$ with k or fewer vertices such that each edge in E has at least one of its endpoints in C.

In Chapter 7 on data reduction and problem kernels we will see simple as well as involved data reduction techniques that can be used to "cut away" easy parts of the input instance, leaving just the problem kernel, which can be considered as the basic cause of the problem's hardness. A problem kernel for VERTEX COVER with only $2k$ vertices can always be achieved in this way. In Chapter 8 we will describe simple and—based on case distinctions—complicated depth-bounded search tree strategies that provide fixed-parameter algorithms solving VERTEX COVER with a combinatorial explosion of c^k with $c < 2$. Dynamic programming, as described in Chapter 9, is another fruitful technique in parameterized algorithm design—in the case of VERTEX COVER it helps to shrinking the search tree size further. VERTEX COVER can be solved efficiently for graphs of bounded treewidth—a deep and central topic of algorithmic graph theory that we will explore to some extent in Chapter 10. Finally, in Chapter 11 we will encounter various further techniques that have mostly arisen very recently and whose development in a sense goes hand in hand with the advancement of fixed-parameter studies for tackling hard combinatorial problems. Here, VERTEX COVER appears when illustrating the prospective iterative compression technique.

Clearly, VERTEX COVER cannot serve as a running example of all the features and particularities of the algorithmic techniques to be described. Nevertheless, in order to try to avoid any form of "over-formalization" we always study concrete algorithmic problems when introducing and discussing these techniques.

Probably the three most elementary and best understood techniques in fixed-parameter algorithmics so far are:

- data reduction and problem kernels (Chapter 7);
- depth-bounded search trees (Chapter 8); and
- dynamic programming (Chapter 9).

Each of these has a number of aspects and tricks that appear differently in context with various problems. Hence each of the corresponding chapters contains applications to several example problems highlighting different aspects. Chapter 10, however, is somewhat different in the sense that it focuses on one single concept—tree decompositions of graphs—which is more demanding and needs more space to be reviewed in at least some depth. Two central aspects to be dealt with here are the construction of tree decompositions of small "width" and their use in solving graph problems. The latter is again based on dynamic programming algorithms whose time and space consumption are both exponential with respect to the width of the given tree decomposition. Finally, Chapter 11 gives an overview of

- color-coding;
- integer linear programming;
- iterative compression;
- greedy localization; and
- graph minor theory,

as further advanced and valuable tools for deciding on fixed-parameter tractability.

Many ideas described in this part are fairly new and/or have been explored only to some limited extent. We hope that it serves as a springboard into discovering new applications and techniques that further advance the field of algorithmics for computationally hard problems.

7

DATA REDUCTION AND PROBLEM KERNELS

Every phone book is sorted. In a sorted structure, by using binary search we can find items in logarithmic time. Sorting is the fundamental algorithmic problem in computer science. Indeed, "when in doubt, sort" is considered as one of the first rules of algorithm design. Thus sorting is a prime example of simplifying or, in this case more specifically, structuring the input by efficient preprocessing. In this chapter we will present a special form of data simplification. Whereas in the case of sorting the simplification of the input consists in restructuring it to make it sorted, the focus in the parameterized complexity context is on simplifying the input by shrinking it in size.

Whereas the sorting problem is a computationally feasible problem, we focus on "intractable" problems. If a computationally hard problem must be solved in practice, one of the first things usually done is to try to perform a reduction of the size of the input data. Many input instances consist of some parts that are relatively easy to cope with and other parts that form the "really hard" core of the problem. Hence, before starting a cost-intensive algorithm to solve the difficult problem, a polynomial-time preprocessing phase is executed in order to shrink the given input data as much (and as fast) as possible.

In Part I we have already seen two examples of such data reduction techniques that can be executed in polynomial time. In Section 1.2, considering a problem from railway optimization, the two simple reductions given by the Train Rule and the Station Rule were presented as efficient (polynomial-time) preprocessing rules with enormous success in empirical studies. A theoretical confirmation of the strength of these rules is still missing today, though. In Section 1.3, considering the communication network problem MULTICUT IN TREES, we discussed several simple data reduction rules. In this case, by way of contrast, the "quality" of data reduction can actually be proven by showing a size-bounded "problem kernel"—this form of "guaranteed quality of data reduction" will be the core concept of this chapter.

Observe the "universal importance" of data reduction by preprocessing. It is not only a ubiquitous topic for the design of efficient fixed-parameter algorithms, but it is of importance for basically *any* method (such as approximation or purely heuristic algorithms) that tries to cope with hard problems. It is important to emphasize, however, that the use of data reduction techniques is not restricted to a preprocessing phase only. On the contrary, there is empirical as well as theoretical evidence that it is beneficial to combine or interleave data reduction techniques with the "main algorithm" for problem solution, achieving significant speedups in this way. Data reduction is one of the most important techniques in designing fixed-parameter algorithms and is also useful for other paradigms in the field of algorithmics for hard problems.

We start with two concrete examples of data reduction. First, consider CNF-SATISFIABILITY, where one is given a Boolean formula F in conjunctive normal form and the task is to decide whether or not F has a satisfying truth assignment (refer to Section 1.1). Clearly, if there are clauses consisting of only one literal then to satisfy F one must satisfy these "unit-clauses" by setting the value of the corresponding variable accordingly. There is no choice here. This can be accomplished by a simple scan through the formula phase, and it may shrink the original input formula considerably, resulting in a reduced formula.

Second, let us return to our running example VERTEX COVER:

Input: A graph $G = (V, E)$ and a nonnegative integer k.
Task: Find a subset of vertices $C \subseteq V$ with k or fewer vertices such that each edge in E has at least one of its endpoints in C.

It is clearly permissible to remove isolated vertices, that is, vertices with no adjacent edges. Moreover, if one is looking for only *one* optimal vertex cover and *not all* of them, then vertices with only one adjacent edge, and thus one adjacent vertex, can easily be dealt with by putting the neighboring vertex into the cover. This is correct because in order to cover the corresponding edge one of the two endpoints *must* be in the vertex cover and a vertex with higher degree has the potential to cover more edges. In the fixed-parameter setting, where we ask for a vertex cover of size at most k, we can further do the following. If there is a vertex of degree at least $k+1$, that is, a vertex with more than k adjacent edges, then, if a vertex cover of size k exists, this particular vertex must be part of it. Otherwise, to cover all its adjacent edges would require all its at least $k + 1$ neighbors, a contradiction. This is known as *Buss's reduction to a problem kernel* for VERTEX COVER. One can easily verify that after performing the above rules, in order to have a VERTEX COVER of size at most k the remaining graph can have at most $k^2 + k$ vertices and at most k^2 edges. For the time being, however, our sole point of interest here is that all three of the above rules (concerning isolated vertices, vertices of degree one, and vertices of degree at least $k + 1$) can be executed in polynomial time. Thus, we may obtain an efficiently executable data reduction. In Section 7.4, we will see that VERTEX COVER even allows for a much more sophisticated and stronger data reduction.

The methodological approach, including various techniques of data reduction or problem kernelization is best learned by a series of concrete examples. This will be the main contents of this chapter. The examples are selected in such a way that it is hoped that they well represent standard techniques that have been employed in this context. So, our extensive discussions are based on diverse problems such as MAXIMUM SATISFIABILITY, CLUSTER EDITING, 3-HITTING SET, VERTEX COVER, and DOMINATING SET IN PLANAR GRAPHS. To facilitate illustration, the majority of the chosen problems are graph problems, although data reduction techniques are of interest and apply to all sorts of hard problems.

In summary, data reduction is a topic with practical importance that cannot be overrated and it belongs as an important key technique in *every* algorithm designer's toolbox. Formal studies of data reduction techniques are still under-represented in the algorithms literature and it will become a growing field of research on its own. In the context of fixed-parameter algorithms, this basically comes down to what is known as reduction to a problem kernel. We formally introduce this notion in the first section.

7.1 Basic definitions and facts

Besides the notion of fixed-parameter tractability itself, the concept of *reduction to a problem kernel* or, equivalently, *kernelization*, leads to the most important definition in this book. It captures data reduction taken from a fixed-parameter complexity point of view. To simplify matters, it is assumed that the parameter is a positive integer. Cases with more than one parameter being a number are each handled analogously.

Definition 7.1 *Let $\mathcal{L}$ be a parameterized problem, that is, $\mathcal{L}$ consists of input pairs (I, k), where I is the problem instance and k is the parameter.* Reduction to a problem kernel *then means to replace instance (I, k) by a "reduced" instance (I', k') (called* problem kernel*) such that*

$$k' \leq k, \qquad |I'| \leq g(k)$$

for some function g only depending on k, and

$$(I, k) \in \mathcal{L} \text{ iff } (I', k') \in \mathcal{L}.$$

Furthermore, the reduction from (I, k) to (I', k') must be computable in polynomial time $T_K(|I|, k)$. The function $g(k)$ is called the size *of the problem kernel.*

Clearly, the aforementioned kernelization for VERTEX COVER due to Buss fits into the given definition.

Two remarks:

- Definition 7.1 requires that $k' \leq k$. In principle, it could even be allowed that $k' = f(k)$ for some arbitrary function f. We are not aware of a concrete, natural parameterized problem with a problem kernel where $k' > k$.
- This remark concerns the type of parameterizations for which we study reduction to a problem kernel. All kernelizations we are aware of refer to parameterized problems where the parameter reflects the value to be optimized in the corresponding optimization problem.

To further substantiate and to state more precisely the last remark, recall the problem MULTICUT IN TREES introduced in Section 1.3:

Input: An undirected tree $T = (V, E)$, $n := |V|$, and a collection H of m pairs of vertices in V, $H = \{(u_i, v_i) \mid u_i, v_i \in V, u_i \neq v_i, 1 \leq i \leq m\}$.
Task: Find a minimum size subset E' of E such that the removal of the edges in E' separates each pair of vertices in H.

In Section 1.3 we discussed the parameterization by $k := |E'|$. In Section 5.1 we briefly introduced another parameterization, namely the maximum number d of demand paths passing through a tree node. By employing dynamic programming techniques also with respect to parameter d we can achieve fixed-parameter tractability. For parameter k, however, with some technical expenditure a problem kernel can be proven. By way of contrast, we see no point in searching for a problem kernel with respect to d. Note that k corresponds to the optimization value in MULTICUT IN TREES, whereas d corresponds to a "structural property" of the input which is basically unrelated to the optimization value. Hence it is not conceivable to get a reduced instance with an upper bound on its size that depends only on the parameter d because arbitrarily large instances with small d but a large solution set exist.

For computationally hard problems, the best one can hope for is that a problem kernel has size linear in k, a so-called *linear problem kernel*. According to Definition 7.1, this formally means that, for a given problem parameter k, the reduced instance I' is of size $O(k)$. For graph problems, however, this is often confused with the property that the reduced graph $G' = (V', E')$ fulfills $|V'| = O(k)$, whereas the total instance size also including the number of edges still may be $O(k^2)$. Nevertheless, one frequently terms this a linear problem kernel. A concrete example of this is given with the problem kernel for VERTEX COVER that contains only $2k$ vertices; see Section 7.4.

One can classify kernelization algorithms or, more specifically, data reduction rules, into two types:

- *parameter-independent* and
- *parameter-dependent* ones.

The distinguishing factor is whether or not the kernelization makes explicit use of the parameter value. For instance, consider VERTEX COVER. The notion of "high-degree vertices" as employed in the kernelization attributed to Buss is parameter-dependent because it needs to know the value of parameter k. By way of contrast, the rule which simply put the neighbor of a degree-one vertex into the vertex cover does not need to know the value of k; it is parameter-independent. Clearly, "parameter-independence" is preferable, but it seems hard to achieve in all cases as we will see in the case studies to follow. For parameter-dependent rules, to find a solution with *optimal* parameter value one has to try several candidates for parameter k in a systematic fashion. For the time being, just note that currently the only known kernelization of MULTICUT IN TREES is based on a mixture of parameter-dependent and parameter-independent data reduction rules.

To get further acquainted also with the pitfalls of problem kernels, let us now have a brief look at a somewhat strange kind of problem kernelization. Consider INDEPENDENT SET IN PLANAR GRAPHS:

> **Input:** A planar graph $G = (V, E)$ and a nonnegative integer k.
> **Task:** Find a subset of vertices $I \subseteq V$ with k or more vertices that form an independent set, that is, I induces an edgeless subgraph of G.

INDEPENDENT SET IN PLANAR GRAPHS has a problem kernel consisting of only $4k$ vertices: due to the famous four-color theorem for planar graphs and the corresponding polynomial-time coloring algorithm one can color the vertices of a given planar graph with four colors such that no two neighboring vertices possess the same color. Hence each "color class" forms an independent set of the graph, and, since we need only four color classes, one of them must contain at least one fourth of all vertices.

Thus the reduction to a problem kernel may simply proceed as follows: if for a given planar graph with n vertices and parameter k it holds that $k \leq \lceil n/4 \rceil$, then answer "yes" and produce an independent set using the polynomial-time coloring algorithm. If $k > \lceil n/4 \rceil$, then $n < 4k$ and, voilà, we have a problem kernel with $4k$ vertices. On the one hand, in a way, this kind of kernelization is not satisfactory because in the second case we did not reduce the size of the input graph at all, but simply made the indirect observation that it must have a big (!) independent set that contains at least one quarter of all graph vertices. This contradicts the common assumption of parameters being "small" and directs our attention to the now more natural parameterization above the guaranteed value $\lceil n/4 \rceil$, see Section 5.2. On the other hand, this example also shows that there may be deep theory behind the construction of problem kernels: here, the famous four-color theorem and the corresponding intricate coloring algorithm.

This problem kernelization for INDEPENDENT SET IN PLANAR GRAPHS based on the four-color theorem also has a "global flavor", whereas the majority of known reductions to a problem kernel work by using "local rules". More specifically, the corresponding data reduction rules—we usually need more than one single rule to prove a problem kernel—explore local structures within an input instance. Based exclusively on this, these rules decide whether the input can be be simplified (or, equivalently, reduced) here. For instance, in the case of VERTEX COVER such a local structure may be a vertex together with all its neighbors. If it has only one neighbor, a very simple data reduction rule says that we should put its neighbor into the vertex cover and remove it from the given instance together with all its incident edges, no matter what the remaining graph looks like.

Finally, let us mention in passing that in parameterized complexity theory it has become a commonplace that "every fixed-parameter tractable problem is kernelizable". To begin with, note that it is obvious that if there is a reduction to a problem kernel for a decidable parameterized problem, then it is fixed-parameter tractable: simply perform a brute-force search algorithm on the remaining problem kernel. The opposite direction is a little less obvious: assume that the given fixed-parameter algorithm has running time $f(k) \cdot n^c$ for some positive constant c. The idea is to run this algorithm on the problem for at most n^{c+1} steps and then to consider the two cases that either the algorithm has finished its task within that time or it has not finished. In the first case, we directly obtain a kernelization algorithm running in polynomial time n^{c+1}, which simply outputs either a trivial "no"-instance or a trivial "yes"-instance. In the second case, we can argue that $n < f(k)$. Thus, our problem kernel is the original input instance itself. We can summarize these arguments as follows.

Proposition 7.2 *A decidable parameterized problem $\mathcal{L}$ is fixed-parameter tractable with respect to parameter k iff there exists a reduction to a problem kernel for $\mathcal{L}$ with respect to k.*

Proof Following the discussion preceding Proposition 7.2, it remains to be shown that $n < f(k)$ in the case that the fixed-parameter algorithm has not finished its task within n^{c+1} steps for a positive constant c. If so, this means that

$$n^{c+1} < f(k) \cdot n^c,$$

which yields $n < f(k)$. □

Clearly, Proposition 7.2 is of no direct practical use and gives only a trivial problem kernel with no algorithmic impact. As a matter of experience, one may conclude that, although it is often not hard to find some simple data reduction rules for a fixed-parameter tractable problem, to prove that they really yield a non-trivial problem kernel in the sense of Definition 7.1 frequently becomes a mathematically challenging task—it is worth the effort! We will elaborate several concrete problem kernelizations in the remainder of this chapter.

7.2 Maximum Satisfiability

To start our series of example kernelizations, we present a simple reduction to a problem kernel for the *NP*-complete Maximum Satisfiability (MaxSat) problem (see also Section 1.1):

> **Input:** A Boolean formula in conjunctive normal form consisting of m clauses and a nonnegative integer k.
> **Task:** Find a truth assignment satisfying at least k clauses.

We represent the Boolean values true and false by 1 and 0, respectively. A *truth assignment* I can be defined as a set of literals that contains no pairs of complementary literals. Then for a variable x we have $I(x) = 1$ iff $x \in I$ and $I(x) = 0$ iff $\bar{x} \in I$. We deal only with propositional formulae in conjunctive normal form. These are often represented in *clause form*, that is, as a set of clauses, where a clause is a set of literals. We represent formulae as *multi-sets* of sets since a formula might contain some identical clauses. For the Satisfiability problem (see Section 1.1) multiple clauses can be eliminated, but this is of course no longer true if we are interested in the number of satisfiable clauses. The formula

$$(x \vee y \vee \bar{z}) \wedge (x \vee y \vee \bar{z}) \wedge (\bar{x} \vee z) \wedge (\bar{y} \vee z)$$

will be represented as

$$\{\{x, y, \bar{z}\}, \{x, y, \bar{z}\}, \{\bar{x}, z\}, \{\bar{y}, z\}\}.$$

Note that the outer curly brackets denote the multi-set and the inner curly brackets denote the sets of literals. The *length of a clause* is its cardinality, and the *length of a formula* is the sum of the lengths of its clauses. To simplify presentation, we assume that each clause contains at most one occurrence of a variable x. Note that a clause where both literals x and $\bar{x}$ occur is trivially always satisfied.

Suppose that we are given an input instance for MAXSAT. The first simple observation is that if $k \leq \lceil m/2 \rceil$, then the desired truth assignment trivially exists: take a random truth assignment. If it satisfies at least k clauses then we are done. Otherwise, "flipping" each bit in this truth assignment to its opposite value yields a new truth assignment that now satisfies at least $\lceil m/2 \rceil$ clauses.

Hence from now on we can assume that $k > \lceil m/2 \rceil$, which implies that $m < 2k$. The next observation gives a problem kernel of quadratic size: partition the clauses of the given formula F into long clauses (that is, clauses containing k or more literals) and into short clauses (that is, clauses containing less than k literals). Thus,

$$F = F_l \wedge F_s,$$

where F_l contains all long clauses and F_s contains all short clauses. Let L be the number of long clauses. If $L \geq k$ then again at least k clauses can be satisfied by picking (in the worst case) in each long clause another variable and setting its value accordingly such that the corresponding clause gets satisfied.

If $L < k$, then an important point is that now it is sufficient to focus attention on the new MAXSAT instance $(F_s, k - L)$, which already gives our problem kernel due to the following observations. Note that (F, k) is a yes-instance of MAXSAT iff $(F_s, k - L)$ is a yes-instance of MAXSAT: the same reasoning as in the preceding paragraph shows that the remaining L large clauses can always be satisfied. This is true because to satisfy $k - L$ clauses at most $k - L$ variables (at most one variable per satisfied clause) are needed. Thus, for the L large clauses at least $k - (k - L) = L$ variables remain to be freely set and the claimed equivalence is shown.

It remains to show that the size of the formula F_s (making the reduced problem instance) is bounded from above by $O(k^2)$: we have that there are $m - L \leq m$ small clauses, each containing at most k literals, and we have that $m < 2k$ as explained before. Hence the total number of literal occurrences in F_s is bounded from above by $2k \cdot k = 2k^2$. This means a quadratic-size problem kernel for MAXSAT with respect to parameter k.

Proposition 7.3 MAXIMUM SATISFIABILITY *has a problem kernel of size* $O(k^2)$, *and it can be found in linear time.*

Proof The above described method and its correctness are clear. With regard to the running time, it is easy to see that the determination of the small and large clauses as well as the determination of an assignment satisfying all large clauses can be done in linear time by simply scanning the given formula. □

In summary, we observe that the given problem kernelization for MAXIMUM SATISFIABILITY makes explicit use of parameter value k to perform the partitioning into a small and a large formula; hence the described data reduction is parameter-dependent. Moreover, the reduction to a problem kernel for MAXSAT has the same unsatisfactory flavor as that which INDEPENDENT SET IN PLANAR GRAPHS (see Section 7.1) had. The point is that again it is the use made of the structural property of the input instance which guarantees "beforehand" that at least half of all clauses are always satisfiable. Hence, similar to the case of INDEPENDENT SET IN PLANAR GRAPHS, parameterizing above guaranteed values would probably be more appropriate here (see Section 5.2). This is also discussed in the related literature, see the bibliographical remarks in Section 7.10.

7.3 Cluster Editing

Our next example of problem kernelization deals with a graph modification problem arising in the field of data clustering. Different from the MAXIMUM SATISFIABILITY case, in this new example we work with with mainly *local* data reduction rules. Again, however, the rules are parameter-dependent. The problem definition is based on the notion of a *similarity graph* whose vertices correspond to data elements and in which there is an edge between two vertices iff the similarity of their corresponding elements exceeds a predefined threshold. The goal is to obtain a *cluster graph* by as few edge modifications (that is, edge deletions or edge additions) as possible; a cluster graph is a graph in which each of the connected components is a clique. Thus we arrive at the *NP*-complete edge modification problem CLUSTER EDITING:

> **Input:** A graph $G = (V, E)$ and a nonnegative integer k.
> **Task:** Find out whether we can transform G, by deleting or adding at most k edges, into a graph that consists of a disjoint union of cliques.

In our data reduction rules we employ two annotations for unordered vertex pairs which are defined as follows.

"*permanent*": In this case, $\{u, v\} \in E$ and it is not allowed to delete $\{u, v\}$;
"*forbidden*": In this case, $\{u, v\} \notin E$ and it is not allowed to add $\{u, v\}$.

Note that whenever we delete an edge $\{u, v\}$ from E, we make vertex pair $\{u, v\}$ *forbidden*, since it would not make sense to reintroduce previously deleted edges. In the same way, whenever we add an edge $\{u, v\}$ to E, we make $\{u, v\}$ *permanent*.

Although the following data reduction rules also *add* edges to the graph, we consider the resulting instances as *simplified*. The reason is that for every added edge, the parameter is decreased by one. In the following rules, it is implicitly assumed that, whenever an edge is added or deleted, parameter k is decreased by one. In the formulation of our rules, we use the following terminology. Given a graph $G = (V, E)$ and a vertex pair $v_i, v_j \in V$, we use *common neighbor* of v_i and v_j to refer to a vertex $z \in V$ with $\{z, v_i\} \in E$ and $\{z, v_j\} \in E$. Similarly, a *non-common neighbor* of v_i and v_j is a vertex $z \in V$ with $z \neq v_i$ and $z \neq v_j$ such that either $\{z, v_i\} \in E$ or $\{z, v_j\} \in E$ but not both.

Rule 1 For every pair of vertices $u, v \in V$:

1. If u and v have more than k common neighbors, then $\{u, v\}$ must be in E and we make the vertex pair $\{u, v\}$ *permanent*. If $\{u, v\} \notin E$, we add it to E.
2. If u and v have more than k non-common neighbors, then $\{u, v\}$ may not be in E and we make $\{u, v\}$ *forbidden*. If $\{u, v\} \in E$, we delete it.
3. If u and v have both more than k common and more than k non-common neighbors, then the given instance has no solution.

Lemma 7.4 *Rule 1 is correct.*

Proof In the following arguments it is useful to employ the easily verified fact (also see Section 8.2) that a graph is a cluster graph iff it contains no induced path of three vertices.
Case 1: Vertices u and v have more than k common neighbors. If we did exclude $\{u, v\}$ from E, then we would have to, for every common neighbor z of u and v, delete $\{u, z\}$, $\{v, z\}$, or both. This, however, would require at least $k + 1$ edge deletions, a contradiction of the fact that at most k edge modifications are allowed.
Case 2: Vertices u and v have more than k non-common neighbors. If we did include $\{u, v\}$ in E, then we would have to, for every non-common neighbor z of u and v, edit one of the edges $\{u, z\}$ and $\{v, z\}$: without loss of generality, let z be a neighbor of u and not a neighbor of v. Then, we would either have to delete $\{u, z\}$ from E or add $\{v, z\}$ to E. With at least $k + 1$ non-common neighbors, this would require at least $k + 1$ edge modifications.
Case 3: Vertices u and v have more than k common neighbors and more than k non-common neighbors. From the proofs for Case 1 and Case 2 it is clear that it would require more than k edge modifications both when including $\{u, v\}$ in E and when excluding $\{u, v\}$ from E. □

Note that Rule 1 applies to every vertex pair $u, v \in V$ for which the number of vertices which are neighbors of u or v (or both) is greater than $2k$.

Rule 2 For every three vertices $u, v, w \in V$:

1. If $\{u, v\}$ has an annotation permanent and $\{u, w\}$ has an annotation *permanent*, then $\{v, w\}$, if not already existing, must be added to E and $\{v, w\}$ is set *permanent*.

2. If $\{u, v\}$ has an annotation *permanent* and $\{u, w\}$ has an annotation *forbidden*, then $\{v, w\}$, if already existing, must be deleted from E and $\{u, w\}$ is set *forbidden*.

The correctness of Rule 2 is obvious. Regarding the running time, one must analyze the interleaved application of Rules 1 and 2 together. Since the analysis is technical and requires the efficient use of additional data structures, we refer to the literature for the proof and just state the result:

Lemma 7.5 *An n-vertex graph can be transformed into a graph which is reduced with respect to Rules 1 and 2 in $O(n^3)$ time.*

We complete our set of reduction rules with the following.

Rule 3 Delete the connected components which are cliques from the graph.

The correctness of Rule 3 is straightforward. Computing the connected components of a graph and checking for cliques can easily be done in linear time. Hence Rule 3 is executable in linear time.

Notably, for the problem kernel size to be shown, Rules 1 and 3 would be sufficient. Rule 2 is also taken into account since it is simple and general and it can be executed in the course of executing Rule 1. Thus, Rules 1 to 3 constitute a small set of data reduction rules yielding a problem kernel with $O(k^2)$ vertices and $O(k^3)$ edges. It is computable in $O(n^3)$ worst-case time.

The proof of the subsequent main theorem neglects the running time analysis as mentioned before.

Theorem 7.6 *For n-vertex graphs that are connected,* CLUSTER EDITING *has a problem kernel with a graph that contains $O(k^2)$ vertices and $O(k^3)$ edges. It can be found in $O(n^3)$ time.*

Proof Let $G = (V, E)$ be a graph which is reduced with respect to Rules 1 to 3. Without loss of generality, we assume that G is connected. Since Rule 3 deletes all isolated cliques from the given graph, G is not a clique and we need at least one edge modification to transform $G = (V, E)$ into a graph $G' = (V, E')$, consisting of disjoint cliques. Let k be the minimum number of required edge modifications, namely k_a edge additions and k_d edge deletions. Under the assumption that G is reduced with respect to Rules 1 to 3, we will show by contradiction that $n \leq (2k + 1) \cdot k$ and that $|E| \leq \binom{2k+1}{2} \cdot k$ as follows.

Assume that $|V| > (2k + 1) \cdot k$. We distinguish two cases, namely the case that $k_a = 0$ and the case that $1 \leq k_a \leq k$. In both cases, we show a contradiction to our assumption that the graph is reduced with respect to Rule 1.

(Case 1): $k_a = 0$. We have k_d edge deletions, $1 \leq k_d = k$, to transform G into G'. Let $V_C \subseteq V$ denote the vertex set of a largest clique in G'. The vertices in V_C also form a clique in G since $k_a = 0$. Since G is connected, at least one vertex $u \in V_C$ is connected to a vertex $v \notin V_C$. We further distinguish two subcases: either v is not connected to any other vertex $u' \in V_C$ with $u' \neq u$ or there is a $u' \in V_C$ with $u' \neq u$ and $\{u', v\} \in E$.

(Case 1.1): Vertex v is not connected to any other vertex $u' \in V_C$ with $u' \neq u$. We can lower-bound the clique size by $|V_C| \geq |V|/(k_d + 1)$: by k_d edge deletions, G is transformed into a graph G' containing at most k_d+1 cliques and, therefore, a largest clique in G' contains at least $|V|/(k_d + 1)$ vertices. First, we assume that $k_d \geq 2$ $^{(*)}$. Using our further assumptions that $|V| > (2k+1) \cdot k$ $^{(**)}$ and $k_d = k$ $^{(***)}$ we obtain

$$|V_C| \geq \frac{|V|}{k_d + 1} \overset{(**)}{>} \frac{(2k+1) \cdot k}{k_d + 1} \overset{(***)}{=} \frac{2k^2 + k}{k+1} = \frac{k^2 + k}{k+1} + \frac{k^2}{k+1} \overset{(*)}{\geq} k + 1.$$

Consequently, $|V_C| \geq k+2$ and u has at least $k+1$ neighbors—all vertices $u' \in V_C$ with $u' \neq u$—which are not neighbors of v. This contradicts the assumption that G is reduced with respect to Rule 1. Second, assuming that $k_d = k = 1$ while $|V| > (2k+1) \cdot k = 3$, G' consists of two cliques; either both contain at least two vertices or one of them contains at least three vertices—both times, Rule 1 would apply, a contradiction.

(Case 1.2): There is a $u' \in V_C$ with $u' \neq u$ and $\{u', v\} \in E$. We can lower-bound the clique size by at least $|V|/k_d$: G is transformed into a graph G' containing at most k_d cliques and, therefore, a largest clique in G' contains at least $|V|/k_d$ vertices. With the assumptions $|V| > (2k+1) \cdot k$ $^{(*)}$ and $k_d = k$ $^{(**)}$, we obtain

$$|V_C| \geq \frac{|V|}{k_d} \overset{(*)}{>} \frac{(2k+1) \cdot k}{k_d} \overset{(**)}{=} 2k + 1.$$

Consequently, V_C contains more than $2k+1$ vertices and at most k many of them are connected to v. Therefore, u has more than $k + 1$ neighbors in this clique which are not neighbors of v, contradicting the assumption that G is reduced with respect to Rule 1.

(Case 2): $1 \leq k_a \leq k$. We know, since $k_a + k_d = k$, that $k_d < k$. Again, let $V_C \subseteq V$ denote the vertex set of a largest clique in G'. Since G' contains at most k_d+1 cliques, we have $|V_C| \geq |V|/(k_d+1)$. With $k_d < k$, this yields $|V_C| \geq |V|/k$ and, using $|V| > (2k + 1) \cdot k$, we obtain

$$|V_C| > 2k + 1. \tag{7.1}$$

Since the vertices of V_C form a clique in G' and at most k many edges are added in the transformation from G to G', in G there are at most k vertex pairs (v_i, v_j) with $i < j$ and $v_i, v_j \in V_C$ which are not connected by an edge. In the following, we show that, under the two assumptions that $|V_C| \geq 2(k+1)$ and that $|\{(v_i, v_j) \mid v_i, v_j \in V_C, i < j, \{v_i, v_j\} \notin E\}| \leq k$, the graph cannot be reduced with respect to Rule 1. To this end, we consider the cases that $k_a = k$ and that $k_a < k$ separately.

(Case 2.1): $k_a = k$. We conclude that $V_C = V$ and G' consists of only one clique. Since $k_a = k \geq 1$, there are $v_i, v_j \in V$ with $i < j$ and $\{v_i, v_j\} \notin E$. Since at most $k - 1$ of the at least $2k$ vertices besides v_i and v_j are not connected to v_i or v_j in G, v_i and v_j have at least $k + 1$ common neighbors in G and Rule 1 would apply, in contradiction to our assumption that G is reduced with respect to Rule 1.

(Case 2.2): $k_a < k$. We can conclude that there are $u \in V_C$ and $v \notin V_C$ such that $\{u, v\} \in E$. Due to inequality (7.1), there are more than $2k + 1$ vertices in V_C. On the one hand, there are at least $(2k + 1) - k_a - 1$ vertices $u' \in V_C$ with $\{u', u\} \in E$. On the other hand, there are at most $k_d - 1$ vertices $u' \in V_C$ with $\{u', v\} \in E$. Consequently, we have at least

$$(2k + 1) - (k_a + k_d) = (2k + 1) - k = k + 1$$

vertices $u' \in V_C$ with $\{u', u\} \in E$ but $\{u', v\} \notin E$. This implies that u has at least $k + 1$ neighbors in G which are not neighbors of v and Rule 1 applies. In both cases, for $k_a = k$ and for $1 \leq k_a < k$, we obtain a contradiction to the assumption that G is reduced since Rule 1 would apply.

Regarding the edge set of the connected component, we infer a contradiction from the assumption that $|E| > \binom{2k+1}{2} \cdot k$ in an analogous way as for the vertex set: again, we let V_C be the vertex set of a largest clique in G' and we distinguish between the cases $k_d = 0$ and $k_d > 0$. If $k_d = 0$, then we can easily derive that $|V_C| > 2k + 1$ and the contradiction follows in analogy to (Case 2.1) above. If $k_d > 0$, we can derive (omitting some details here) that $|V_C| \geq 2k + 1$. Then the contradiction follows in analogy to (Case 2.2) above.

Summarizing, the reduced graph contains at most $2k^2 + k$ vertices and at most $\binom{2k+1}{2}k = 2k^3 + k^2$ edges; otherwise, no solution exists. □

The proof of Theorem 7.6 illustrates that, although the data reduction may be fairly simple, the proof of the upper bound on the size of the problem kernel may become demanding. This is an observation that can often be made.

7.4 Vertex Cover

At the beginning of this chapter we discussed the size-$O(k^2)$ problem kernel due to Buss, which is based on a simple observation concerning high-degree vertices. There are several kernelization techniques with much improved bounds on the kernel size and which are based on maximum matching and linear programming techniques. Two of them are directly based on a theorem of George L. Nemhauser and William T. Trotter from 1975. This was developed in the context of approximation algorithms, but turns out to be very useful when designing fixed-parameter algorithms for VERTEX COVER.

7.4.1 *Kernelization based on matching*

We start with an algorithmic approach employing basic combinatorial arguments not relying on integer or linear programming. As we will show, the described method guarantees a "kernelized graph" where the number of vertices is bounded from above by twice the size of an optimal vertex cover of this graph.

Theorem 7.7 *For an n-vertex graph $G = (V, E)$ with m edges, we can compute two disjoint sets $C_0 \subseteq V$ and $V_0 \subseteq V$ in $O(\sqrt{n} \cdot m)$ time, such that the following three properties hold:*

Input: $G = (V, E)$
Output: C_0 and V_0.
Phase 1:
Construct the bipartite graph $B = (V, V', E_B)$
where $E_B := \{\{x, y'\}, \{x', y\} \mid \{x, y\} \in E\}$
and V' is a copy of V.
Phase 2:
Let C_B be an optimal vertex cover of B determined
by computing a maximum matching using
standard methods in $O(\sqrt{n}m)$ time.
$C_0 := \{x \in V \mid x \in C_B$ and $x' \in C_B\}$.
$V_0 := \{x \in V \mid$ either $x \in C_B$ or $x' \in C_B\}$.

FIG. 7.1. An algorithm to compute the sets C_0 and V_0 in Theorem 7.7. A *maximum matching* is a maximum cardinality set of edges in a graph such that no two edges share an endpoint. It is well known in matching theory that for bipartite graphs the size of a maximum matching coincides with the size of a minimum vertex cover of that graph (König's Minimax Theorem).

1. *There is a minimum-size vertex cover of G which comprises C_0.*
2. *The induced subgraph $G[V_0]$ has a minimum vertex cover of size at least $|V_0|/2$.*
3. *If $D \subseteq V_0$ is a vertex cover of the induced subgraph $G[V_0]$, then $C := D \cup C_0$ is a vertex cover of G.*

Proof The algorithm given in Figure 7.1 computes the sets C_0 and V_0 employing the bipartite graph $B = (V, V', E_B)$ where

$$E_B := \{\{x, y'\}, \{x', y\} \mid \{x, y\} \in E\}$$

and V' simply is a copy of V. For $x \in V$, its copy is $x' \in V'$. Moreover, C_B denotes a minimum-size vertex cover of B.

To prove the validity of the three claims in the theorem, we further need to introduce the third "remaining set".

$$\begin{aligned} I_0 &:= \{x \in V \mid x \notin C_B \text{ and } x' \notin C_B\} \\ &= V \setminus (V_0 \cup C_0). \end{aligned}$$

Note that I_0 forms an independent set in G. Moreover, for every vertex in I_0 all its neighbors are in C_0.

1. **$D \cup C_0$ is a vertex cover of G:** For $\{x, y\} \in E$ we have to show that $x \in C$ or $y \in C$. We distinguish four cases.
   Case 1: If $x \in I_0$, that is, $x, x' \notin C_B$, then it must hold that $y, y' \in C_B$, and, thus, $y \in C_0$.
   Case 2: The case $y \in I_0$ is completely analogous to Case 1.

Case 3: $x \in C_0$ or $y \in C_0$ is trivial.

Case 4: If $x, y \in V_0$, then $x \in D$ or $y \in D$ by definition of D.

2. **There is a minimum-size vertex cover of G comprising C_0:** Let S be any minimum-size vertex cover of G. We show how to use S to construct a minimum-size vertex cover of G that comprises C_0. To this end, define $S_V := S \cap V_0$, $S_C := S \cap C_0$, $S_I := S \cap I_0$ and $\bar{S}_I := I_0 \setminus S_I$. By $S'_C \subseteq V'$ we denote the copy of S_C. We make use of the following auxiliary claim.
Claim: $\hat{C}_B := (V \setminus \bar{S}_I) \cup S'_C$ is a vertex cover of the bipartite graph B.
Using the Claim (which we prove afterwards), next we will show $|C_0| \leq |S \setminus S_V|$ which clearly implies $|C_0 \cup S_V| \leq |S|$. Since S_V is a vertex cover of $G[V_0]$, with the already proven first point of the theorem we can conclude that $C_0 \cup S_V$ yields an optimal vertex cover.

$$\begin{aligned}
|V_0| + 2|C_0| &= |V_0 \cup C_0 \cup C'_0| \\
&= |C_B| \quad \text{[Definition of } V_0 \text{ and } C_0] \\
&\leq |\hat{C}_B| \quad \text{[Optimality of } C_B] \\
&= |V \setminus \bar{S}_I| + |S'_C| \quad \text{[Claim]} \\
&= |V_0 \cup C_0 \cup I_0 \setminus (I_0 \setminus S_I)| + |S'_C| \\
&= |V_0| + |C_0| + |S_I| + |S_C| \\
\Rightarrow \quad |C_0| &\leq |S_I| + |S_C| \\
&= |S_I| + |S_C| + |S_V| - |S_V| \\
&= |S| - |S_V|.
\end{aligned}$$

It remains to show the above claim. Let $\{x, y'\} \in E_B$. We have to show that $x \in \hat{C}_B$ or $y' \in \hat{C}_B$.

Case 1: If $x \notin \bar{S}_I$, then $x \in \hat{C}_B$ according to the definition of $\hat{C}_B$.

Case 2: If $x \in \bar{S}_I$, then $x \in I_0$, and, thus, $x \notin C_0$ and $x \notin S$. Then, $y \in S$ and $y \in C_0$. Thus, $y \in S \cap C_0 = S_C$ and $y' \in S'_C$.

3. **$G[V_0]$ has a minimum vertex cover of size at least $|V_0|/2$:** Suppose S_0 is a minimum vertex cover of $G[V_0]$. According to the first point of the theorem, $S_0 \cup C_0$ is a vertex cover of G. Then, according to the definition of the bipartite graph B the set $S_0 \cup S'_0 \cup C_0 \cup C'_0$ is a vertex cover of B. Hence:

$$\begin{aligned}
|V_0| + 2|C_0| &= |C_B| \quad \text{[Definition of } V_0 \text{ and } C_0] \\
&\leq |C_0 \cup C'_0 \cup S_0 \cup S'_0| \quad \text{[Optimality of } C_B] \\
&= 2|C_0| + 2|S_0|.
\end{aligned}$$

This implies $|V_0| \leq 2|S_0|$.

□

Theorem 7.7 is the key to a problem kernel for VERTEX COVER which is formed by a graph that has at least half of its vertices in a minimum-size vertex cover.

Theorem 7.8 *Let* $(G = (V, E), k)$ *be an input instance of* VERTEX COVER. *In* $O(k \cdot |V| + k^3)$ *time we can compute a reduced instance* $(G' = (V', E'), k')$ *with* $|V'| \leq 2k$ *and* $k' \leq k$ *such that* G *admits a vertex cover of size* k *iff* G' *admits a vertex cover of size* k'.

Proof We begin by using Buss's reduction to a problem kernel as sketched at the beginning of the chapter to get a reduced instance containing at most $O(k^2)$ vertices and edges with a parameter value $k'' \leq k$. Buss's kernelization takes $O(k|V|)$ time. Then, Phase 2 of the algorithm described in the proof of Theorem 7.7—which basically consists of a maximum matching computation in a bipartite graph containing $O(k^2)$ vertices and $O(k^2)$ edges—can be done in time $O(k^3)$. From the maximum matching in linear time we get the set C_0 of vertices that have to be in the vertex cover, and we define $k' := k'' - |C_0|$. We also get the set V_0 that induces the subgraph $G[V_0]$, the problem kernel G'. Observe that due to Theorem 7.7 we directly know that if $|V_0| > 2k'$, then there is no vertex cover of size k for the original graph G. Otherwise, the remaining vertices for a minimum vertex cover of G can be found by searching for a minimum vertex cover of G'. □

There are several important observations to be made with respect to Theorem 7.8.

- According to the current state of knowledge, the size bound $2k$ for the number of vertices in the reduced graph is the best one may hope for because a problem kernel with $(2 - \varepsilon) \cdot k$ vertices with constant $\varepsilon > 0$ would imply a factor $(2 - \varepsilon)$ polynomial-time approximation algorithm for VERTEX COVER—simply put all vertices of the reduced graph into the vertex cover which then is by a factor at most $2 - \epsilon$ larger than a minimum one. This, however, is a longstanding open question in approximation theory and an answer to it would mean a major breakthrough in approximation algorithms for VERTEX COVER.
- In contrast with the simpler Buss kernelization, after performing the above Nemhauser–Trotter reduction to a problem kernel, at first glance we cannot expect to find *all* vertex covers up to size k. Theorem 7.7 only leads to (at least) one particular minimum vertex cover, excluding others from further consideration. Recent research shows, however, how to modify these results in order to obtain all optimal solutions; see the bibliographical remarks.
- Theorem 7.7 can be generalized to find minimum *weight* vertex covers, where vertices have a positive real weight. This is useful for fixed-parameter algorithms for WEIGHTED VERTEX COVER.

- The Nemhauser–Trotter kernelization itself does not make explicit use of the parameter value k when computing the problem kernel graph $G[V_0]$—it is an example of a parameter-independent data reduction.

7.4.2 *Kernelization based on linear programming*

Having described in detail a purely combinatorial approach to a $2k$-vertices problem kernel for VERTEX COVER, we now investigate an alternative route to this result. It is rooted in (integer) linear programming techniques, from where the original work of Nemhauser and Trotter also originated. Without delving into the details of the theory behind it, we focus attention on the easy to grasp and immediately clear fundamental ideas of this approach. All is based on the following elementary observation. The optimization version of VERTEX COVER can be stated as a simple integer linear program: associate with each graph vertex v a 0/1-variable x_v. Then, $x_v = 1$ means that v is in the vertex cover set and $x_v = 0$ means that it is not. Clearly, for a given graph $G = (V, E)$ VERTEX COVER can be expressed as the following optimization problem

1. Minimize $\sum_{v \in V} x_v$ where
2. $x_u + x_v \geq 1$ is satisfied for all $\{u, v\} \in E$.

Since integer linear programming is generally intractable—the corresponding decision problem is *NP*-complete—we relax the integer programming formulation to polynomial-time solvable linear programming. Without changing the above two points, the only difference here is that we do not require a solution based on 0/1-variables x_v but we allow x_v to take values in the real-valued interval $0 \leq x_v \leq 1$. The value of the objective function under this "relaxation" then gives a lower bound for the size of an optimal vertex cover.

Based on the relaxation to linear programming, however, we may obtain a small problem kernel for VERTEX COVER. To this end, define

$$C_0 := \{v \in V \mid x_v > 0.5\},$$
$$V_0 := \{v \in V \mid x_v = 0.5\},$$
$$I_0 := \{v \in V \mid x_v < 0.5\}.$$

The following theorem results in a $2k$-vertices problem kernel in the same way that Theorem 7.7 does.

Theorem 7.9 *Let $(G = (V, E), k)$ be a* VERTEX COVER *instance. For C_0, V_0, and I_0 as defined above, there is a minimum-size vertex cover S with $C_0 \subseteq S$ and $S \cap I_0 = \emptyset$. In addition, V_0 induces the problem kernel $(G[V_0], k - |C_0|)$ with $|V_0| \leq 2k$.*

Proof Consider an optimal vertex cover set S and define $\bar{S}_C := C_0 \setminus S$ and $S_I := S \cap I_0$. We show that $S' := (S \setminus S_I) \cup \bar{S}_C$ is an optimal vertex cover as well. Clearly, according to the linear programming formulation we must have that $N(I_0) \subseteq C_0$. With this, $|\bar{S}_C| \geq |S_I|$ because otherwise we could replace S_I

in S with $\bar{S}_C$, obtaining a smaller vertex cover set, contradicting the optimality of S. Moreover, $|\bar{S}_C| \leq |S_I|$ as can be seen as follows. Consider

$$\epsilon := \min\{x_v - 0.5 \mid v \in C_0\}$$

and

1. replace the values x_u with $x_u + \epsilon$ for all $u \in S_I$; and
2. replace the values x_v with $x_v - \epsilon$ for all $v \in \bar{S}_C$.

If $|\bar{S}_C| > |S_I|$ then such a modification would give a better objective value for the linear program, contradicting the optimality of the given solution. Altogether, hence, $|\bar{S}_C| = |S_I|$, and we can replace S_I with $\bar{S}_C$ in S because $N(I_0) \subseteq C_0$.

To derive the bound $|V_0| \leq 2k$, notice that since the value of the objective function for the linear program is a lower bound for the objective function of the integer linear program, the size of any optimal cover of $G[V_0]$ is bounded from below by $\sum_{v \in V_0} x_v = |V_0|/2$. Hence we conclude that if $|V_0| \geq 2k$ (or even $|V_0| \geq 2(k - |C_0|)$), there is no solution to the given VERTEX COVER instance. This finishes the proof. □

We end this section with the remark that the running time of the kernelization associated with Theorem 7.9 is essentially that needed for linear programming, and hence is polynomial.

7.4.3 *Kernelization based on crown structures*

Recall that vertices of degree one can be easily deleted—we can simply decide to take their neighboring vertices into the vertex cover. This simple observation, in a sense, finds its generalization in the *crown reduction* rules which have very recently appeared in the literature. In this sense, degree-one vertices lead to the simplest crowns.

Let I be an independent set in the given graph $G = (V, E)$. Assume that there is a maximum matching in the *bipartite* graph induced by I and $N(I)$ that contains $|N(I)|$ edges. Clearly then at least $|N(I)|$ vertices from $I \cup N(I)$ must appear in any vertex cover of G. Hence there is a minimum-size vertex cover that contains all the vertices in $N(I)$ and none of the vertices in I. We may delete all of $I \cup N(I)$ from G. This is the "crown reduction rule", which obviously generalizes the "degree-one vertices rule" as discussed above. Moreover, the crown reduction rule is related to the Nemhauser–Trotter kernelization in the sense that matching again plays a key role.

In what follows, we formalize the concept of crown structures and show that their use again leads to a problem kernel with a linear number of vertices.

Definition 7.10 *A* crown *of a graph $G = (V, E)$ consists of $H \subseteq V$ and $I \subseteq V$ with $H \cap I = \emptyset$ such that the following conditions hold:*

1. *$H = N(I)$,*
2. *I forms an independent set, and*

3. *the edges connecting H and I contain a matching in which all elements of H are matched.*

It is easy to observe that, given a crown H and I, we can reduce the graph.

Lemma 7.11 *If G is a graph with a crown H and I, there exists a minimum-size vertex cover of G that contains all of H and none of I.*

Proof Since there is a matching of the edges between H and I, any vertex cover must contain at least one vertex from each matched edge. Thus the matching will require at least $|H|$ vertices in the vertex cover. This minimum number can be realized by selecting H to be in the vertex cover: vertices from H can be used to cover edges that do not connect I and H, while this is not true for vertices in I. Thus including the vertices from H does not increase, and may decrease, the size of the vertex cover compared with including vertices from I. Therefore there is a minimum-size vertex cover that contains all vertices in H and none of the vertices in I. □

Two questions remain to be answered:

1. how to find crowns efficiently, that is, in polynomial time; and
2. what is the size of the problem kernel that can be obtained via crown reductions?

Both questions are answered in the following, starting with an algorithm to compute a crown in an arbitrary input graph. The algorithm is given in Figure 7.2.

The next lemma shows that the algorithm from Figure 7.2 correctly produces a crown. For the definitions of the subsequently used vertex sets I' and I_0 and matching M_2 see Figure 7.2.

Lemma 7.12 *The algorithm from Figure 7.2 finds a crown as long as at least one of the following two conditions is met.*

1. *Every vertex in $N(I')$ is matched by M_2, or*
2. *the set I_0 is nonempty.*

Proof We only sketch the idea of proof. The correctness of the claim is straightforward if the first condition is met. In addition, the set I is clearly an independent set—otherwise, M_1 would not be a maximal matching. Moreover, it is clear from the algorithm's steps that $H = N(I)$. It remains to be shown that the third condition in Definition 7.10 is fulfilled. Suppose that it is not and assume that there exists an element $h \in H$ unmatched by M_2. It is not very hard to verify that by the construction of H and I from I_0 we could then determine a matching M_2' that matches h and $|M_2'| > |M_2|$, a contradiction of the fact that the matching M_2 is of maximum size. □

Now we have all the prerequisites at hand to show another problem kernelization for Vertex Cover which, according to worst-case analysis, may be slightly larger than those given before.

Input: $G = (V, E)$
Output: Crown H and I.
Step 1:
Find a *maximal* matching M_1 of G,
and let I' be the set of all unmatched vertices.
Step 2:
Find a *maximum* matching M_2 in the bipartite graph
induced by the edges between I' and $N(I')$.
Step 3:
If every vertex in $N(I')$ is matched by M_2, then
$H := N(I')$ and $I := I'$ form a crown, and we are done.
Otherwise:
Step 4:
Let I_0 be the set of vertices in I' that are unmatched by M_2.
Step 5:
Start with $i := 0$ and repeat the following two steps
until $I_i = I_{i-1}$.
$H_i := N(I_i)$.
$I_{i+1} := I_i \cup N_{M_2}(H_i)$.
Step 6:
$H := H_i$ and $I := I_i$ form a crown.

FIG. 7.2. An algorithm to compute a crown H and I in a graph. Here, $N_{M_2}(H_i)$ denotes the neighbors of H_i that are connected to H_i by edges contained in the matching M_2.

Theorem 7.13 *Let $(G = (V, E), k)$ be a* VERTEX COVER *instance where both matchings M_1 and M_2 from the algorithm in Figure 7.2 shall have size at most k. Then there is a crown H and I such that the reduced instance $(G[V_0], k - |H|)$ with $V_0 := V \setminus (H \cup I)$ is a problem kernel with $|V_0| \leq 3k$.*

Proof The size of M_1 is at most k, hence it consists of at most $2k$ vertices and the set I' in the algorithm contains at least $|V| - 2k$ vertices. The size of M_2 is at most k, hence at most k vertices in I' are matched by M_2. In other words, at least $|V| - 3k$ vertices contained in I' are unmatched by M_2. All these vertices are included in $I_0 \subseteq I$ in Step 4 of the algorithm. Hence, $|V \setminus (H \cup I)| \leq 3k$.
□

Observe that if one of the above matchings M_1 and M_2 in G is of size greater than k, then the VERTEX COVER instance (G, k) clearly has no solution and we are done. Thus Theorem 7.13 covers all the cases of interest. The running time of the kernelization, dominated by the complexity of the algorithm computing the maximum bipartite matching M_2, is clearly polynomial.

7.4.4 *Comparison and discussion*

There is no other problem in the literature whose accessibility to kernelization has been as well investigated as VERTEX COVER. We have presented three approaches to obtain a problem kernel with a number of vertices linear in the parameter k denoting the size of the desired vertex cover. Roughly speaking, although completely different at first sight, they have common roots. Empirical investigations indicate that kernelization based on linear programming tends to be slowest. The determination of crown structures and their usefulness for other (graph) problems is currently under heavy investigation. Among others, applications have been discovered for packing and coloring problems; refer to the bibliographical remarks.

Although a problem kernel with $2k$ vertices appears to be optimal according to the current state of the art, the best one—a size bound $(2-\epsilon)\cdot k$ would give a ratio-$(2-\epsilon)$ polynomial-time approximation algorithm—at least from a practical point of view, might not be the final word. All our analysis is clearly related to a worst-case scenario and simple data reduction rules which may be ineffective with respect to improving the worst-case upper bounds may be highly effective in practical applications, or they may further simplify proofs and algorithms for known worst-case bounds.

One such efficient data reduction rule is called "folding" (also known as "struction" in the literature) and it was developed to get rid of degree-two vertices—in addition to the trivial handling of degree-one vertices—in reduced VERTEX COVER instances. The point is that each degree-two vertex in a graph together with its two neighbors can be melted into one super-vertex: consider a degree-two vertex x with its two neighbors u and v. Without loss of generality, u and v are not connected because if $\{u, v\}$ exists, then we can always choose u and v to be in a vertex cover without loosing the guarantee to find an optimal solution. Then delete x, u, and v from the graph and insert a new vertex x' with $N(x') := N(u) \cup N(v) \setminus \{x, u, v\}$. The decisive observation then is that the original graph has a vertex cover of size k iff the reduced graph has a vertex cover of size $k-1$. Moreover, the vertex cover of the original graph can easily be reconstructed from the vertex cover of the reduced graph. Thus the "combinatorially explosive" part of the search for a minimum vertex cover can be restricted to graphs with minimum vertex degree three. See the exercises for more on this.

7.5 3-Hitting Set

In this section we show how to generalize beyond Buss's reduction to a problem kernel for VERTEX COVER to solve the more general d-HITTING SET problem for an integer $d \geq 3$. Simply speaking, this lifts the vertex covering issue from graphs to hypergraphs. We focus on the 3-HITTING SET (3HS) problem, the generalization of VERTEX COVER to "hypergraphs with size-three edges":

Input: A collection $\mathcal{C}$ of subsets of size at most three of a finite set S and a nonnegative integer k.
Task: Find a subset $S' \subseteq S$ with $|S'| \leq k$ such that S' contains at least one element from each subset in $\mathcal{C}$.

VERTEX COVER is the same as 2-HITTING SET.

Our kernelization extends ideas behind the VERTEX COVER problem kernel due to Buss as sketched in the introduction to Chapter 7. In this way, we easily obtain a problem kernel for 3HS consisting of $O(k^3)$ sets, that is, $|\mathcal{C}| = O(k^3)$. The data reduction strategy presented is parameter-dependent. We remark that it can be generalized to achieve an upper bound $O(k^d)$ for d-HITTING SET with $d > 3$. The running time increases, though.

Similar to Buss's kernelization for VERTEX COVER the kernelization for 3HS employs the idea that we *must* put "high-degree elements" into the hitting set.

Theorem 7.14 3-HITTING SET *has a problem kernel with* $|\mathcal{C}| = O(k^3)$, *and it can be found in linear time.*

Proof First, let us consider two fixed elements $x, y \in S$:
Claim 1: For an instance $(\mathcal{C}, k)$ we can find an instance $(\mathcal{C}', k)$ in linear time so that $(\mathcal{C}, k) \in$ 3HS iff $(\mathcal{C}', k) \in$ 3HS. Moreover, there can be at most k size-three subsets in the collection $\mathcal{C}'$ that contain both x and y.
Claim 1 is seen as follows. Assume that there are more than k subsets containing x and y. Since each set appears only once in $\mathcal{C}$ this implies that there are more than k different "third" elements in the corresponding sets. Hence to cover these more than k different sets with at most k elements from the base set S, we *must* bring at least one of x and y into our hitting set S'. This means, however, that all sets containing both x and y can be replaced by the single set $\{x, y\}$. This proves Claim 1.

Second, we consider the case where there is only one fixed element $x \in S$:
Claim 2: For an instance $(\mathcal{C}, k)$ we can find an instance $(\mathcal{C}', k')$ in linear time so that $(\mathcal{C}, k) \in$ 3HS iff $(\mathcal{C}', k') \in$ 3HS where $k' \leq k$. Moreover, there can be at most k^2 size-three subsets in the collection $\mathcal{C}'$ that contain x.
Claim 2 is seen as follows. Assume that there are more than k^2 subsets containing x. By Claim 1 we can assume that x can occur in a subset together with another element y at most k times. Hence, if there were more than k^2 subsets containing x, these could not be covered by any $S' \subseteq S$ with $|S'| \leq k$ without taking x. Thus, x must be in S' and the corresponding sets can be deleted. This proves Claim 2.

Now, from Claim 2 we can conclude that each element x from S can occur in at most k^2 subsets in C because, otherwise, x *had* to be in the hitting set S'. Clearly, since $|S'| \leq k$ this means (provided that $\mathcal{C}$ has a hitting set of size $\leq k$) that $\mathcal{C}$ can consist of at most $k \cdot k^2 = k^3$ size three subsets. This gives the desired upper bound on the size of the problem kernel.

Finally, we remark that we can easily count in how many sets each element occurs and throw away in linear time all elements (and their corresponding subsets) occurring in more than k^2 subsets. □

7.6 Dominating Set in Planar Graphs

In Section 7.4 we became acquainted with a problem kernel for VERTEX COVER consisting of only $2k$ vertices. Specializing this result to planar graphs, we can even speak of a linear-size problem kernel because for planar graphs the number of edges is linearly bounded by the number of vertices. We know so far of only a few problem kernels of linear size—a further example is that for DOMINATING SET IN PLANAR GRAPHS.

> **Input:** A planar graph $G = (V, E)$ and a nonnegative integer k.
> **Task:** Find a subset $S \subseteq V$ with at most k vertices such that every vertex $v \in V$ is contained in S or has at least one neighbor in S.

The kernelization for DOMINATING SET IN PLANAR GRAPHS consists of two main parts—the algorithmic side with the actual data reduction rules and the mathematical side with the proof of correctness and the analysis of the problem kernel size. Because of the significant technical machinery involved, we will essentially focus on the algorithmic side.

Two data reduction rules are used to prove the linear-size problem kernel. Both reduction rules are based on the same principle: explore the local neighborhood of one vertex or the neighborhood of two vertices and try to replace it by a simpler structure. In what follows, the minimum k such that the graph G has a size-k dominating set is called the *domination number* of G, denoted by $\gamma(G)$. We begin with the simpler rule which refers to the local neighborhood of a single vertex.

7.6.1 *The neighborhood of a single vertex*

Consider a vertex $v \in V$ of the given graph $G = (V, E)$, and partition the neighbors $N(v)$ of v into three different sets $N_1(v)$, $N_2(v)$, and $N_3(v)$ depending on what the neighborhood structure of these vertices looks like. More specifically, setting $N[v] := N(v) \cup \{v\}$, we define

$$\begin{aligned} N_1(v) &:= \{u \in N(v) \mid (N(u) \setminus N[v]) \neq \emptyset\}, \\ N_2(v) &:= \{u \in N(v) \setminus N_1(v) \mid (N(u) \cap N_1(v)) \neq \emptyset\}, \\ N_3(v) &:= N(v) \setminus (N_1(v) \cup N_2(v)). \end{aligned}$$

An example which illustrates the partitioning of $N(v)$ into the subsets $N_1(v)$, $N_2(v)$, and $N_3(v)$ is given in the left-hand diagram of Figure 7.3. A helpful intuitive interpretation is that of

- considering vertices in $N_1(v)$ as *exits* because they have direct connections to the world outside the closed neighborhood of v;

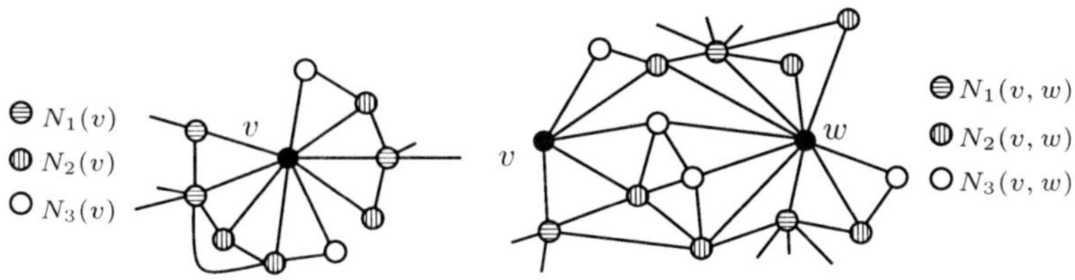

FIG. 7.3. The left-hand side shows the partitioning of the neighborhood of a single vertex v. The right-hand side shows the partitioning of a neighborhood $N(v,w)$ of two vertices v and w. Since, in the left-hand figure, $N_3(v) \neq \emptyset$, reduction Rule 1 applies. In the right-hand figure, since $N_3(v,w)$ cannot be dominated by a single vertex at all, Case 2 of Rule 2 applies.

- considering vertices in $N_2(v)$ as *guards* because they have direct connections to exits; and
- considering vertices in $N_3(v)$ as *prisoners* because they have absolutely no chance to see the world outside $N[v]$, the closed neighborhood of v.

Based on the above definitions we obtain the first reduction rule.

Rule 1 *If $N_3(v) \neq \emptyset$ for some vertex v, then*

- *remove $N_2(v)$ and $N_3(v)$ from G, and*
- *add a new vertex v' with the edge $\{v, v'\}$ to G.*

Note that the vertex v' is used as a "gadget vertex" that forces us to take v into an optimal dominating set in the reduced graph. More specifically, if finally in the reduced graph we find an optimal dominating set containing v', we can safely replace it by v without destroying the optimality of the solution.

Lemma 7.15 *Let $G = (V,E)$ be a graph and let $G' = (V', E')$ be the resulting graph after having applied Rule 1 to G. Then $\gamma(G) = \gamma(G')$.*

Proof Consider a vertex $v \in V$ such that $N_3(v) \neq \emptyset$. The vertices in $N_3(v)$ can only be dominated by either v or by vertices in $N_2(v) \cup N_3(v)$. But, clearly, $N(w) \subseteq N(v)$ for every $w \in N_2(v) \cup N_3(v)$. This shows that an optimal way to dominate $N_3(v)$ is given by taking v into the dominating set. This is simulated by the "gadget" $\{v, v'\}$ in G'. It is safe to remove $N_2(v) \cup N_3(v)$, since $N(N_2(v) \cup N_3(v)) \subseteq N(v)$; in other words, since the vertices that could be dominated by vertices from $N_2(v) \cup N_3(v)$ are already dominated by v. Hence, $\gamma(G') = \gamma(G)$. □

Lemma 7.16 *Rule 1 can be carried out in $O(n)$ time for planar graphs and in $O(n^3)$ time for general graphs.*

Proof We first discuss the planar case. To carry out Rule 1, for each vertex v of the given planar graph G we have to determine the neighbor sets $N_1(v)$, $N_2(v)$, and $N_3(v)$. By definition of these sets, one easily observes that it is sufficient to consider the subgraph G that is induced by all vertices that are connected to v by a path of length at most two. To do so, we employ a "partial" depth-first search tree of depth two, rooted at v. More precisely, this means that we explore all vertices at distance one from v (that is, connected to v by an edge in G) and some vertices at distance two from G (details to follow). We perform two phases.

In phase 1, constructing the search tree, we determine the vertices from $N_1(v)$. Each vertex of the first level (that is, distance one from the root v) of the search tree that has a neighbor at the second level of the search tree belongs to $N_1(v)$. Observe that it is enough to stop the expansion of a vertex from the first level as soon as its *first* neighbor in the second level is encountered. Hence, denoting the degree of v by $\deg(v)$, phase 1 takes $O(\deg(v))$ time because there clearly are at most $2 \cdot \deg(v)$ tree edges and at most $O(\deg(v))$ non-tree edges to be explored. The latter holds true since these non-tree edges all belong to the subgraph of G induced by $N[v]$. Since this graph is planar and $|N[v]| = \deg(v) + 1$, the claim follows.

In phase 2, it remains to determine the sets $N_2(v)$ and $N_3(v)$. To get $N_2(v)$, one basically has to go through all vertices from the first level of the above search tree that are not already marked as being in $N_1(v)$ but have at least one neighbor in $N_1(v)$. All this can be done within the planar graph induced by $N[v]$, using the already marked $N_1(v)$-vertices, in $O(\deg(v))$ time. Finally, $N_3(v)$ simply consists of vertices from the first level that are neither marked being in $N_1(v)$ nor marked being in $N_2(v)$. In summary, this shows that for vertex v the sets $N_1(v)$, $N_2(v)$, and $N_3(v)$ can be constructed in $O(\deg(v))$ time.

Having determined these three sets, the sizes of which are all bounded by $\deg(v)$, it is clear that the possible removal of vertices from $N_2(v)$ and $N_3(v)$ and the addition of a vertex and an edge as required by Rule 1 all can be done in $O(\deg(v))$ time. Finally, it remains to analyze the overall complexity of this procedure when going through all n vertices of $G = (V, E)$. But this is easy. The running time can be bounded from above by

$$\sum_{v \in V} O(\deg(v)) = O(m) = O(n).$$

Since G is planar, we have the $O(n)$ bound, that is, the whole reduction takes linear time.

For general graphs, the method described above leads to a worst-case cubic time implementation of Rule 1. Here, one ends up with the sum

$$\sum_{v \in V} O((\deg(v))^2) = O(n^3).$$

Note that the size of the graph that is induced by the neighborhood $N[v]$ again is relevant for the time needed to determine the sets $N_1(v)$, $N_2(v)$, and $N_3(v)$. For general graphs, this neighborhood may contain $O(\deg(v)^2)$ many edges. □

7.6.2 *The neighborhood of a pair of vertices*

Similar to Rule 1, we explore the neighborhood set $N(v,w) := N(v) \cup N(w)$ of two vertices $v, w \in V$. Analogously, we now partition $N(v,w)$ into three disjoint subsets $N_1(v,w)$, $N_2(v,w)$, and $N_3(v,w)$. Setting $N[v,w] := N[v] \cup N[w]$, we define

$$\begin{aligned} N_1(v,w) &:= \{u \in N(v,w) \mid (N(u) \setminus N[v,w]) \neq \emptyset\}, \\ N_2(v,w) &:= \{u \in N(v,w) \setminus N_1(v,w) \mid (N(u) \cap N_1(v,w)) \neq \emptyset\}, \\ N_3(v,w) &:= N(v,w) \setminus (N_1(v,w) \cup N_2(v,w)). \end{aligned}$$

The right-hand diagram of Figure 7.3 shows an example which illustrates the partitioning of $N(v,w)$ into the subsets $N_1(v,w)$, $N_2(v,w)$, and $N_3(v,w)$.

Our second reduction rule—compared to Rule 1—is slightly more complicated.

Rule 2 *Consider $v, w \in V$ ($v \neq w$) and suppose that $N_3(v,w) \neq \emptyset$. Suppose that $N_3(v,w)$ cannot be dominated by a single vertex from $N_2(v,w) \cup N_3(v,w)$.*

Case 1 *If $N_3(v,w)$ can be dominated by a single vertex from $\{v,w\}$:*

(1.1) *If $N_3(v,w) \subseteq N(v)$ as well as $N_3(v,w) \subseteq N(w)$:*
- *remove $N_3(v,w)$ and $N_2(v,w) \cap N(v) \cap N(w)$ from G, and*
- *add two new vertices z, z' and edges $\{v,z\}, \{w,z\}, \{v,z'\}, \{w,z'\}$ to G.*

(1.2) *If $N_3(v,w) \subseteq N(v)$ but not $N_3(v,w) \subseteq N(w)$:*
- *remove $N_3(v,w)$ and $N_2(v,w) \cap N(v)$ from G, and*
- *add a new vertex v' and the edge $\{v,v'\}$ to G.*

(1.3) *If $N_3(v,w) \subseteq N(w)$ but not $N_3(v,w) \subseteq N(v)$:*
- *remove $N_3(v,w)$ and $N_2(v,w) \cap N(w)$ from G, and*
- *add a new vertex w' and the edge $\{w,w'\}$ to G.*

Case 2 *If $N_3(v,w)$ cannot be dominated by a single vertex from $\{v,w\}$:*
- *remove $N_3(v,w)$ and $N_2(v,w)$ from G, and*
- *add two new vertices v', w' and edges $\{v,v'\}$, $\{w,w'\}$ to G.*

Clearly, Cases (1.2) and (1.3) are symmetric to each other. Again, the newly added vertices v' and w' of degree one act as gadgets that enforce us to take v or w into an optimal dominating set. A special situation is given in Case (1.1). Here, the gadget added to the graph G simulates that at least one of the vertices v or w has to be taken into an optimal dominating set.

Lemma 7.17 *Let $G = (V,E)$ be a graph and let $G' = (V',E')$ be the resulting graph after having applied Rule 2 to G. Then $\gamma(G) = \gamma(G')$.*

Proof Similar to the proof of Lemma 7.15, vertices from $N_3(v,w)$ can only be dominated by vertices from $M := \{v,w\} \cup N_2(v,w) \cup N_3(v,w)$. All cases in Rule 2 are based on the fact that $N_3(v,w)$ needs to be dominated. All cases only apply if there is not a *single* vertex in $N_2(v,w) \cup N_3(v,w)$ which dominates $N_3(v,w)$.

We first of all discuss the correctness of Case (1.2) (and similarly the symmetric Case (1.3)): if v dominates $N_3(v,w)$ (and w does not), then it is optimal to take v into the dominating set—and at the same time still leave the option of taking vertex w. It is never better to take any combination of two vertices $\{x,y\}$ from the set $M \setminus \{v\}$. It may happen that we still have to take w to obtain a minimum dominating set, but in any case the vertex set $\{v,w\}$ dominates at least as many vertices as $\{x,y\}$. The "gadget edge" $\{v,v'\}$ simulates the effect of taking v. It is safe to remove $R := (N_2(v,w) \cap N(v)) \cup N_3(v,w)$ since, by taking v into the dominating set, all vertices in R are already dominated and since, as discussed above, it is always at least as good to take $\{v,w\}$ into a minimum dominating set as to take any other of the vertices from M.

In the situation of Case (1.1), we can dominate $N_3(v,w)$ by either v or w. Since we cannot decide at this point which of these vertices should be chosen to be in the dominating set, we use the "gadget" with vertices z and z' which simulates a choice between v or w. In any case, however, it is at least as good to take one of the vertices v and w (maybe both) than to take any other two vertices from M. The argument for this is similar to the one for Case (1.2). The removal of $N_3(v,w) \cup (N_2(v,w) \cap N(v) \cap N(w))$ is safe by a similar argument to the one that justified the removal of R in Case (1.2).

In Case 2, we need at least two vertices to dominate $N_3(v,w)$. Since $N(v,w) \supseteq N(x,y)$ for all pairs $x,y \in M$, it is optimal to take v and w into the dominating set, simulated by the gadgets $\{v,v'\}$ and $\{w,w'\}$. As in the previous cases, removing $N_3(v,w) \cup N_2(v,w)$ is safe since these vertices are already dominated and since these vertices need not be used for an optimal dominating set. □

It is easy to see that applying the reduction rules to planar graphs always results in a planar graph again. This is due to the fact that the removal of vertices and edges does not affect planarity and the gadget vertices (and edges) that are introduced by Rules 1 and 2 can clearly be drawn without causing edge crossings. Here, only Case (1.1) of Rule 2 needs a little care: since $N_3(v,w) \subseteq N(v)$ as well as $N_3(v,w) \subseteq N(w)$, the removal of $N_3(v,w)$ provides "space" for the (clearly planar) gadget drawn between v and w without any edge crossings.

Lemma 7.18 *Rule 2 can be carried out in time $O(n^2)$ for planar graphs and in time $O(n^4)$ for general graphs.*

Proof To prove the time bounds for Rule 2, basically the same ideas as for Rule 1 apply (compare the proof of Lemma 7.16). Instead of a depth-two search tree, one now has to argue on a search tree where the levels indicate the minimum of the distances to vertex v and w. Hence we associate the vertices v and w with the root of this search tree. The first level consists of all vertices that lie in $N(v,w)$ (that is, at distance one from either of the vertices v or w). Determining the subset $N_3(v,w)$ means to check whether some vertex on the first level has a neighbor on the second level. We do the same kind of construction as in Lemma 7.16. The running time is again determined by the size of the subgraph induced by the vertices that correspond to the root and the first level of this search tree, that is, by $G[N[v,w]]$ in this case. For planar graphs, we have $|G[N[v,w]]| = O(\deg(v) + \deg(w))$. Hence, we get $\sum_{v,w \in V} O(\deg(v) + \deg(w))$ as an upper bound on the overall running time in the case of planar graphs. Making use of the fact that $\sum_{v \in V} \deg(v) = O(n)$ for planar graphs, this is upper-bounded by

$$O\left(\sum_{v,w \in V} \deg(v) + \sum_{v,w \in V} \deg(w)\right) = O(n^2).$$

In the case of general graphs, we have $|G[N[v,w]]| = O\left((\deg(v) + \deg(w))^2\right)$, which trivially yields the upper bound

$$\sum_{v,w \in V} O\left((\deg(v) + \deg(w))^2\right) = O(n^4)$$

for the overall running time. □

We remark that the running times given in Lemmas 7.16 and 7.18 are pure worst-case estimates and the algorithms seem to be much faster in experimental studies than would be expected from the given worst-case bounds. In particular, for practical purposes it is important to see that Rule 2 can only be applied for vertex pairs that are at distance at most three from each other.

7.6.3 *Reduced graphs and the problem kernel*

We say that an application of a reduction rule leaves the graph *unchanged* if the "new" graph after applying the rule is isomorphic to the old one. Clearly, we are only interested in applications of the reduction rules that *change* the graph: Let $G = (V, E)$ be a graph such that both the application of Rule 1 and the application of Rule 2 leave the graph unchanged. Then we say that G is *reduced* with respect to these rules.

Observing that the (successful) application of any reduction rule always "shrinks" the given graph implies that there can only be $O(m)$ (where $m := |E|$) successful applications of reduction rules. This leads to the following.

Lemma 7.19 *A graph G can be transformed into a reduced graph G' with $\gamma(G) = \gamma(G')$ in time $O(n^3)$ in the planar case and in time $O(n^6)$ in the general case.*

Proof We prove the general statement that for a graph with m edges there can be at most $O(m)$ successful applications of reduction rules. The decisive claim we show is that after one application of Rule 1 or Rule 2 that changes the graph, the resulting graph has at most the same number of vertices, but at least one edge fewer than before the application of the rule.

Note that it is easy to verify that the total number of vertices never increases by applying the reduction rules. Now we go through Rule 1 and the various subcases of Rule 2, checking the validity of our claim. As to Rule 1, a change only occurs if there is more than one vertex affected by the rule—this means that more than one vertex and at least two edges are removed, whereas one vertex and one edge are newly introduced by the gadget.

Cases (1.2) and (1.3) of Rule 2 trivially fulfill the claim since only one gadget vertex and one gadget edge are introduced, but at least two $N_3(v, w)$ vertices together with at least two incident edges are deleted. The validity of Case 2 of Rule 2 also follows easily because clearly the rule never adds more than it deletes—at least two vertices together with their edges are removed. If a change takes place, however, more edges will be removed.

Finally, concerning Case (1.1) of Rule 2 we can observe that, although the gadget introduces two more vertices and four more edges, at least the same number of vertices and more than four edges are deleted. This is true because if this case applies, then at least two $N_3(v, w)$ vertices with edges to v as well as w each must exist. These and at least one additional edge will be deleted if a change takes place (otherwise, there were no changes).

This concludes the proof of the claim and the theorem follows by Lemmas 7.16 and 7.18 noting that $m = O(n)$ for planar graphs and $m = O(n^2)$ for general graphs. □

The kernelization procedure implied by the above reduction rules allows to prove the following main result. Herein, $\gamma(G)$ denotes the size of an optimal dominating set of graph G.

Theorem 7.20 *For a planar graph $G = (V, E)$ which is reduced with respect to Rules 1 and 2, we get $|V| = O(\gamma(G))$, that is,* DOMINATING SET IN PLANAR GRAPHS *admits a linear problem kernel.*

To present the proof of correctness of Theorem 7.20 is beyond the scope of this book. We refer to the literature for a complete exposition.

7.7 On lower bounds for problem kernels

Lower bounds are very hard to achieve in computational complexity analysis. Not surprisingly, relatively little is known about lower bounds on kernel sizes. There are two exceptions, though. First, one may make use of known lower bounds on the polynomial-time approximability of certain problems. Recent years have brought several deep results in this direction. Thus, if we know that a certain problem "cannot" be approximated better than a factor c, then we may often conclude that there is no kernel smaller than $c \cdot k$ where parameter k refers to the size of the solution set. More specifically, we can make the following concrete statement about problem kernels for VERTEX COVER with parameter k denoting the size of the desired vertex cover:

Theorem 7.21 VERTEX COVER *has no problem kernel formed by a graph with $1.36k$ vertices unless $P = NP$.*

Proof Assume that VERTEX COVER has a problem kernel with $1.36k$ vertices. Then this set of vertices yields a ratio-(1.36) polynomial-time approximation of an optimal solution. By deep results from approximation theory (based on the famous "PCP theorem") this is not possible unless $P = NP$ □

Based on the above approach, we may obtain further lower bounds for problem kernels employing deep results from the theory of polynomial-time approximation. We are not aware of any problem where the lower bound and the upper bound on the problem kernel are as close as for VERTEX COVER.

Now we discuss a fairly easy approach, based on elementary calculations and yielding surprising new results. The key observation made use of is based on "parametric duality". Again, we illustrate matters using VERTEX COVER. Recall that an n-vertex graph G has a vertex cover of size k iff G has an independent set of size $n - k$. In this sense, the parameters "size of a vertex cover" and "size of an independent set" are duals of each other. Moreover, let us focus attention on planar graphs. The Nemhauser–Trotter problem kernel for VERTEX COVER consisting of $2k$ vertices clearly also applies to the special case of planar graphs. In particular, however, due to the four-color theorem and the corresponding polynomial-time coloring algorithm, we also know that there is a size-$(4k)$ problem kernel for INDEPENDENT SET IN PLANAR GRAPHS. This observation can be used for a "two-sided" algorithmic attack on VERTEX COVER IN PLANAR GRAPHS, yielding the subsequent theorem. Herein, observe that the first approach exhibited with Theorem 7.21 does not work for VERTEX COVER IN PLANAR GRAPHS. The reason is that there is no constant-ratio lower bound for its polynomial-time approximability because there exists a polynomial-time approximation scheme (PTAS) with approximation ratio $(1 + \epsilon)$ for arbitrarily small $\epsilon > 0$.

Theorem 7.22 *For any $\epsilon > 0$,* VERTEX COVER IN PLANAR GRAPHS *has no problem kernel formed by a planar graph with $(4/3-\epsilon)\cdot k$ vertices unless $P = NP$.*

Proof Consider the polynomial-time kernelization for VERTEX COVER's dual problem INDEPENDENT SET IN PLANAR GRAPHS—now referred to as "4C"—with upper bound $4k' = 4(n - k)$ for the number of vertices in the problem kernel. Assume that there exists a size-$((4/3 - \epsilon) \cdot k)$ problem kernel for VERTEX COVER IN PLANAR GRAPHS. Call the corresponding polynomial-time kernelization algorithm "VK". Let (G, k) be an input instance of VERTEX COVER IN PLANAR GRAPHS and perform the following steps.

if $k \leq \frac{4}{4/3-\epsilon+4} \cdot |V|$
  then run algorithm VK on (G, k)
  else run algorithm 4C on $(G, |V| - k)$
fi

Let $G' = (V', E')$ be the graph of the "reduced instance" computed in the above way. Now, if $k \leq \frac{4}{4/3-\epsilon+4} \cdot |V|$ then

$$|V'| \leq \frac{4 \cdot (4/3 - \epsilon)}{4/3 - \epsilon + 4} \cdot |V| < |V|.$$

Otherwise, we conclude

$$|V'| \leq 4 \cdot (|V| - k) < 4 \cdot \left(\frac{4/3 - \epsilon}{4/3 - \epsilon + 4} \right) \cdot |V| < |V|.$$

Thus, in both cases we get $|V'| < |V|$. Clearly, after a linear number of steps we will arrive at a graph with a constant number of vertices—the VERTEX COVER problem can be trivially solved here. Altogether, this would yield a polynomial-time algorithm for the *NP*-complete VERTEX COVER problem, implying $P = NP$. □

We remark that in the same way we can show that there is no $(2-\epsilon)k$ vertices problem kernel for INDEPENDENT SET IN PLANAR GRAPHS unless $P = NP$. The above method can be generalized to other problems with "linear problem kernel sizes" for both the "primal" and the "dual" problem, one further example being DOMINATING SET IN PLANAR GRAPHS. The approach seems to be tailored for linear bounds, however; we are not aware of any nonlinear lower bound for problem kernels. Much remains to be done in the field for lower bounds for problem kernels.

7.8 Summary and concluding remarks

In this chapter we have seen several concrete examples of how to derive data reduction rules that yield a reduction to a problem kernel.

- In the case of MAXIMUM SATISFIABILITY we discussed a parameter-dependent kernelization which is based on a fairly simple observation. Moreover, this kernelization is algorithmically completely trivial and its practical use seems limited— a particular reason for this being the fact that no small parameter values are to be expected. In this sense the case of MAXIMUM SATISFIABILITY is related to INDEPENDENT SET IN PLANAR GRAPHS where a parameterization above guaranteed values would seem more appropriate—and leads to significantly harder algorithmic problems.
- With a parameter-dependent reduction to a problem-kernel for the CLUSTER EDITING problem, we encountered an example where the reasoning is similar in spirit to the obvious rule concerning Buss's reduction to a problem kernel for VERTEX COVER. Nevertheless, although the rules appear very natural, to prove their effectiveness—that is, to show a size bound on the problem kernel—is not easy.
- VERTEX COVER is the problem that appears in most places of this book. Also with respect to reduction to a problem kernel it plays a major role since we achieve an in a sense "optimal" problem kernel size and there are several ways to do so. The fact that there is more than one way to achieve this result and also how it is achieved by interesting relations to matching theory and linear programming shows that data reduction often is a technically demanding topic worth thorough research. In particular, the idea behind crown reduction rules has now been applied to several other problems besides VERTEX COVER.

- 3-HITTING SET is the VERTEX COVER problem adapted to hypergraphs with hyperedges built between at most three vertices. The parameter-dependent data reduction technique employed here resembles Buss's reduction to a problem kernel for VERTEX COVER. It remains open to investigate whether the more advanced kernelization techniques for VERTEX COVER can also be extended to this case, achieving a problem kernel of size $o(k^3)$. It would also be interesting to see whether the VERTEX COVER kernelization may help for the one of 3-HITTING SET. With DOMINATING SET IN PLANAR GRAPHS we presented "locality-based", parameter-independent data reduction rules. Whereas their correctness is relatively easy to show, the corresponding linear-size upper bound on the problem kernel requires a significant technical expenditure. Although being elementary in the sense that it does not use any deep graph-theoretical concepts or results, the lengthy and technical proof is omitted.
- The issue of (relative) lower bounds for problem kernel sizes, on the one hand, exhibits close connections to the theory of approximation algorithms. On the other hand, it demonstrates how in a surprising way upper bounds for a problem and its "dual" can lead to lower bounds for the very same problem as well.

When considering all the above-mentioned problems, note that the parameter is always related to the "size" of the solution we seek. This seems to be the natural thing in the case of problem kernelization—"structural parameters" such as the treewidth of graphs (see Chapter 10) seem to be less adequate for that purpose.

In summary, the design and analysis of good kernelization algorithms clearly is among the most important and practically most relevant contributions to fixed-parameter algorithmics for hard problems. The reason for that is the ubiquitous need for efficient preprocessing procedures in the algorithmics for hard problems. In addition, as the Nemhauser and Trotter problem kernel for VERTEX COVER and the lower bounds issue exhibit, there are close connections to the theory of approximation algorithms. These should be further pursued in future investigations. All in all, a good problem kernelization for a combinatorially hard parameterized problem is among the best and most valuable objectives from a practical as well as a theoretical view that a designer of fixed-parameter algorithms can achieve.

7.9 Exercises

1. Show that an exhaustive application of the data reduction rules for MULTICUT IN TREES presented in Section 1.3, namely the rules Idle Edge, Unit Path, Dominated Edge, and Dominated Path, leads to a reduced instance where there is either no demand path remaining or there is at least one demand path of length exactly two.
2. Show that the CLOSEST STRING problem as introduced in Section 5.3 has a problem kernel of size $k \cdot d$.

3. We subsequently propose several data reduction rules for MAXIMUM SATISFIABILITY. Decide on the correctness of these rules by giving a short argument each time.
   (a) If formula F contains a clause with only one literal, then set the corresponding variable accordingly and decrease the parameter k by one.
   (b) If a variable x occurs only positively in F, then set x to true, decrease parameter k by the number of clauses satisfied this way, and delete these clauses.
   (c) If formula F contains clauses (x) and $(\overline{x})$, then delete both clauses and decrease the parameter k by one.
   (d) If the variables x, y, and z occur in F exclusively in the subformula $(x \vee y) \wedge (\overline{y} \vee z) \wedge (\overline{x} \wedge \overline{z})$, then delete all three clauses and decrease parameter k by three.
   (e) If variable x occurs in F exclusively in the subformula $(x \vee y) \wedge (y \vee z) \wedge (\overline{x})$, then replace x by y and leave parameter k unchanged.
   (f) If variable x occurs in F exclusively in the subformula $(x \vee y) \wedge (y \vee z) \wedge (\overline{x})$, then replace x by $\overline{y}$, decrease parameter k by one, and delete the first clause.
4. Prove the correctness of the "folding reduction rule" for VERTEX COVER as described in Section 7.4.4.
5. Consider the following facility location problem where one is given n points in the Euclidean plane. The task is to select k of these as transmission stations such that each of the n points is within the radius of at least one sending station. Here, each transmission station shall have transmission radius 2 and two sending stations have to have distance at least 1 from each other. Find a reduction to a problem kernel with respect to parameter k.
6. The following problem, SORTING BY REVERSALS, originates in computational biology and occurs in the context of genome rearrangements. Consider a set of genes $G = \{1, 2, \ldots, n\}$. We denote strings over G where every gene occurs exactly once by

$$\pi = 0, e_1, e_2, \ldots, e_n, n+1$$

with two extra symbols 0 and $n+1$ to mark the start and the end of the string. A *reversal* $\mathrm{Rev}(\pi, i, j)$, $1 \leq i < j \leq n$, is an operation which transforms π into π' such that the substring from e_i to e_j gets reversed; that is, $\mathrm{Rev}(\pi, i, j)$ transforms

$$\pi = 0, e_1, e_2, \ldots, e_{i-1}, e_i, e_{i+1} \ldots e_{j-1}, e_j, e_{j+1} \ldots, e_n, n+1$$

into

$$\pi' = 0, e_1, e_2, \ldots, e_{i-1}, e_j, e_{j-1} \ldots e_{i+1}, e_i, e_{j+1} \ldots, e_n, n+1.$$

The problem SORTING BY REVERSALS is then defined as follows:
Input: Two strings π_1 and π_2 over $G = \{1, 2, \ldots, n\}$ and a nonnegative

integer d.
Task: Find a sequence of at most d reversals that transforms π_1 into π_2.
Hint: it is known that it can be assumed that a minimum number of reversals transforming π_1 into π_2 does not "destroy" a substring or its reverse that is part of both π_1 and π_2.
Show that SORTING BY REVERSALS can be reduced to a problem kernel consisting of only $O(d)$ genes.

7. Consider the following problem, MATRIX DOMINATION:
 Input: An $n \times n$-matrix M with entries 0 or 1 and a nonnegative integer k.
 Task: Find a set C of at most k nonzero-entries such that all other nonzero entries are in the same row or in the same column with at least one entry from C.
 Show a reduction to a problem kernel for MATRIX DOMINATION.
8. In Section 7.6 two data reduction rules for DOMINATING SET are introduced. Show that both rules are "orthogonal" to each other by
 (a) giving a graph that is reduced with respect to the first rule but can be further reduced by the second rule, and
 (b) by giving a graph that is reduced with respect to the second rule but can be further reduced by the first rule.

 Now consider the second rule and more precisely its subcase (1.2). Construct a graph which fulfills the prerequisites of this case and in which one of the two vertices v and w is not contained in any optimal dominating set.

7.10 Bibliographical remarks

Preprocessing of hard problems has been around since the very beginnings of algorithm research. It seems impossible to trace it back to one particular piece of research. Besides the exact, fixed-parameter algorithms that we study here, data reduction is important in all sorts of heuristic methods for combinatorial optimization. Note that data reduction techniques in the literature are often considered as *preprocessing* methods. It seems clear, however, that it is beneficial to combine and interleave them with the "main algorithm" such that data reduction techniques may appear in all phases of algorithm design. Downey and Fellows (1999) formalized data reduction for parameterized complexity purposes by introducing the concept of reduction to a problem kernel.

"When in doubt, sort" (referring to our starting motivating example in this chapter) is a quote from the algorithm design manual by Skiena (1998). The discussed data reduction for the railway optimization problem is due to Weihe (1998) and Weihe (2000); also see Mecke and Wagner (2004) and Nemhauser and Wolsey (1988). The technically costly proof for a problem kernel for MULTICUT IN TREES can be found in Guo and Niedermeier (2005*b*).

The simple data reduction for VERTEX COVER is attributed to Buss and appears in Buss and Goldsmith (1993) in Downey and Fellows (1999). The four-color theorem for planar graphs is due to Appel and Haken (1977*a*) and Appel and Haken (1977*b*); refer to Robertson *et al.* (1997) for later improvements. The polynomial-time algorithm to four-color a graph is given in Robertson *et al.* (1996). The observation that every fixed-parameter tractable problem is kernelizable goes back to Cai *et al.* (1997) (see also Alber (2003) for a recent presentation).

Studying the parameterized and exact complexity of MAXIMUM SATISFIABILITY was initiated by Mahajan and Raman (1999), where also the quadratic-size problem kernel as presented here was proven. Moreover, they introduced the study of parameterizing above guaranteed values here. Further research was pursued in Bansal and Raman (1999), Chen and Kanj (2004), and Niedermeier and Rossmanith (2000*b*). In particular, Niedermeier and Rossmanith (2000*b*) contains several simple data reduction rules for MAXIMUM SATISFIABILITY. Related studies for the still *NP*-complete case MAXIMUM-2-SATISFIABILITY appear in Gramm *et al.* (2003*a*).

The kernelization for CLUSTER EDITING appears in Gramm *et al.* (2005). The problem itself was studied in Shamir *et al.* (2004) under different complexity aspects. CLUSTER EDITING is also an important special case of the "correlation clustering" scenario as studied in Bansal *et al.* (2004). Its applications mainly lie in data clustering in fields such as computational biology or machine learning. Its *NP*-completeness is (implicitly) proven in Křivánek and Morávek (1986). CLUSTER EDITING falls into the category of graph modification problems whose parameterized complexity is studied in a broader context by Cai (1996). Finally, CLUSTER EDITING is also the central problem in the automated generation of search tree algorithms as pursued in Gramm *et al.* (2004).

The fundamental result concerning the size-$(2k)$ problem kernel for VERTEX COVER is due to Nemhauser and Trotter (1975) (see also Bar-Yehuda and Even (1985) and Khuller (2002)). Its applicability for parameterized complexity was observed in Chen *et al.* (2001). The folding technique is also described there. The other routes to a linear problem kernel for VERTEX COVER are described in Abu-Khzam *et al.* (2004), Fellows (2003*a*), and Langston and Suters (2005). The generalization of the described Nemhauser-Trotter kernelization in order to enumerate all optimal solutions is investigated in Chlebík and Chlebíková (2004). The hardness of approximating VERTEX COVER better than ratio 2 is studied in Khot and Regev (2003).

The size-$O(k^3)$ problem kernel for 3-HITTING SET is shown in Niedermeier and Rossmanith (2003*a*). Also the more general d-HITTING SET case for $d > 3$ is discussed there.

The linear-size problem kernel for DOMINATING SET IN PLANAR GRAPHS appears in Alber *et al.* (2004). Follow-up work with better upper bounds on the kernel size or generalizations to non-planar graphs appears in Chen *et al.* (2005) and Fomin and Thilikos (2004*a*). Some additional experimental work concerning these reduction rules is undertaken in Alber *et al.* (2003). As shown there, the data reduction rules exhibit good performance on other sparse but non-planar graphs.

Theorem 7.22 is due to Chen *et al.* (2005). The lower bound 1.36 for the approximation ratio of Vertex Cover appears in Dinur and Safra (2002).

To learn more about Sorting by Reversals (see exercises), refer to Hannenhalli and Pevzner (1996). Its *NP*-completeness is shown in Caprara (1999).

8

DEPTH-BOUNDED SEARCH TREES

When studying data reduction and problem kernels in Chapter 7 the focus was on polynomial-time (pre)processing of the input. The basic idea is to "cut away easy parts" of the given input instance, leaving behind the "really hard problem kernel". For *NP*-hard problems one usually cannot avoid using some exponential-time method to finally determine an optimal solution. A standard way to explore the huge search space related to a computationally hard problem is to perform a systematic exhaustive search. This can be organized in a tree-like fashion, which is the subject of this chapter. In the fixed-parameter context, the depths of these search trees are usually bounded from above by some numbers depending on the parameter values. Bounded search trees lie at the heart of many efficient fixed-parameter algorithms today.

The basic idea behind a systematic search by way of depth-bounded search trees is as follows: in polynomial time find a "small subset" of the input instance such that at least one element of this subset is part of an optimal solution to the problem. For instance, in the case of VERTEX COVER this "small subset" can be chosen as a two-element set consisting of the two endpoints of an edge—one of these two vertices must be part of the vertex cover. This leads to the previously mentioned (see Chapter 4) search tree of size $O(2^k)$, where the parameter k denotes the size of the vertex cover. In principle, a depth-bounded search tree coincides with the tree of recursive calls of the corresponding recursive algorithm, where the depth of the recursion is upper-bounded by the parameter value.

In the case of 3-HITTING SET, we have the same observation with size-three sets. Hence we obtain a size-$O(3^k)$ search tree here, where parameter k denotes the size of the hitting set. The "art" of constructing search trees lies in detecting more clever—and usually more complicated—ways of "defining" these small subsets. Both VERTEX COVER and 3-HITTING SET are subject to an elaborate search tree machinery leading to significantly reduced search tree sizes. The problematic nature of search trees, as we will see in the course of this section, may come from the fact that many small size search trees are based on using numerous case distinctions. These case distinctions require a complicated analysis and correctness proof and, when implemented, may cause administrative overhead. Thus the theoretically best search tree with the smallest worst-case size bound may not always be the best one in practice. Implementation and experiments must have the final say in this respect.

Before we deal with several examples of more or less intricate bounded search trees used in fixed-parameter algorithmics, let us first consider a very simple search tree, which, to the best of our knowledge, is the smallest in size known for that particular problem. Consider the INDEPENDENT SET problem:

> **Input:** A graph $G = (V, E)$ and a nonnegative integer k.
> **Task:** Find a subset $I \subseteq V$ with k or more vertices that form an independent set, that is, I induces an edgeless subgraph of G.

For general graphs, INDEPENDENT SET is $W[1]$-complete with respect to parameter k—there is no hope of fixed-parameter tractability. If INDEPENDENT SET is restricted to the class of planar graphs—where it nevertheless remains NP-complete—fixed-parameter tractability can be easily attained. There is an important property of planar graphs directly following form the well-known Euler formula that says that an n-vertex planar graph has at most $3n - 6$ edges: in every planar graph there is at least one vertex of degree five or smaller. Using this knowledge, our search strategy to find a size-k independent set in planar graphs is as follows. Pick a vertex v in G which has minimum degree (bounded by five) and branch into at most six cases. Either put v into the independent set or one of its, at most, five neighbors. In each branching case, delete the corresponding vertex together with all its adjacent edges and vertices from G. Thus, obtain a smaller graph G' in each case, and recursively search for an independent set of size $k - 1$ in each of these G'. This is correct because from the set $\{v\} \cup N(v)$, that is, the closed neighborhood of v, at least one vertex *must* be in a maximum independent set. Since the parameter in each branch decreases by one, we thus obtain a search tree of size $O(6^k)$. Using a suitable edge list representation of the $O(n)$ graph vertices and edges, picking a vertex and generating G' can easily be done in linear time $O(n)$. Altogether, we have that INDEPENDENT SET IN PLANAR GRAPHS can be solved in $O(6^k \cdot n)$ time, where k is the size of the independent set we search for.

Sometimes, such a tempting simple argument as that for INDEPENDENT SET IN PLANAR GRAPHS above can go wrong. Consider the DOMINATING SET problem restricted to planar graphs.

> **Input:** A planar graph $G = (V, E)$ and a nonnegative integer k.
> **Task:** Find a subset $S \subseteq V$ with at most k vertices such that every vertex $v \in V$ is contained in S or has at least one neighbor in S.

For general graphs, in a parameterized sense, DOMINATING SET is even "harder" than INDEPENDENT SET; that is, it is $W[2]$-complete. Can we use a similar argument as for INDEPENDENT SET in order to show fixed-parameter tractability of DOMINATING SET on planar graphs? Unfortunately, it is much harder to prove the existence of a bounded search tree here. The problem is the following. Assume that we want to argue along the same lines as we did for INDEPENDENT SET, that is, choosing a vertex of minimum degree, then branching on it by recursively solving the problem in each branch. It is true that for such a minimum-degree vertex v again either itself or one of its neighbors must be in the dominating set. A vertex that has become dominated, however, in the course of the algorithm can only be discarded from further consideration when all its neighbors are dominated as well. The point is that it could still be necessary to incorporate an already dominated vertex into the dominating set because it is needed to optimally dominate other vertices. To circumvent this problem, in Section 8.6 we will formulate a more general, "annotated" version of DOMINATING SET, where there are two kinds of vertices in our graph. The technical effort required to solve this problem increases significantly (see Section 8.6). Notably, the difficulty mainly comes from the mathematical analysis of correctness; the algorithm presented is still fairly easy.

Let us end this introductory part by returning to the VERTEX COVER problem and taking a first glimpse at how to obtain a search tree size $o(2^k)$ by means of case distinctions. In particular, the subsequent consideration motivates a closer look at how to determine upper bounds on search tree sizes. There is a fairly simple "degree-branching strategy" that beats the trivial $O(2^k)$-bound for VERTEX COVER search trees. Here, let V' be the vertex cover of size at most k that we are searching for:

1. If there is a vertex of degree one, then put its neighbor into V'.
2. If there is a vertex v of degree two, then either put v into V' together with all neighbors of its neighbors or put both neighbors of v into V'.
3. If there is a vertex v of degree at least three, then put either v or all its neighbors into V'.

Actually, we could even avoid the second step when making use of the folding of degree-two vertices as discussed at the end of Section 7.4. Since we have a different focus here, for the time being we omit these considerations. It is not hard to see that the above search strategy always leads to an optimal solution. This is simply based on the fact that to cover the edges adjacent to v either v or all its neighbors must be in the vertex cover—the second step needs a little thought, though: basically, the second step is based on branching to put either v or both its neighbors into the vertex cover. In the first branch, however, we can argue that we may even bring all the neighbors of v's neighbors into the vertex cover because of the following. Assume that there would exist a minimum vertex cover containing v and one of its neighbors. Then changing this set by replacing v by its second neighbor clearly yields a minimum vertex cover as well. This will be found in the second branching case. Hence, if there should be a vertex cover *smaller* than the one that contains both v's neighbors then it must contain v and it *must not* contain its neighbors. This implies that all neighbors of v's neighbors have to be part of this vertex cover as well. In the algorithm's second and third steps the search branches into two cases each time. In the second step each branch puts at least two vertices into V'. In the third step the first branch puts one vertex into V' and the second branch puts at least three vertices into V'.

This branching process is recursively repeated until an optimal solution is found. If the solution has size k, the corresponding search tree has size bounded from above by $O(1.47^k)$. How do we obtain this bound? The recursive construction of the search tree makes it possible to analyze its size with the help of simple recurrences which upper-bound the search tree sizes. These recurrences can be solved using standard mathematical tools—see Section 8.1 for more on that.

In what follows, we will start with basic facts about depth-bounded search trees. In particular, we discuss how to determine search tree sizes by solving recurrences with standard mathematical tools. After that we will study various concrete problems and the corresponding search tree strategies. Emphasis is laid on the fact that in some cases it is not "clear" how the depth-bounded search tree paradigm can be applied, whereas in other cases the challenge is to shrink the size of the search trees as much as possible. The latter often involves complicated case distinctions. To this end, frequently the combination with data reduction rules is very fruitful. Moreover, the interleaving of search trees with repeated application of data reduction may further accelerate the solution finding process; see Section 8.7. Finally, we discuss how to replace the error-prone and messy "case distinction business" in the construction of small search trees by a mechanized approach using the help of computers; see Section 8.8. The specific problems we study include CLUSTER EDITING, VERTEX COVER, HITTING SET, CLOSEST STRING, and DOMINATING SET IN PLANAR GRAPHS.

8.1 Basic definitions and facts

The central theme in this chapter is the development of "small" search trees that lead to efficient fixed-parameter algorithms. These search trees are nothing but the trees of recursive calls of the underlying algorithms. Depending on the structure of the recursion, determining an upper bound on the search tree size will require more or less mathematical effort. The determination of the sizes of the above depth-bounded search trees for VERTEX COVER, 3-HITTING SET, and INDEPENDENT SET IN PLANAR GRAPHS was trivial: we branched into either two, three, or six cases, in each of which the parameter value could be decreased by exactly one. This is due to the fact that in each case we selected exactly one element for inclusion into the solution set to be constructed. Thus we had extremely regular branchings, and since the parameter value (and thus the depth of the search tree) was bounded by k, we ended up with search tree sizes $O(2^k)$, $O(3^k)$, and $O(6^k)$ in the respective problems. The upper bounds for VERTEX COVER and 3-HITTING SET, however, can be significantly improved. These improvements on 2^k and 3^k, respectively, rely heavily on more complicated branching strategies with numerous case distinctions. In particular, in one branching step often more than one element is selected for inclusion into the desired set. In addition, the numbers of selected elements may differ in different branches. This leads to more complicated recursive algorithms, and to estimate the worst-case sizes of the corresponding depth-bounded search trees requires rigorous mathematical analysis. Fortunately, the necessary tools are readily available and are easy to use. This is explained next.

Search tree algorithms work in a recursive manner. The number of recursion calls is the number of nodes in the according tree. This number is governed by linear recurrences with constant coefficients. It is well known how to solve these, and the asymptotic solution is determined by the roots of the so-called *characteristic polynomial.* If the algorithm solves a problem of size n and calls itself recursively for problems of sizes $n-d_1, \ldots, n-d_i$, then $(d_1, \ldots, d_i)$ is called the *branching vector* of this recursion. It corresponds to the recurrence

$$T_n = T_{n-d_1} + \cdots + T_{n-d_i}. \tag{8.1}$$

Observe that in recurrence (8.1) we actually give a recurrence to estimate the number of leaves of a search tree. This is sufficient because for non-degenerate trees (that is, all inner nodes have at least two children), as we always consider here, the leaves make more than half of the total number of tree nodes. In addition, also note that we assume that $T_0 = T_1 = \ldots = T_{d_i-1} = 1$ for the termination of the recursion. Again, the validity of this is clear: we assume that constant-size problems can be solved by one final recursive call. To solve the recurrence, now the characteristic polynomial comes into play. The characteristic polynomial of recurrence (8.1) is

$$z^d - z^{d-d_1} - \cdots - z^{d-d_i}, \tag{8.2}$$

where $d := \max\{d_1, \ldots, d_i\}$.

Now, we describe a simple scheme how to solve recurrence equations in the form of (8.1). Let us assume that recurrence (8.1) has a solution of the form $T_n = \alpha^n$ for some real or even complex number α. Setting $n := d$ and plugging this into recurrence (8.1) we obtain

$$\alpha^d = \alpha^{d-d_1} + \ldots + \alpha^{d-d_i},$$

that is, α is a zero of the characteristic polynomial (8.2). In the same way, it is also not hard to see the converse direction if α is a zero of the characteristic polynomial then α^n is a solution of the recurrence (8.1). Moreover, with some more effort one can show that if α is a jth zero of the characteristic polynomial (8.2) then $n^l \cdot \alpha^n$ for all $0 \leq l < j$ is a solution of recurrence (8.1). This finally leads to the following result, which we state without proof here.

Proposition 8.1 *A depth-bounded search tree with branching vector* $(d_1, \ldots, d_i)$ *and its root labeled with parameter value* n *has size* $n^{O(1)} \cdot |\alpha|^n$, *where* α *is the zero of the corresponding characteristic polynomial.*

The base α of the exponentially growing function in Proposition 8.1 is called the *branching number.* Obviously, the ordering of the entries of a branching vector plays no role in the mathematical analysis; it may only reflect the order of the recursive calls corresponding to each vector entry. We mention in passing that in the examples in the remainder of this chapter the zero with the largest absolute value is always a single root, so the polynomial multiplicative factor can be replaced by a constant.

In most concrete applications we will have *sets* of recurrence equations, each describing a particular case of the recursive branching. Clearly, an upper worst-case bound for the overall search tree size then simply derives by analyzing every recurrence equation separately and then simply take the maximum branching number as the base for the exponential term to bound the search tree size from above. Thus, in the examples to follow, the size of the search tree is $O(\alpha^k)$, where k is the parameter and α is the biggest branching number that will occur. For instance, in an intricate search tree algorithm for VERTEX COVER, among others, the branching vectors $(3, 5, 7)$, $(4, 5, 8, 9, 9)$, and $(3, 5, 8, 8)$ occur. Here, it is no longer obvious which one the branching with the largest branching number is. By solving the corresponding recurrences—which can be done using any standard computer algebra system or by simply determining the zeros of the characteristic polynomials—we obtain the respective (approximate) branching numbers 1.273739, 1.290649, and 1.291743. The last number gives the worst case and the corresponding search tree algorithm indeed has the worst-case bound $O(1.291743^k)$ on its search tree size.

One more thing to learn from the above brief description is that all the bounds for search tree sizes we give are *worst-case* estimates. They are based on the determination of worst-case branching vectors and branching numbers. It is quite conceivable that the average-case sizes of the constructed search trees are significantly smaller in many cases. Since average-case complexity analysis is a very elusive matter, the normal way to find out the typical average-case and practical behaviour of a search tree algorithm is through implementation and experiments. Experience tells us that additional heuristic improvements can often be incorporated and the resulting algorithm then frequently performs much better than would be expected from sole consideration of the proven worst-case bounds. Finally, "iterative branching" is a worthwhile thing to do as a means of perhaps further improved upper bounds and branching strategies. Roughly speaking, the idea is to pick special branches iteratively in a skilful way in order to keep some kind of invariant concerning "nice" branching situations which lead to good branching vectors.

8.2 Cluster Editing

Our first more demanding example concerning the application of the depth-bounded search tree paradigm is taken from the field of graph modification problems. We describe a recursive algorithm for CLUSTER EDITING that follows the bounded search tree paradigm. Recall that in Section 7.3 we have seen a reduction to a problem kernel for this problem. The problem is as follows:

Input: A graph $G = (V, E)$ and a nonnegative integer k.
Task: Find out whether we can transform G, by deleting or adding at most k edges, into a graph that consists of a disjoint union of cliques.

The basic idea underlying the depth-bounded search tree is to identify a "conflict triple" consisting of three vertices and to branch into subcases to repair this "conflict" by adding or deleting edges between the three considered vertices. Thus we invoke recursive calls on instances which are simplified in the sense that the value of the parameter is decreased by at least one. Central to the branching strategy is the following observation. In fact, our branching strategy will make use of a *forbidden subgraph characterization* of graphs that are disjoint unions of cliques.

Lemma 8.2 *A graph $G = (V, E)$ consists of disjoint cliques iff there are no three distinct vertices $u, v, w \in V$ with $\{u, v\} \in E$, $\{u, w\} \in E$, but $\{v, w\} \notin E$.*

Proof If a graph is a collection of disjoint cliques then it clearly contains no path induced by three vertices, proving the "if"-direction. For the reverse direction, if there are $u, v, w \in V$ with $\{u, v\}, \{u, w\} \in E$ but $\{v, w\} \notin E$ then the connected component containing u, v, and w cannot be a clique, which is in contradiction to the fact that G is a collection of cliques. □

Lemma 8.2 says that a graph is a disjoint union of cliques iff it contains no P_3 (a path of three vertices) as an induced subgraph. Lemma 8.2 implies that, if a given graph does not consist of disjoint cliques, then we can find a conflict triple of vertices between which we must either insert or delete an edge in order to transform the graph into disjoint cliques. In the following, we describe the recursive procedure that results from this observation. Inputs are a graph $G = (V, E)$ and a nonnegative integer k, and the procedure reports, as its output, whether G can be transformed into a union of disjoint cliques by deleting and adding at most k edges.

- If the graph G is already a union of disjoint cliques, then we are done: report the solution and return.
- Otherwise, if $k \leq 0$, then we cannot find a solution in this branch of the search tree: return.
- Otherwise, identify $u, v, w \in V$ with $\{u, v\} \in E$, $\{u, w\} \in E$, but $\{v, w\} \notin E$ (they exist with Lemma 8.2). Recursively call the branching procedure on the following three instances consisting of graphs $G' = (V, E')$ with nonnegative integer k' as specified below:
  (B1) $E' := E \setminus \{\{u, v\}\}$ and $k' := k - 1$.
  (B2) $E' := E \setminus \{\{u, w\}\}$ and $k' := k - 1$.
  (B3) $E' := E \cup \{\{v, w\}\}$ and $k' := k - 1$.

Proposition 8.3 *There is a size-$O(3^k)$ search tree for* Cluster Editing.

Proof The recursive procedure suggested above is obviously correct and directly implies a search tree of size $O(3^k)$. □

Note that so far we have ignored the overall running time behind the above search tree algorithm. It is easy to see, however, that multiplying the search tree size $O(3^k)$ by a polynomial factor clearly bounds the running time from above. In fact, a cubic factor suffices.

The "art of case distinction"—still comparatively modest in case of Cluster Editing— now leads us to a search tree of size $O(2.27^k)$. The preceding branching strategy can be easily improved as described in the following. We still identify

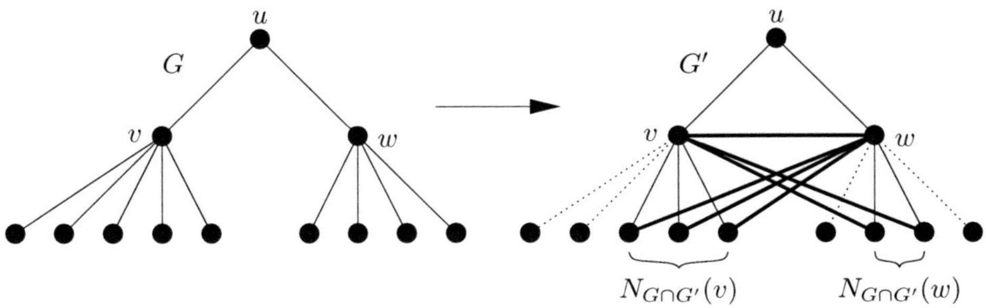

FIG. 8.1. In case (C1), adding edge $\{v, w\}$ does not need to be considered. Here, G is the given graph and G' is a clustering solution of G by adding edge $\{v, w\}$. The dashed lines denote edges being deleted to transform G into G', and the bold lines denote edges being added. Observe that the drawing only shows that parts of the graphs (in particular, edges) which are relevant for our argument.

a conflict triple of vertices, that is, $u, v, w \in V$ with $\{u, v\} \in E$, $\{u, w\} \in E$, but $\{v, w\} \notin E$. Based on a case distinction, we provide for every possible situation additional branching steps. The amortized analysis of successive branching steps, then, yields the better worst-case bound on the search tree size. To this end, in the same way as in Section 7.3, we make use of two annotations for unordered vertex pairs:

"*permanent*": In this case, $\{u, v\} \in E$ and it is not allowed to delete $\{u, v\}$;

"*forbidden*": In this case, $\{u, v\} \notin E$ and it is not allowed to add $\{u, v\}$.

Clearly, if an edge $\{u, v\}$ is deleted, then the vertex pair is made *forbidden*. If an edge $\{u, v\}$ is added, then the vertex pair is made *permanent*.

We start by distinguishing three main situations that may apply when considering the conflict triple u, v, w:

(C1) Vertices v and w do not share a common neighbor, that is, $\forall x \in V, x \neq u : \{v, x\} \notin E$ or $\{w, x\} \notin E$.

(C2) Vertices v and w have a common neighbor $x \neq u$ and $\{u, x\} \in E$.

(C3) Vertices v and w have a common neighbor $x \neq u$ and $\{u, x\} \notin E$.

Regarding case (C1), the following lemma shows that a branching into two cases (B1) and (B2) as described in the preceding algorithm suffices.

Lemma 8.4 *Given a graph $G = (V, E)$, a nonnegative integer k and $u, v, w \in V$ with $\{u, v\} \in E$, $\{u, w\} \in E$, but $\{v, w\} \notin E$, then if v and w do not share a common neighbor besides u, then branching case* (B3) *cannot yield a better solution than cases* (B1) *and* (B2), *and it can therefore be omitted.*

Proof Consider a clustering solution G' for G where we did add $\{v, w\}$ (see Figure 8.1 for an example). We use $N_{G\cap G'}(v)$ to denote the set of vertices

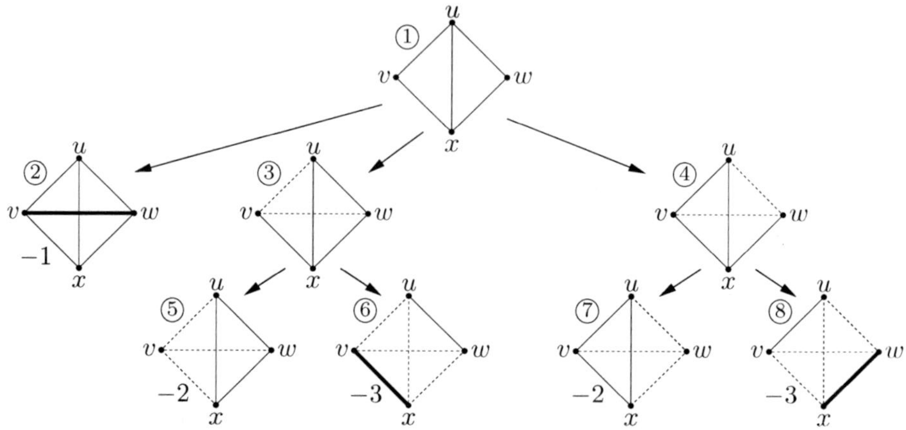

FIG. 8.2. Branching for case (C2). Bold lines denote permanent edges and dashed lines forbidden edges.

which are neighbors of v in G and in G'. Without loss of generality, assume that $|N_{G\cap G'}(w)| \leq |N_{G\cap G'}(v)|$. We then construct a new graph G'' from G' by deleting all edges adjacent to w. It is clear that G'' is also a clustering solution for G. We compare the cost of the transformation $G \to G''$ to that of the transformation $G \to G'$:

- -1 for not adding $\{v, w\}$,
- $+1$ for deleting $\{u, w\}$,
- $-|N_{G\cap G'}(v)|$ for not adding all edges $\{w, x\}$, $x \in N_{G\cap G'}(v)$,
- $+|N_{G\cap G'}(w)|$ for deleting all edges $\{w, x\}$, $x \in N_{G\cap G'}(w)$.

Here we have omitted possible vertices which are neighbors of w in G' but not neighbors of w in G because they would only increase the cost of transformation $G \to G'$.

In summary, the cost of $G \to G''$ is not higher than the cost of $G \to G'$, that is, we do not need more edge additions and deletions to obtain G'' from G than to obtain G' from G. □

Lemma 8.4 shows that in case (C1) a branching into two cases is sufficient, namely to recursively consider graphs $G_1 = (V, E \setminus \{\{u, v\}\})$ and $G_2 = (V, E \setminus \{\{u, w\}\})$, each time decreasing the parameter value by one.

For case (C2), we change the order of the basic branching. In the first branch, we add edge $\{v, w\}$. In the second and third branches, we delete edges $\{u, v\}$ and $\{u, w\}$, as illustrated by Figure 8.2.

- Add $\{v, w\}$ as labeled by ② in Figure 8.2. The cost of this branch is 1.
- Make $\{v, w\}$ *forbidden* and delete $\{u, v\}$, as labeled by ③. This creates the new conflict triple u, v, x. To resolve this conflict, we make a second

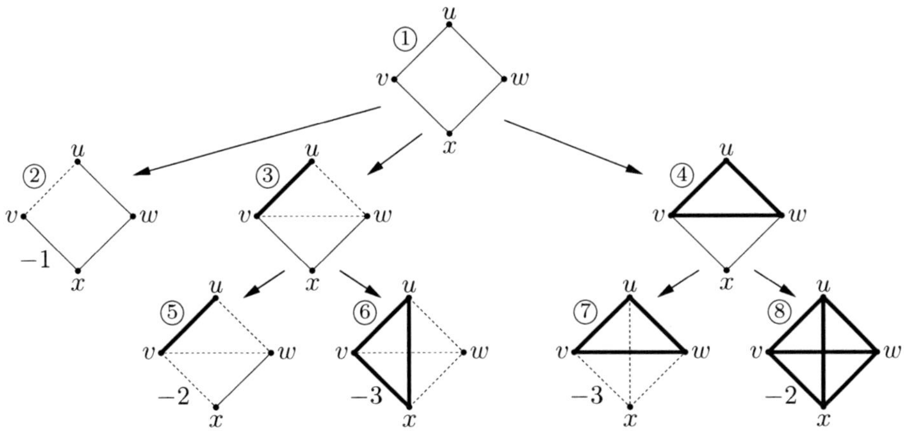

FIG. 8.3. Branching for case (C3).

branching. Since adding $\{u, v\}$ is forbidden, there are only two branches to consider:

- ∗ Delete $\{v, x\}$, as labeled by ⑤. The cost is 2.
- ∗ Make $\{v, x\}$ *permanent* and delete $\{u, x\}$. With reduction Rule 2 from Section 7.3, we then delete $\{w, x\}$, too, as labeled by ⑥. The cost is 3.

- Make $\{v, w\}$ *forbidden* and delete $\{u, w\}$ (④). This case is symmetric to the previous one, so we have two branches with costs 2 and 3, respectively.

In summary, the branching vector for case (C2) is $(1, 2, 3, 2, 3)$.

For case (C3), we perform a branching as illustrated by Figure 8.3:

- Delete $\{u, v\}$, as labeled by ②. The cost of this branch is 1.
- Make $\{u, v\}$ *permanent* and delete $\{u, w\}$, as labeled by ③. With Rule 2, we can additionally make $\{v, w\}$ *forbidden*. We then identify a new conflict triple u, v, x. Not being allowed to delete $\{u, v\}$, we can make a 2-branching to resolve the conflict:
  - ∗ Delete $\{v, x\}$, as labeled by ⑤. The cost is 2.
  - ∗ Make $\{v, x\}$ *permanent*. This implies $\{u, x\}$ needs to be added and $\{w, x\}$ needs to be deleted due to reduction rule 2, as labeled by ⑥. The cost is 3.
- Make $\{u, v\}$ and $\{u, w\}$ *permanent* and add $\{v, w\}$, as labeled by ④. Vertices u, w, and x form a conflict triple. To solve this conflict without deleting $\{u, w\}$, we make a branching into two cases:
  - ∗ Delete $\{w, x\}$ as labeled by ⑦. We then also need to delete $\{v, x\}$. The cost is 3. Additionally, we can make $\{u, x\}$ *forbidden*.
  - ∗ Add $\{u, x\}$, as labeled by ⑧. The cost is 2. Additionally, we can make $\{u, x\}$ and $\{v, x\}$ *permanent*.

It follows that the branching vector for case (C3) is $(1, 2, 3, 3, 2)$.

In summary, this leads to a refinement of the branching with a worst-case branching vector of $(1, 2, 2, 3, 3)$, yielding branching number 2.27. Since the recursive algorithm stops whenever the parameter value has reached 0 or below, we obtain a search tree size of $O(2.27^k)$. This gives the following result.

Theorem 8.5 *There is a size-$O(2.27^k)$ search tree for* CLUSTER EDITING.

In fact, combining the reduction to a problem kernel from Section 7.3 with the search tree shown here and applying the interleaving technique as discussed in Section 8.7, one can show that CLUSTER EDITING can be solved in $O(2.27^k + |V|^3)$ time. Note, however, that already data reduction Rule 2 from Section 7.3 has played a decisive role in the above search tree strategy. Thus it gives a first example of the fact that data reduction rules are frequently also of central importance in the development of branching strategies, and that they improve their behaviour in the sense of upper bounds on the search tree size. A further reduction of the search tree size is possible based on computer-assisted analysis. This will be explored in Section 8.8.

8.3 Vertex Cover

Recall the problem definition:

> **Input:** A graph $G = (V, E)$ and a nonnegative integer k.
> **Task:** Find a subset $C \subseteq V$ with k or fewer vertices such that each edge in E has at least one of its endpoints in C.

VERTEX COVER has a long history of more and more improved search tree sizes, the state of the art now being $O(1.28^k)$. Here we give a complete and self-contained description of a search tree of size $O(1.33^k)$, which should be seen as a compromise between reasonable complexity of description and a good worst-case upper bound. Other than in the case of CLUSTER EDITING we do not employ a forbidden subgraph characterization here, but our branching strategy is that of "degree-branching". Based on the various vertex degrees occurring in the input graph, we distinguish several cases for recursive calls. More specifically, here we consider degrees from one up to at least five. Note that in the more complex case distinctions leading to better upper bounds on the search tree size vertices of degree five and six are considered separately. In what follows, without loss of generality, we assume that the graph is connected because connected components can be handled completely independently of each other.

The algorithm finds recursively an optimal vertex cover as follows. Given a graph G, we choose several subgraphs $G_1, \ldots, G_i$ and compute optimal vertex covers for all of them. From these we can construct an optimal vertex cover for G. For example, let x be some vertex of G and let G_1 be the subgraph that results from G by deleting x and all incident edges. A vertex cover of G_1, together with x, then is a vertex cover of G. Moreover, if there is an optimal vertex cover for G that contains x, then we can construct an optimal vertex cover from an optimal vertex cover of G_1. Otherwise, if no optimal vertex cover of G contains x, then it must contain all neighbors of x. Hence, let G_2 be the graph that results from G by deleting all neighbors of x. Again, we can construct a vertex cover of G by taking a vertex cover of G_2 and adding all neighbors of x. If we start from optimal vertex covers for G_1 and G_2, then one of the resulting covers for G must be optimal, since either x or its neighbors must be part of any vertex cover. We say we *branch according to x and $N(x)$*, where $N(x)$ denotes the neighbors of x. In the first branch, x will be part of the vertex cover and in the second branch $N(x)$ will be part of it. The vertex cover constructed grows in size with each step. Since its size cannot exceed k, the algorithm terminates, implicitly having constructed a search tree of depth upper-bounded by k.

In principle this is the way our algorithm works, but we choose the subgraphs $G_1,\ldots,G_i$ in a more complicated way and branch according to much more complicated sets. Subsequently, we call a graph *regular* if all its vertices have exactly the same degree. If the graph is connected, the rules for how to choose these branching sets are as follows.

1. If there is a vertex x with degree one, then choose $N(x)$ to be in the vertex cover. There is no other branch since there is always an optimal vertex cover that contains $N(x)$ and does not contain x.
2. If there is a vertex x with degree at least five, then branch according to x and $N(x)$.
3. If 1. and 2. do not apply and if the graph is regular, then choose any vertex x and branch according to x and $N(x)$. (This can happen at most three times in each path of the search tree and increases its size at most by a small constant factor.)
4. If there are no vertices with degree one or at least five but there is a vertex with degree two, then proceed as shown in the subsequent case distinction.
5. If 1.–4. do not apply and if there is a vertex with degree three then proceed as shown in the subsequent case distinction.

Before going into the details of the subcases concerning the vertices of degree two and three, let us first verify that the above case distinction is complete, that is, that it covers all cases that may occur. The only subtle point herein is to ask why we do not have to study vertices of degree four in a separate case: after the fourth case, we know that the graph *has* to consist of at least one degree-three and at least one degree-four vertex. Hence the fifth case *always* applies and there is no need for extra consideration of degree-four vertices, because this situation will be subsumed by one of the given cases. It thus remains to give a closer description of the two cases concerning degree-two and degree-three vertices. We study several subcases in the first and second cases.

Degree-two vertices. Let x be a degree-two vertex with two neighbors a and b. We distinguish three cases.

1. If there is an edge between a and b then bring both a and b into the vertex cover. No branching is necessary because x, a, and b form a triangle—thus,

two of the three vertices must be chosen anyway—and a and b cover a superset of edges covered by any other "two out of three" combination.

2. If there is no edge between a and b but a and b both have degree two with a common neighbor c different from x, then bring x and c into the vertex cover. No branching is necessary because x and c cover a superset of the edges covered by choosing any other pair of vertices from the set $\{x, a, b, c\}$.
3. Otherwise, we know that $|N(a) \cup N(b)| \geq 3$ and we can branch according to $N(v)$ and $N(a) \cup N(b)$. At least one of these two branches leads to an optimal solution: the first branch deals with the case that x is not part of an optimal vertex cover, and hence all its neighbors must be. The second branch then of course means that x is in an optimal vertex cover. Then, however, we can infer even more. The central point is that it is not necessary to search for an optimal vertex cover containing x and a or x and b. The reason is that choosing a and b must also then give an optimal vertex cover. Hence, the only possibility that choosing x could lead to a *smaller* vertex cover than choosing a and b already does may occur when we choose x and neither a nor b. This implies that we can choose $N(a) \cup N(b)$ in the second branch. The corresponding branching vector is at least $(2, 3)$ because $|N(a) \cup N(b)| \geq 3$. The branching number then is smaller than 1.33.

Degree-three vertices. Let x be a degree-three vertex with its three neighbors a, b, and c. We distinguish four subcases where one of these cannot occur.

1. Assume that x is part of a triangle, for instance, let $\{x, a, b\}$ be the triangle (but there can be more triangles). Then we can branch according to $N(x)$ and $N(c)$. If x is not part of the vertex cover, then $N(x)$ is. If x is part of the vertex cover, then either a or b is in order to cover all triangle edges. If c is also in the cover, then two neighbors of x are and the set $N(x)$ covers a superset of the edges. Hence, to get a smaller vertex cover than by choosing $N(x)$, the only chance is to choose $N(c)$ which includes x. The branching vector is at least $(3, 3)$. The branching number is smaller than 1.27.
2. Assume that x is part of a cycle of length four consisting of a, x, b, and d. Here d is a new vertex. Then we can branch according to $N(x)$ and $\{x, d\}$. If x is not part of the vertex cover, then $N(x)$ is. If x is part of the vertex cover, then not choosing d would mean that we would have to choose both a and b together with x. This, however, cannot give a smaller vertex cover than choosing $N(x)$ already does. Hence we can confine the second branch to picking $\{x, d\}$. The branching vector is as least $(3, 2)$. The branching number then is smaller than 1.33.
3. Assume that there is no edge between a, b, and c, and without loss of generality $N(a) = \{x, a_1, a_2, a_3\}$ with $|N(a)| = 4$. Then we branch according to $N(x)$ and $N(a)$ and $\{a\} \cup N(b) \cup N(c)$. If x is not part of the vertex cover, then $N(x)$ is. If x is part of the vertex cover, then further distinguish two cases. If a is not part of the vertex cover, then $N(a)$ is. Thus it remains to

study what happens when both x and a are chosen. As in previous cases, we can infer that then it is of no interest to choose additionally b or c. This leads to the choice of $\{a\} \cup N(b) \cup N(c)$ which contains a. The branching vector is at least $(3, 4, 6)$ because according to the previous cases we can conclude that $N(b) \cap N(c) = \{x\}$, and hence $|N(b) \cup N(c)| \geq 5$, implying that $|\{a\} \cup N(b) \cup N(c)| \geq 6$ because $a \notin N(b)$ and $a \notin N(c)$. The branching number is smaller than 1.31.

4. The situation that seems to remain now is that we have to consider a degree-three vertex all of whose neighbors have degree three. Since, however, in the course of the search tree algorithm the degree of every graph vertex is a monotonically decreasing function and since regular graphs are handled in the third main case, we must always be able to find a degree-three vertex with a degree-four neighbor here and the preceding case applies.

This concludes the case distinction for degree-three vertices and the whole search tree algorithm. Observe that the above example shows that one can avoid some case distinctions by "discussing away" certain situations—the example here was the observation concerning regular graphs. Often, and in particular in the case of VERTEX COVER, one can somewhat improve the branchings by studying what happens after having applied a certain branching. It may happen that afterwards a favorable situation with a particularly good branching vector always occurs that can be made use of to improve the upper bounds. The idea is to merge two subsequent recursive calls into one big one. In the same direction we find the idea of "iterative branching" where one tries to guide the selection of branching cases and which one to apply with the goal of always generating new situations where somehow better branchings than the worst case can be applied. See the literature for more on this.

To summarize, the presented case distinction led us to branching numbers 1.33, 1.27, and 1.31. We can infer an upper worst-case bound on the search tree as follows.

Theorem 8.6 *There is a size-$O(1.33^k)$ search tree for* VERTEX COVER.

In other words, Theorem 8.6 says that VERTEX COVER can be solved with a branching algorithm as described above that performs $O(1.33^k)$ recursive calls. The currently best fixed-parameter algorithms for VERTEX COVER—which are based on depth-bounded search trees—have a search tree size of $O(1.28^k)$.

8.4 Hitting Set

The next example is a depth-bounded search tree for d-HITTING SET for some fixed positive integer d. Here the determination of the size of the search tree is a little more involved than usual. Recall from Section 7.5 that we already know that 3HS has a problem kernel of size $O(k^3)$. The definition of d-HITTING SET for $d \geq 3$ is as follows.

Input: A collection $\mathcal{C}$ of subsets of size at most d of a finite set S and a nonnegative integer k.
Task: Find a subset $S' \subseteq S$ with $|S'| \leq k$ such that S' contains at least one element from each subset in $\mathcal{C}$.

We assume that no subset occurs more than once within the collection. Equally, d-HITTING SET can be seen as a vertex cover problem for *hypergraphs:* Interpret the elements of S as vertices and interpret the size-d subsets as hyperedges. Thus a hyperedge may join three vertices. Then, 3-HITTING SET requires the *covering* of all these three-element sets (hyperedges) by elements (vertices) completely analogously to VERTEX COVER.

A special property of the subsequent mathematical analysis of the search tree size is that in contrast to previous estimates of search tree sizes, we use a *system* of recurrence equations.

Before we describe the search tree approach, note that there are some special cases, which are always considered first. For instance, if there is a singleton $\{x\}$ in our collection, we must clearly put x into our hitting set. A simple but important concept is that of *domination*. An element x is dominated by an element y if, whenever x occurs in a set of the collection, y occurs in this set as well. In this case, we can delete x from the sets without repercussion.

The trivial search tree algorithm for d-HITTING SET tries all d possibilities for a set of d elements and yields a search tree of size $O(d^k)$. Our algorithm is better, having a search tree of size $O(\alpha^k)$, where

$$\alpha = \frac{d-1}{2} + \frac{d-1}{2}\sqrt{1 + \frac{4}{(d-1)^2}} = d - 1 + \frac{1}{d-1} + O(d^{-3}) = d - 1 + O(d^{-1}).$$

The algorithm proceeds as follows.

1. Eliminate all dominated elements.
2. Choose some set $s = \{x_1, x_2, \dots, x_d\}$.
3. Branch according to the following possibilities:
   (a) Choose x_1 for the hitting set, or
   (b) x_1 is not in the hitting set but x_i is for $i = 2, \dots, d$.

This makes d branches in total. If T_k is the number of leaves in a branching tree with upper bound k on its depth, then the first branch has at most T_{k-1} leaves. Let B_k be the number of leaves in a branching tree where there is at least one set of size $d-1$ or smaller. For each $i = 2, \dots, d$, there is some set s' in the given collection such that $x_1 \in s'$ but $x_i \notin s'$. Therefore, the size of s' is at most $d-1$ after excluding x_1 from and including x_i into the hitting set. Altogether we get

$$T_k \leq T_{k-1} + (d-1)B_{k-1}.$$

If there is already a set with at most $d-1$ elements, we can play the same game and get

$$B_k \leq T_{k-1} + (d-2)B_{k-1}.$$

The branching number of this recursion for T_k is α as given above.

d	3	4	5	6	7	8	9	10	20	50	100
T_k	2.41^k	3.30^k	4.23^k	5.19^k	6.16^k	7.14^k	8.12^k	9.11^k	19.05^k	49.02^k	99.01^k

Table 8.1 *Search tree sizes for d-Hitting Set.*

Table 8.1 shows the resulting search tree sizes for several values of d. Note that α is always smaller than $d - 1 + (d - 1)^{-1}$. Summarizing, we obtain the following result.

Theorem 8.7 *There is a size-T_k search tree for d-*HITTING SET *where T_k is specified as in Table 8.1. Each search tree node can be processed in time linear in the input size.*

The above simple search tree strategy can be improved by case distinguishing, giving the currently best search tree size of $O(2.18^k)$ for 3-HITTING SET.

Observe, however, that the general HITTING SET problem (unbounded subset size) is $W[2]$-complete (see Section 13.3.2), so it seems hopeless to try to show fixed-parameter tractability for the general problem with unbounded value of d.

8.5 Closest String

This example presents a depth-bounded search tree where case distinguishing is *not* the point. It deals with a string problem with applications in coding theory and computational molecular biology. We encountered it already in Section 5.3 when giving an example of a problem with more than one reasonable parameterization. Here, we will concentrate on one of these parameterizations. The problem is named CLOSEST STRING or sometimes also CONSENSUS STRING or CENTER STRING.

> **Input:** A set of k strings $s_1, \ldots, s_k$ over alphabet Σ of length L each, and a nonnegative integer d.
> **Task:** Find a string s such that $d_H(s, s_i) \leq d$ for all $i = 1, \ldots, k$.

Here, $d_H(s, s_i)$ denotes the Hamming distance between strings s and s_i. We present a fixed-parameter algorithm with respect to the distance parameter d. To start with, we need to introduce some notation and state some easy observations.

Given a set of k strings of length L, we can think of these strings as a $k \times L$ character matrix. The *columns* of a CLOSEST STRING instance are the columns of this matrix. With the following observation we find that it is sufficient to solve instances containing at most kd columns. We call a column *dirty* iff it contains at least two different symbols from alphabet Σ. Clearly, all the work in solving CLOSEST STRING concentrates on the dirty columns of the input instance, because otherwise the desired center string simply carries the same letter at the corresponding position as all input strings do.

It is easy to verify the following simple lemma using just the definition of CLOSEST STRING.

Lemma 8.8 *Given a* CLOSEST STRING *instance with k length-L strings and distance parameter d. If the corresponding $k \times L$ matrix has more than kd dirty columns, then there is no solution to this instance.*

Lemma 8.8 gives a simple reduction to a problem kernel. Notably, it is not with respect to the parameter d alone, but needs to incorporate both parameters k *and* d. A nontrivial reduction to a problem kernel only with respect to parameter d is not known. Note, however, that Lemma 8.8 has no effect on the upper size bound on the subsequently shown depth-bounded search tree, but it only may influence the polynomial factors in the running time when really implementing the algorithm.

Whereas the existence of a depth-bounded search tree for the preceding problems was more or less obvious and the central challenge was to shrink it as much as possible, in the case of CLOSEST STRING the point now is to *detect* that there indeed is a solution using a depth-bounded search tree. Other than for the preceding problems, the derivation of the upper bound for the search tree size then will be straightforward.

The central idea is to make use of the fact that the desired closest string can differ in at most d positions from each input string. Hence, if the candidate center string differs from an input string in too many positions, then one of these positions has to be changed in the candidate string such that it coincides with this input string at the respective position. In addition, as a starting candidate center string one may choose any of the given input strings, that is, without loss of generality string s_1.

In Figure 8.4, we outline a recursive algorithm solving CLOSEST STRING. It yields a search tree algorithm obeying the depth bound d. For the correctness of the algorithm, we need the following simple observation.

Lemma 8.9 *Let $S = \{s_1, s_2, \ldots, s_k\}$ be a set of strings and let d be a nonnegative integer. If there are $i, j \in \{1, \ldots, k\}$ with $d_H(s_i, s_j) > 2d$, then there is no string s with $\max_{i=1,\ldots,k} d_H(s, s_i) \leq d$.*

Proof The Hamming distance clearly satisfies the triangle inequality, that is,

$$d_H(q, r) \leq d_H(q, t) + d_H(t, r)$$

for arbitrary strings q, r, and t. If $d_H(s_i, s_j) > 2d$, we therefore know that $d_H(s, s_i) + d_H(s, s_j) > 2d$ for any string s. Since Hamming distances are nonnegative, it follows that $d_H(s, s_i) > d$ or $d_H(s, s_j) > d$ (or both). □

The following fixed-parameter algorithm for CLOSEST STRING is based on a size-$O(d^d)$ search tree which is implicitly constructed by the algorithm in Figure 8.4.

Theorem 8.10 *There is a size-$O(d^d)$ search tree for* CLOSEST STRING.

Proof Figure 8.4 presents the recursive procedure *CSd* which after a "successful" reduction to a problem kernel—that is, Lemma 8.9 does not exclude the

Recursive procedure $CSd(s, \Delta d)$:
Global variables: Set of strings $S = \{s_1, s_2, \ldots, s_k\}$, nonnegative integer d.
Input: Candidate string s and integer Δd.
Output: A string $\hat{s}$ with $\max_{i=1,\ldots,k} d_H(\hat{s}, s_i) \leq d$ and $d_H(\hat{s}, s) \leq \Delta d$,
if it exists, and "not found," otherwise.
Method:
(D0) **if** $\Delta d < 0$ **then return** "not found";
(D1) **if** $d_H(s, s_i) > d + \Delta d$ for some $i \in \{1, \ldots, k\}$ **then return** "not found";
(D2) **if** $d_H(s, s_i) \leq d$ for all $i = 1, \ldots, k$ **then return** s;
(D3) choose any $i \in \{1, \ldots, k\}$ such that $d_H(s, s_i) > d$:
    $P := \{\, p \mid s[p] \neq s_i[p] \,\}$;
    choose any $P' \subseteq P$ with $|P'| = d + 1$;
    for all $p \in P'$ **do**
        $s' := s$;
        $s'[p] := s_i[p]$;
        $s_{ret} := CSd(s', \Delta d - 1)$;
        if $s_{ret} \neq$ "not found" **then return** s_{ret};
(D4) **return** "not found"

FIG. 8.4. **Algorithm CS-D**. Inputs are a CLOSEST STRING instance consisting of a set of strings $S = \{s_1, s_2, \ldots, s_k\}$ of length L each, and a nonnegative integer d. The recursion is invoked with $CSd(s_1, d)$. Instead of s_1, we could choose an arbitrary element from S here.

existence of a solution—is invoked by the call $CSd(s_1, d)$. Referring to this by "Algorithm CS-D," we subsequently analyze the size of the tree of recursive calls and prove that it correctly solves CLOSEST STRING.

Search tree size. Consider the recursive part of the algorithm. Parameter Δd is initialized to d. Every recursive call decreases Δd by one. The algorithm stops when $\Delta d < 0$. Therefore, the algorithm builds a search tree of height at most d. In one step of the recursion, the algorithm chooses, given the current candidate string s, a string s_i such that $d_H(s, s_i) > d$. It explores $d+1$ subcases for positions in which s and s_i disagree. This yields an upper bound of $O((d+1)^d)$ on the search tree size.

Correctness. What must be shown is that Algorithm CS-D always finds a string $\hat{s}$ with $\max_{i=1,\ldots,k} d_H(\hat{s}, s_i) \leq d$, if one exists. Here, we explicitly show only the correctness of the first recursive step where s_1 is the candidate string; the correctness of the algorithm then follows with a straightforward inductive application of the argument.

In the situation that s_1 satisfies $\max_{i=1,\ldots,k} d_H(s_1, s_i) \leq d$, we already have a solution, namely s_1. If s_1 is not a solution but there exists a desired closest string for this instance with distance value d, then there must be a string s_i, $i = 2, \ldots, k$, such that $d_H(s_1, s_i) > d$. For branching, we consider the positions where s_1 and s_i differ, that is,

$$P := \{ p \mid s_1[p] \neq s_i[p] \}.$$

According to Lemma 8.9, we may assume that $|P| \leq 2d$. Algorithm CS-D successively creates subcases for $d+1$ positions p from P in order to create a new candidate by altering the respective position p from $s_1[p]$ to $s_i[p]$. Assume that $\hat{s}$ is a desired solution. A change as described is correct if we choose a position p from a set P_1 of positions defined as

$$P_1 := \{ p \mid s_1[p] \neq s_i[p] \wedge s_i[p] = \hat{s}[p] \}.$$

We show that at least one of the $d+1$ moves is a correct one. To this end, we observe that $P = P_1 \cup P_2$ for

$$P_2 := \{ p \mid s_1[p] \neq s_i[p] \wedge s_i[p] \neq \hat{s}[p] \}.$$

Since $d_H(\hat{s}, s_i) \leq d$ we can conclude that $|P_2| \leq d$. Therefore, at least one of our $d+1$ subcases will try a position from P_1. An inductive application of this argument shows that Algorithm CS-D finds a closest string for this instance, if one exists. Note that in deeper levels of the search tree we may allow alteration of the candidate s in only that position p in which $s[p] = s_1[p]$: having started with s_1 as the candidate string, every position p with $s[p] \neq s_1[p]$ has been previously altered. It does not make sense to alter the candidate twice in one position. In addition, with Δd we store how many positions we may still change, that is, how deep the search tree may still be explored.

Regarding instruction (D1), we can analogously to Lemma 8.9 observe that it is correct to omit branches where the candidate string s satisfies $d_H(s, s_i) > d + \Delta d$ for some string s_i of the given strings $s_1, \ldots, s_k$: assume that there is a solution $\hat{s'}$. The solution $\hat{s'}$ can differ from every s_i in at most d positions. Due to the triangle inequality, $\hat{s'}$ would differ from s in more than Δd positions, contradicting the assumption that $\hat{s'}$ is a solution. □

Combining the search tree from Theorem 8.10 with the reduction to a problem kernel (as a preprocessing phase) from Lemma 8.8, with little effort one achieves the following result.

Corollary 8.11 Closest String *can be solved in* $O(kL + kd \cdot d^d)$ *time.*

With Algorithm CS-D, we can find a solution if one exists. One may easily observe that we find *all* solutions if the given distance parameter d is minimum. We do not necessarily find all solutions to a given instance when d is not optimal. Moreover, if there are two strings s_i, s_j with $d_H(s_i, s_j) = 2d$ then we can use a special strategy with improved time bounds: we know that a solution must differ from both s_i and s_j in d positions. Thus, search a solution by trying all ways to partition the set of positions p with $s_i[p] \neq s_j[p]$ into two sets of size d each. In the candidate string, we give to one set of positions the characters of s_i, to the second set the characters of s_j. For each candidate, we only have to check whether it is a solution or not. It is not hard to verify that if such a special instance has a solution then we will find it—actually all of them—using this approach. This yields an exponential part limited by $O(2^{2d}) = O(4^d)$.

It is open to give a depth-bounded search tree to prove fixed-parameter tractability with respect to parameter k, the number of input strings. In Section 11.2, however, we will see that a deep result from *integer linear programming* theory implies that CLOSEST STRING is fixed-parameter tractable with respect to parameter k as well.

8.6 Dominating Set in Planar Graphs

DOMINATING SET IN PLANAR GRAPHS is *the* example of this book for a depth-bounded search tree whose existence is not obvious at all. This problem also represents cases where the search tree algorithm does *not* employ many case distinctions, and hence is comparatively easy. By way of contrast, the proof of correctness of the underlying branching strategy—more precisely, to show that one always branches into only a constant number of subcases—is hard.

Recall from the beginning of the chapter the problems we face with handling DOMINATING SET instead of INDEPENDENT SET. The difficulties described there lead us to the study of a more general version of DOMINATING SET, that is, ANNOTATED DOMINATING SET:

> **Input:** A graph $G = (B \uplus W, E)$ with its vertices colored either black or white and a nonnegative integer k.
> **Task:** Find a subset $S \subseteq B \uplus W$ with k or fewer vertices such that each vertex in B is contained in S or has at least one neighbor in S. In other words, find a set of at most k vertices (which may be either black or white) that dominates the set of black vertices.

Then, in each inner node of the search tree we may branch according to a low-degree *black* vertex. More specifically, we have to decide whether v shall be dominated by itself or by one of its neighbors. Restricting ANNOTATED DOMINATING SET to planar graphs we can guarantee the existence of a vertex $u \in B \uplus W$ with $\deg(u) \leq 5$. However, this vertex needs not necessarily be black. As a consequence, a direct $O(6^k \cdot n)$ search tree algorithm for (ANNOTATED) DOMINATING SET IN PLANAR GRAPHS by analogy with the one for INDEPENDENT SET IN PLANAR GRAPHS seems out of reach.

In what follows, we sketch a recursive search strategy for (ANNOTATED) DOMINATING SET on planar graphs with search tree size $O(8^k)$. To this end, we provide a set of simple data reduction rules and then use a depth-bounded search tree in which we are constantly simplifying the instance according to the data reduction rules. The branching in the search tree will be done with respect to low-degree vertices. More specifically, we will be able to guarantee that we can always find a black vertex v with degree at most seven and branch on v by either putting v or one of v's at most seven neighbors into the dominating set.

This yields eight recursive calls to branch into and the search tree size $O(8^k)$ follows. Here, the central technical obstacle that has to be surmounted is to prove that whenever we want to do a branching on a minimum-degree black vertex then a degree-at-most-seven black vertex always actually exists. That is why we introduce the set of data reduction rules mentioned above. They continuously generate a reduced graph which always possesses such a degree-seven black vertex. This is shown by involved technical arguments related to the Euler formula for planar graphs, demonstrating that planar *reduced* black-and-white graphs always contain a vertex of degree at most seven. The technically demanding proof of this property is out of the scope of this book.

Without loss of generality, in what follows we restrict attention to *connected* planar graphs, that is, connected graphs that admit crossing-free embeddings in the plane.

8.6.1 *Data reduction rules*

We consider the following data reduction rules for simplifying instances of ANNOTATED DOMINATING SET IN PLANAR GRAPHS. In developing the search tree, we will always assume that we are branching from a reduced instance—thus, we are constantly simplifying the instance according to the data reduction rules. When a vertex v is placed into the dominating set D by such a rule, the target size k for D is decremented to $k - 1$ and the neighbors of v become colored white. This is because the neighbors no longer need to be dominated—they already are—but they may still serve as "dominating vertices". The list of seven data reduction rules reads as follows.

D1 Delete edges between white vertices.

D2 Delete degree-one white vertices.

D3 If there is a degree-one black vertex v with neighbor u (either black or white), then delete v, place u into the dominating set, and decrement parameter k by one.

D4 If there is a white vertex v of degree two with two black neighbors v_1 and v_2 connected by an edge $\{v_1, v_2\}$, then delete v.

D5 If there is a white vertex v of degree two with black neighbors v_1 and v_2, and if there are a black vertex v_3 and edges $\{v_1, v_3\}$ and $\{v_2, v_3\}$, then delete v.

D6 If there is a white vertex v of degree two with black neighbors v_1 and v_2, and if there is a white vertex v_3 and edges $\{v_1, v_3\}$ and $\{v_2, v_3\}$, then delete v.

D7 If there is a white vertex v of degree three with black neighbors v_1, v_2, and v_3, and additionally existing edges $\{v_1, v_2\}$ and $\{v_2, v_3\}$, then delete v.

Let us call a set of data reduction rules *sound* if whenever (G, k) is some problem instance and the new instance (G', k') is obtained from (G, k) by applying at least one of the data reduction rules, then (G, k) is a yes-instance iff (G', k') is a yes-instance. Besides soundness, we demand that the solution for (G, k) can be easily (re-)constructed having the solution for (G', k'). The following is easily shown by a simple case analysis.

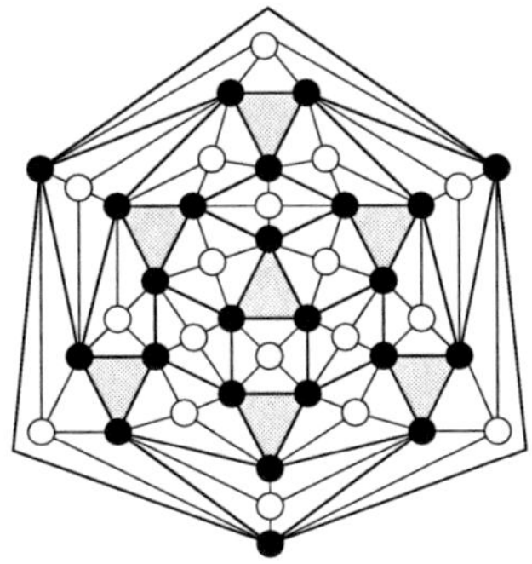

FIG. 8.5. A reduced black-and-white graph where all black vertices have degree exactly seven.

Lemma 8.12 *The transformation rules are sound.*

Not only are they sound, the data reduction rules can also be efficiently executed leading to a reduced graph; that is, a graph where none of the rules applies any more.

Lemma 8.13 *Applying transformation rules* **D1**–**D7**, *a given black-and-white graph* $G = (B \uplus W, E)$ *can be transformed into a reduced black-and-white graph* $G' = (B' \uplus W', E')$ *in* $O(n^2)$ *time, where* $n := |B \uplus W|$.

Proof The claim is easy to see if we perform the data reduction rules in the following order: first apply rule **D1**, and then visit every white vertex, checking whether rules **D4**–**D7** can be applied. Finally, carry out rules **D2** and **D3**. □

8.6.2 *Main result and some remarks*

Based on the above data reduction rules, the following technical lemma can be established. Its lengthy proof, which relies on "Euler-like considerations", although more or less elementary, is beyond the scope of this book.

Lemma 8.14 *If* $G = (B \uplus W, E)$ *is a planar black-and-white graph that is reduced then there exists a black vertex* $u \in B$ *with degree at most seven.*

Figure 8.5 shows a reduced black-and-white graph all whose black vertices have degree exactly equal to seven.

Interestingly, there exists an infinite set of planar reduced black-and-white graphs with the property that all black vertices have degree exactly seven. Hence, in this limited sense, the upper bound provided in Lemma 8.14 is optimal since these examples provide a matching lower bound. Note, however, that it is completely open to prove—if, after all, possible—the existence of a family of graphs which *keeps* this property after every possible branching one is allowed to perform. Moreover, it is open whether extending the given set of data reduction rules may lead to reduced graphs with a guaranteed existence of a black vertex with degree smaller than seven. In summary, we obtain the following main result.

Theorem 8.15 *There is a size-$O(8^k)$ search tree for* (ANNOTATED) DOMINATING SET IN PLANAR GRAPHS.

Proof Use Lemma 8.14 for the construction of a search tree as described in the beginning of the section by branching according to the closed neighborhood of a small-degree black vertex. The size of this neighborhood is bounded from above by eight. □

Observing that to perform the data reduction in each node of the search tree according to Lemma 8.13 can be done in $O(n^2)$ time leads to the following.

Corollary 8.16 (ANNOTATED) DOMINATING SET IN PLANAR GRAPHS *can be solved in $O(8^k \cdot n^2)$ time.*

We remark that by slightly changing the above reduction rules and doing a more refined analysis, the quadratic time factor $O(n^2)$ can be improved to a linear one.

The above sketched data reduction rules lead to the first search tree algorithm with "parameterized bound" on the search tree size for DOMINATING SET IN PLANAR GRAPHS. Unfortunately, the proof of correctness has become fairly technical. Since the "optimality" of Lemma 8.14 only holds with respect to the particular set of transformation rules given above, it remains challenging to improve Lemma 8.14 by adding further, more involved data reduction rules. Moreover, a generalization of the above considerations is possible and yields analogous bounded search tree algorithms for DOMINATING SET on graphs of bounded genus. Finally, we stress that the above algorithm is fairly easy to implement.

It is worth noting that compared with the data reduction rules for DOMINATING SET IN PLANAR GRAPHS given in Section 7.6, the data reduction rules described here are different in purpose. Whereas the rules given in Section 7.6 serve to prove a linear-size problem kernel, the rules from this section are designed to generate an "advantageous branching situation". Whether there is any provable benefit in the worst case by combining both rules remains open. In implementations, however, it is strongly recommended to make combined use of them.

8.7 Interleaving search trees and kernelization

In the preceding parts of this chapter we have seen several different problem types in which the bounded search tree paradigm applies. In Chapter 7 we encountered several examples of the reduction to a problem kernel paradigm. One may say that these two paradigms form the pillars of "feasible fixed-parameter tractability". One obvious way to combine these two methods is, as already indicated, to perform a two-phase attack. Start by preprocessing the given input instance by performing a reduction to a problem kernel, and then systematically process the generated problem kernel using depth-bounded search trees. What we show next is that to do a kernelization repeatedly during the course of the search tree algorithm may provably further accelerate the solution finding process.

8.7.1 *Basic methodology*

In the following, we will deal with a large class of fixed-parameter algorithms. Let us state the conditions that these algorithms must meet: they must be fixed-parameter algorithms that consist of two parts, a *reduction to a problem kernel* and a *depth-bounded search tree*. Let (I, k) be an instance of a parameterized problem to be solved. Reduction to a problem kernel shall take $P(|I|)$ steps and result in an instance of size at most $q(k)$, where both P and q are polynomially bounded. The expansion of a node in the search tree takes $R(|I|)$ steps, which must also be bounded by some polynomial. The search tree size is $O(\alpha^k)$. The overall time complexity of the algorithm is then

$$O(P(|I|) + R(q(k)) \cdot \alpha^k).$$

In the following we show how to modify the second stage of the algorithm in order to improve the time complexity to

$$O(P(|I|) + \alpha^k).$$

To this end, assume that a branching step produces i subcases with branching vector $(d_1, d_2, \ldots, d_i)$. Furthermore, let $I_1, I_2, \ldots, I_i$ be the new instances of the problem generated thereby, with new parameter values $k - d_1, k - d_2, \ldots, k - d_i$. Generally, we now use the following algorithmic steps to expand a node (I, k) in the search tree:

> **if** $|I| > c \cdot q(k)$ **then** $(I, k) := (I', k')$ where (I', k') forms a problem kernel **fi**;
> replace (I, k) with $(I_1, k - d_1), (I_2, k - d_2), \ldots, (I_i, k - d_i)$.

In other words, we enrich the ordinary branching as executed in the second line with a conditional kernelization as executed in the first line. Here $c \geq 1$ is a constant that can be chosen with the aim of further optimizing the running time. There is a tradeoff in choosing c: the optimal choice depends on the implementation of the algorithm but in the end it affects only the constant factor in the overall time complexity. Therefore we neglect optimizing c here.

A closer look shows that we in fact seem to *increase* the time needed to expand a node in the search tree. This is generally speaking true: sometimes we apply reduction to a problem kernel prior to branching into recursive calls. These additional kernelizations, however, also *decrease* the instance size in the "middle" of the search tree. Since the time for branching is bounded polynomially in the *instance size*, this also helps to *decrease* the time to expand a node. It proves to be the case that, while we waste time near the root of the search tree, we gain much more time near the leaves.

In order to mathematically analyze the running time of the above approach, we describe the time needed to expand a node (I, k) and all its descendants by a recurrence equation. Let T_k denote an upper bound on the *time* to process (I, k). The following recurrence holds for T_k:

$$T_k = T_{k-d_1} + T_{k-d_2} + \cdots + T_{k-d_i} + O(P(q(k)) + R(q(k))).$$

The time to expand (I, k) itself is at most $O(P(q(k)) + R(q(k)))$ because $|I| = O(q(k))$ since $|I| > c \cdot q(k)$ is prevented by the conditional kernelization. In order to solve this non-homogeneous linear recurrence we need a special solution. To get its general solution we add the general solution of the corresponding homogeneous recurrence

$$T_k = T_{k-d_1} + T_{k-d_2} + \cdots + T_{k-d_i}.$$

We already know that all solutions of this homogeneous recurrence are bounded by $O(\alpha^k)$. Consequently, we only need to find a small special solution of the non-homogeneous recurrence. In our case the inhomogeneity is a polynomial. Therefore, there exists a special solution that is also a polynomial in k. It is easy to construct such a special solution explicitly. There is always a polynomial solution that has the same degree as the inhomogeneity p. If r is a polynomial special solution then

$$r(k) - \sum_{j=1}^{i} r(k - d_j) = p(k),$$

and the highest degree monomials on the left side cannot cancel each other. All solutions of T_k are therefore bounded by $O(\alpha^k)$.

In order to illustrate this, let us consider the following concrete recurrence equation.

$$T_k = 2T_{k-1} + C \cdot k^2 + D \cdot k + E,$$

where C, D and E are constants that depend on the implementation of the algorithm. The initial conditions are simple, say, $T_0 = 0$. The general solution of the homogeneous recurrence is $\lambda 2^k$ for $\lambda \in \mathbb{R}$. Since it is a recurrence of first order, the dimension of its space of solutions is one, too.

One may easily check that

$$T_k = -Ck^2 - (4C + D)k - (6C + 2D + E)$$

is a special solution. The general solution is then

$$\lambda 2^k - Ck^2 - (4C + D)k - (6C + 2D + E)$$

and the solution for $T_0 = 0$ is

$$T_k = (6C + 2D + E) \cdot 2^k - Ck^2 - (4C + D)k - (6C + 2D + E).$$

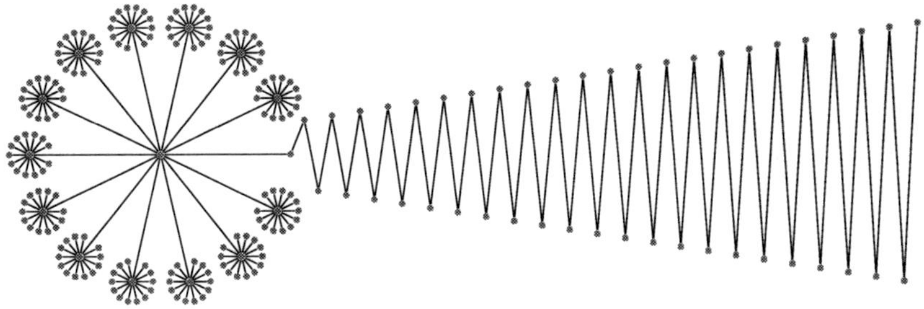

FIG. 8.6. A no-instance of VERTEX COVER. The graph shown is the $k = 15$ member of a family of graphs G_k. The graph G_k consists of a tree with degree $k-1$ and depth 2 to which a path with $3k+1$ vertices is attached (called the *tail*). It is easy to see that the smallest vertex cover for G_k has size $\frac{5}{2}k - \frac{3}{2}$, and therefore the family of inputs (G_k, k) with $k \geq 2$ for VERTEX COVER are all no-instances.

8.7.2 *Interleaving is necessary*

From an algorithmic point of view, the interleaving of kernelization and branching described above is necessary: we show that an improved analysis alone cannot achieve the speedup of the last section. That is, the interleaving of the reduction to a problem kernel and the bounded search tree really is needed to get the claimed improvements. Without modification, the algorithms in general have a running time of $\Omega(P(|I|) + R(f(k))\alpha^k)$. As an example, we can use VERTEX COVER and we assume that we use the trivial size-2^k search tree algorithm.

Look at Figure 8.6 for a definition of a family of graphs defined for odd k. There is no solution of size at most k, since the optimal vertex cover has size $\frac{5}{2}k - \frac{3}{2}$ (in the head $k-2$ vertices and half the vertices of the tail). The graph contains exactly $(k-1)(k-2)+1$ vertices in the head and $3k+1$ vertices in the tail (altogether k^2+4). Trying to apply Buss's reduction to a problem kernel based on the selection of degree-$\geq k$ vertices (see beginning of Chapter 7) at the very beginning does not affect this graph since the degree of every vertex is at most k. Now, assume that the unmodified algorithm chooses edges from right to left. This leads to a search tree of size 2^k, the largest possible. While the algorithm examines this graph, it removes vertices and edges but the *head* remains unchanged. Consequently, instances have size $\Theta(k^2)$ during *each* branching step. The overall time complexity is therefore the worst possible—$\Theta(k^2 2^k)$. Of course, a better time complexity can also be achieved by changing the order of choosing edges. Nevertheless, the time bound is $\Theta(k^2 2^k)$ in the worst case.

When the described interleaving methodology is applied, the running time is decreased tremendously. After the *second* edge is removed and parameter k is decreased by two, the whole head will be removed from the graph.

The presented interleaving technique applies in numerous settings such as 3-Hitting Set and Maximum Satisfiability. It thus belongs in the toolbox for the development of efficient fixed-parameter algorithms. In this context, it is important to note that the achieved improvements when replacing

$$O(\alpha^k \cdot q(k) + p(|I|))$$

by

$$O(\alpha^k + p(|I|))$$

are *not* due to asymptotic tricks, but that $q(k)$ can be replaced by a *small* constant. And the improvement really matters. For 3-Hitting Set, simply compare a $O(2.18^k \cdot k^3 + n)$ time (without interleaving, employing the search tree of size 2.18^k (Section 8.4) and the problem kernel of size $O(k^3)$ (Section 7.5)) with a $O(2.18^k + n)$ time (with interleaving) algorithm.

One must be careful when trying to apply the interleaving technique, however. It is tempting to apply it to Dominating Set in Planar Graphs, for which we know a linear-size problem kernel (see Section 7.6) and a size-8^k search tree (see Section 8.6). We cannot directly apply interleaving because the search tree works with black-and-white graphs, whereas the problem kernelization was only designed for non-colored graphs.

In summary, as a rule, the potential for improvement due to interleaving increases the larger the problem kernel of the underlying parameterized problem is. It probably almost always pays off in practice when perhaps not applied at every search tree node but in a regular manner after some branchings. In this way, the additional administrative overhead can be compensated. The best tradeoff, in the end, must be determined empirically.

8.8 Automated search tree generation and analysis

We now have seen several examples of depth-bounded search trees, each with a somewhat different focus. As with the search trees related to Cluster Editing, Vertex Cover, and 3-Hitting Set (see Sections 8.2, 8.3, and 8.4), however, they all have in common that they were based on fairly extensive case distinctions. As case distinguishing is a tedious and error-prone task, the question arises whether this sort of design of depth-bounded search trees can be automated. The potential benefits of such an approach would be

- rapid development, and
- improved upper bounds

due to computer assistance. In this section we briefly sketch an approach to how to make use of computers in search tree generation and analysis. We study an *NP*-complete special case of the Cluster Editing problem from Section 8.2, the so-called Cluster Deletion.

Search tree algorithms basically consist of a set of branching rules. Branching rules are usually based on local substructures. For graph problems, these can be induced subgraphs having up to s vertices for a constant integer s; we refer to graphs having s vertices as *size-s graphs*. Then each branching rule specifies the branching for a particular local substructure. The idea behind the automation approach is roughly described as follows:

1. For constant s, enumerate all "relevant" subgraphs of size s such that every input instance of the given graph problem has s vertices inducing at least one of the enumerated subgraphs.
2. For every local substructure enumerated in Step 1, check all possible branching rules for this local substructure and select the one corresponding to the *best*, that is, smallest, branching number. The set of all these best branching rules then defines our search tree algorithm.
3. Determine the worst-case branching rule among the branching rules stored in Step 2, because this branching rule yields a worst-case bound on the search tree size of the generated search tree algorithm.

Note that in both Step 1 and Step 2 we usually make use of further *problem-specific rules*: for example, in Step 1, problem-specific rules can determine input instances which do not need to be considered in our enumeration, such as instances which can be solved in polynomial time, instances which can be simplified due to data reduction rules, or instances for which we can use a clever manually developed branching rule; instances with one of these properties are referred to as "trivial instances". Here we restrict ourselves to describing a general framework, indicating where problem-specific rules may apply. The problem-specific rules corresponding to particular graph modification problems are then given in the following sections.

In the following we discuss Steps 2 and 1, respectively, in more detail. We will use CLUSTER DELETION as a running example. CLUSTER DELETION is an edge deletion problem in which the forbidden induced subgraph is a P_3, that is, a path consisting of three vertices. CLUSTER DELETION is defined as follows.

> **Input:** A graph $G = (V, E)$ and a nonnegative integer k.
> **Task:** Find out whether we can transform G, by deleting at most k edges, into a graph that consists of a disjoint union of cliques.

Obviously, there is a trivial size-$O(2^k)$ search tree for CLUSTER DELETION.

We outline a general framework to generate, given a size-s graph $G_s = (V_s, E_s)$ for constant s, an "optimal" branching rule for G_s. To compute a search tree branching rule we again use a search tree to explore the space of possible branching rules. This search tree is referred to as a *meta search tree*. A central reference point herein is a meta search tree that systematically searches through a space of possible branchings for a G_s and then chooses a best one.

(1) Branching rules. A branching rule for G_s specifies a set of "simplified" (to be made precise in the next paragraph) graphs $G_{s,1}, G_{s,2}, \ldots, G_{s,r}$. When invoking the branching rule, one would replace, for every $G_{s,i}$, $1 \leq i \leq r$, G_s

by $G_{s,i}$ and invoke the search tree procedure recursively on the thereby generated instances. By definition, the branching rule has to satisfy the following property: a "solution" is an optimal solution for G_s iff it can be inferred from a "best" one among the optimal solutions for all $G_{s,i}$, $1 \leq i \leq r$, produced by the branching rule. This is referred to by saying that the branching rule is *complete*. The branching is done with respect to the vertex pairs of G_s since we obtain a solution graph by deleting edges.

(2) Annotations. A "simplified" graph $G_{s,i}$, $1 \leq i \leq r$, is obtained from G_s by assigning labels to a subset of vertex pairs in G_s. The labels for a vertex pair $u, v \in V_s$ can be chosen as *permanent* (that is, the corresponding edge is in the solution graph to be constructed) or *forbidden* (that is, the corresponding edge is not in the solution graph to be constructed). All vertex pairs sharing no edge are initially assigned the label *forbidden* since edges cannot be added. An *annotation* then is a mapping π from the vertex pairs to either *permanent*, *forbidden*, or simply *undefined*. The latter simply means that no label is assigned. By G_s with annotation π we then refer to the graph obtained from G_s by deleting $\{u, v\} \in E_s$ if π assigns the label *forbidden* to (u, v). Thus an annotation specifies, in particular, a set of edges to be deleted from the input graph. In this way, an annotation can be used to specify one branch of a branching rule.

(3) Representation of branching rules. A branching rule for G_s with annotation π can be represented by a set A of annotations for G_s such that, for every $\pi' \in A$, π' refines π. Here, an annotation π' *refines* an annotation π iff for every vertex pair p, it holds that

$$\pi(p) \neq \textit{undefined} \implies \pi'(p) = \pi(p).$$

Then, every $\pi' \in A$ specifies one branch of the branching rule. A set A of annotations has to satisfy the following three conditions in order to specify a branching rule:

- The branching rule is complete.
- Every annotation decreases the parameter "number of edge deletions".
- The annotated vertex pairs do not form a P_3, that is, there are no $u, v, w \in V_s$ with $\pi(u, v) = \pi(v, w) = \textit{permanent}$ and $\pi(u, w) = \textit{forbidden}$.

(4) Problem-specific rules that refine annotations. To obtain non-trivial bounds it is necessary to have a set of problem-specific *reduction rules*. In our terminology, a reduction rule specifies how to refine a given annotation π to π' such that an optimal solution for the input graph with annotation π' is also an optimal solution for the input graph with annotation π. For CLUSTER DELETION, we have the following simple data reduction rule: given a graph $G = (V, E)$ with annotation π, if there are three pairwise distinct vertices $u, v, w \in V$ with $\pi(u, v) = \pi(v, w) = \textit{permanent}$, then we can replace π by an annotation π' which refines π by setting $\pi'(u, w) := \textit{permanent}$. Analogously, if $\pi(u, v) = \textit{permanent}$ and $\pi(v, w) = \textit{forbidden}$, then $\pi'(u, w) := \textit{forbidden}$.

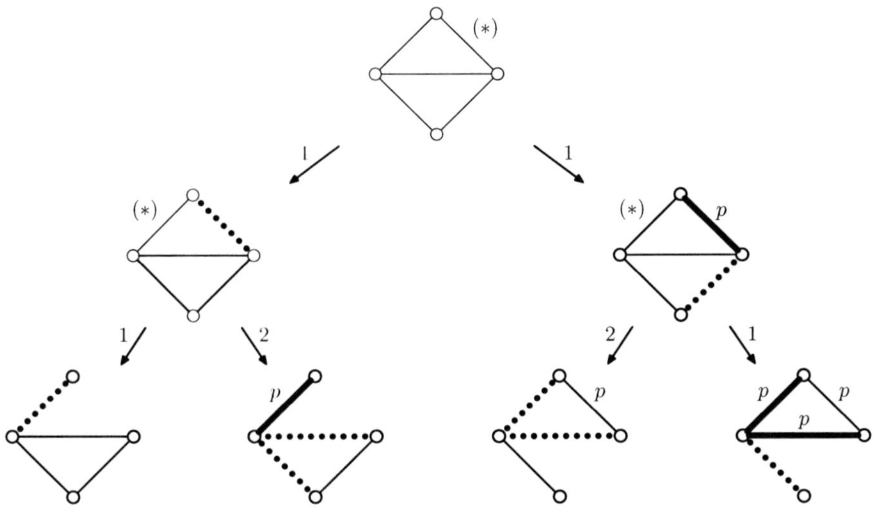

FIG. 8.7. CLUSTER DELETION. Illustration of a candidate for branching in the case of an induced subgraph in form as drawn in the above root. At the root we have a 4-vertex input graph having no labels. *Arrows* indicate the branching steps of the meta search tree. We only display branches of the meta search tree which contribute to the computed branching rule. The vertex pair on which we branch is indicated by $(*)$. Permanent edges are indicated by p, vertex pairs sharing no edge are implicitly forbidden (*bold* or *dotted lines* indicate when a vertex pair is newly set to *permanent* or *forbidden*, respectively). Besides the vertex pair on which we branch, additional vertex pairs are set to *permanent* or *forbidden* due to the problem-specific reduction rule explained in (4). The numbers at the arrows indicate the number of edges deleted in the respective branching step. Thus the resulting branching rule is determined by the leaves of this tree and the corresponding branching vector is $(2, 3, 3, 2)$. Alternative possibilities to branch can be systematically generated by the meta search tree by choosing other edges to branch on, perhaps leading to more favorable branchings with better branching vectors.

(5) Meta search tree. Procedure `compute_br()`, for graph G_s, has as input an annotation π. It returns a set of possible branching rules with respect to G_s and annotation π, that is, a set $B = \{A_1, \ldots, A_r\}$ of annotation sets such that, for every $1 \leq i \leq r$ and every $\pi' \in A_i$, π' refines π. Each $\pi' \in A_i$ represents one branch of the branching rule given by A_i.

For instance, Figure 8.7 illustrates the development of a branching into four subcases—yielding branching vector $(2, 3, 3, 2)$ in the case of a four-vertex input graph with annotation π that assigns to all vertex pairs the label "undefined". This particular branching is derived from starting the branching with respect to one particular edge (marked $(*)$). Other branchings are obtained analogously by means of the meta-search tree where the point is to find a particular edge to branch on. Moreover, using the meta search tree one can systematically search through a whole space of possible branchings and then choose the best one. To do so more efficiently than by pure brute-force search a few further tricks can be applied—see the literature cited in the bibliographical remarks.

To further ameliorate the efficiency of the branching strategy, one can use the following problem-specific rules that improve the enumeration of graphs. Given a constant integer s, we can, firstly, assume that every connected component in a given instance has at least s vertices; every connected component of the input graph having less than s vertices can be processed in constant time. Therefore, we can restrict the enumeration to connected size-s graphs. Secondly, we know that a non-trivial CLUSTER DELETION instance contains a P_3 as a vertex-induced subgraph, since otherwise the input graph is already a solution. Therefore we can restrict the enumeration to connected size-s graphs having a P_3 as a vertex-induced subgraph. Using some more observations, the enumeration can be further sped up by restricting the number of graphs that have to be considered.

With this automated approach one can develop a depth-bounded search tree for CLUSTER DELETION with size $O(1.77^k)$ and a size-$O(1.92^k)$ search tree for CLUSTER EDITING, improving on the previous purely "human-made" bound of $O(2.27^k)$ from Section 8.2.

The idea sketched above and other similar ones (see bibliographical remarks) generalize to other graph problems and also (MAXIMUM) SATISFIABILITY problems. We have concentrated on the CLUSTER DELETION problem. This framework is usable for graph problems in general, observing two main issues where changes must be made depending on the problem considered:

- In CLUSTER DELETION, the branching objects are vertex pairs and the possible labels are "*permanent*" and "*forbidden*". In general, the branching objects and an appropriate set of labels for them are determined by the considered graph problem. For example, in VERTEX COVER, the objects to branch on are vertices instead of edges, and thus the labels would be assigned to the vertices. The labels would be, for example, "is in the vertex cover" and "is not in the vertex cover". Depending on the problem, additional auxiliary labels might be helpful, for example, reflecting the vertex degree.

- The reduction rules are problem-specific. To design an appropriate set of reduction rules working on local substructures is probably the most challenging part when applying the framework to a new problem and for the development of practical search tree algorithms in general.

Until now only a few problems are available where the best known "human upper bound" could be improved by an "automated upper bound" as is the case for CLUSTER DELETION. The reason is that there often are special tricks and data reduction arguments in these better human-made branching algorithms which so far have withstood an automated approach because of their special features. It is unclear now how far automated search tree generation can lead us. Since it is a very new field one may hope for further significant advances to be made. Note that even if the "automated setting" does not always lead to the best known worst-case bounds, however, it might be still considered scientific progress since it usually significantly reduces the "proof complexity" of the corresponding search trees when compared to the hand-made, often highly complicated, case distinctions. In this sense, the sketched framework helps to reveal the (usually) few real "core rules" that lie at the very heart of successfully attacking combinatorially hard problems. This may lead to a better understanding of the problems considered and may smooth the way for new approaches in deriving smaller and smaller search tree sizes. To make this whole line of attack a complete success, many things remain to be done. Obviously, trying to develop new problem-specific rules may lead to improved search tree size bounds in this scenario. It remains for future work to extend the framework in order to translate the computed case distinctions directly into "executable search tree algorithm code" and to test algorithms thus implemented empirically. Finally, a challenge could also be to use the automated framework in order to derive analytical proofs for search tree sizes, that is, proofs and case distinctions that are "compact enough" such that they can again be verified by hand/human brains in reasonable time.

8.9 Summary and concluding remarks

The development of depth-bounded search tree fixed-parameter algorithms offers various challenges. Sometimes the more demanding part is to prove the correctness and the search tree size of the proposed method (examples in this chapter are given with CLOSEST STRING (Section 8.5) and DOMINATING SET IN PLANAR GRAPHS (Section 8.6)), and sometimes the more challenging part is to design and cleverly organize intricate case distinctions leading to good worst-case bounds for the search tree size (CLUSTER EDITING, VERTEX COVER, and 3-HITTING SET). We have seen in Section 8.8 that the latter type of search tree algorithm leads to the question of automating the design and analysis of branching strategies as employed in search tree case distinctions. Little research has been pursued in this direction and more insights should be attainable here. Moreover, a related question arises that should also be considered in future research. Given some complicated branching strategy employed by a depth-bounded search tree algorithm, how can one achieve the same or "nearly the same" upper bounds on the search tree size but with a significantly simplified case distinction? One may consider this as a kind of "re-engineering" of case distinctions and it might have particular importance for the practical side of search tree algorithms, trying to minimize the implementation overhead caused by extensive case distinctions. A lesson to learn from Section 8.7 is that it is highly recommended to interleave search trees with reduction to a problem kernel—there are provable benefits with respect to saving polynomial factors in running time.

Other points when considering the practical sides of search tree algorithms are as follows. Search tree algorithms ...

- ... build one of the most important techniques for coping with the really hard kernel of a problem;
- ... can often be further accelerated by incorporating standard heuristic techniques such as branch-and-bound;
- ... in applications frequently have a few cases of their case distinction that occur very often and the remaining cases occur very seldom, so more systematic studies in this direction are of great interest;
- ... as a rule are easily run on parallel machines because of the straightforward load balancing which is directly implied by the construction process of search trees;
- ... may be combined with polynomial-time approximation algorithms by stopping the recursive search "near" the leaves and running an approximation algorithm instead, thus getting improved approximation factors at the cost of affordable exponential running times.

When dealing with depth-bounded search tree algorithms, it might sometimes appear unsatisfactory that there are no "clear" lower bounds on the search tree size. For instance, the upper bounds on the search tree sizes for VERTEX COVER have been continually improved in a series of papers—and the same holds for MAXIMUM SATISFIABILITY and several other problems. Thus, the impression might be that, at the cost of further extending (which often means further complicating) the case distinctions there will always be some (tiny) progress achievable. Note, however, that it is not always only a matter of refining case distinctions in order to get the search tree size down—sometimes elegant, practically relevant techniques such as the Nemhauser–Trotter problem kernel reduction for VERTEX COVER have resulted from these efforts (see Section 7.4). One might conclude that the community has to decide whether there is enough innovation in a newly proposed search tree—elegance is a thing that matters here. Unfortunately, proving concrete lower bounds for the search tree size seems out of reach of current possibilities—as does the analogous upper bound.

8.10 Exercises

1. Show that DOMINATING SET for graphs with maximum vertex degree bounded from above by a constant is fixed-parameter tractable.
2. Design a depth-bounded search tree for MAXIMUM SATISFIABILITY, using the number of clauses to be satisfied as the bound on the depth of the search tree.
3. Consider the following special structure occurring in a search tree algorithm for VERTEX COVER: a subgraph formed by a cycle of five vertices, and all vertices on the cycle have degree exactly three. At least one vertex has a neighbor with degree four. What is a good branching strategy in this case,

what is your branching vector, and what is the corresponding branching number?

4. Derive a search tree for 3-HITTING SET which improves the size bound $O(2.41^k)$ from Section 8.4.
5. Prove Lemma 8.8.
6. Prove Lemma 8.12.
7. Give all details for a proof of Lemma 8.13.
8. Show that the following graph modification problem is fixed-parameter tractable:
   Input: A bipartite graph $G = (V_1, V_2, E)$ and a nonnegative integer k.
   Task: Find out whether G can be transformed into a new graph G' by adding or deleting (both is allowed) at most k edges such that each connected component of G' is a *biclique*.
   Here a biclique is a complete bipartite graph.
9. In the area of "reconfigurable VLSI" the following problem occurs. Given a rectangular $m \times n$-field in which some of the $m \cdot n$ cells may be faulty. One is looking for a minimal number of rows and columns of this field such that all faults are covered by these. In "parameterized terms" this is to ask whether all faults can be covered by choosing at most k_1 rows and at most k_2 columns.
   (a) Model this problem as a graph problem.
   (b) Show that this problem is fixed-parameter tractable with respect to the combined parameter (k_1, k_2), that is, design a solving algorithm running in $f(k_1, k_2) \cdot (m + n)^{O(1)}$ time.
10. A graph is *chordal* if it contains no induced cycle of length at least 4. Consider the following graph modification problem MINIMUM FILL-IN.
    Input: A graph $G = (V, E)$ and a positive integer k.
    Task: Find out whether G can be made chordal by inserting at most k edges.
    Show that MINIMUM FILL-IN is fixed-parameter tractable with respect to parameter k.
11. Describe modifications to be done in order to generalize the automated analysis described in Section 8.8 from CLUSTER DELETION to CLUSTER EDITING.

8.11 Bibliographical remarks

Depth-bounded search trees are one of the most fundamental concepts in the design of fixed-parameter algorithms. In the literature one often finds synonymous concepts such as splitting algorithms or branching algorithms. Whereas Downey and Fellows (1999) went through this topic fairly quickly, here we have spent more time on them because there appear to be numerous facets of depth-bounded search trees that deserve further investigation. The basic idea behind search trees in our sense can be found in earlier books on algorithms and complexity such as Mehlhorn (1984). The first problem with an extensive list of more and more refined depth-bounded search trees was VERTEX COVER, significant contributions appearing in (chronologically ordered) Balasubramanian *et al.* (1998), Downey *et al.* (1999), Niedermeier and Rossmanith (1999), Chen *et al.* (2001), Niedermeier and Rossmanith (2003*b*), Chandran and Grandoni (2005), and Chen *et al.* (2005).

The paper by Kullmann (1999) provides a detailed description of how to analyze search tree sizes using recurrence equations. Refer to standard mathematical textbooks for more on solving recurrences as appearing in this book. For a more refined mathematical analysis of search tree sizes multivariate recurrences often come into play—Eppstein (2004) shows how to treat these.

The search tree for CLUSTER EDITING appears in Gramm *et al.* (2005)—with later improvement concerning the search tree size using the automated approach sketched in Section 8.8 (Gramm *et al.*, 2004). Its *NP*-completeness follows from work of Křivánek and Morávek (1986)—independently proven in Shamir *et al.* (2004). It is also studied under the name CORRELATION CLUSTERING (Bansal *et al.*, 2004). A more general perspective on the fixed-parameter tractability of graph modification problems such as CLUSTER EDITING is due to Cai (1996).

The VERTEX COVER search tree machinery has been mainly developed in Balasubramanian *et al.* (1998), Chen *et al.* (2001), Chen *et al.* (2005), and Niedermeier and Rossmanith (1999). We give a simplified version with worse worst-case bounds here. By combining search trees with dynamic programming—the idea goes back to Robson (1986)—the worst-case search tree sizes can be lowered slightly further at the cost of exponential space usage (Chandran and Grandoni, 2005; Niedermeier and Rossmanith, 2003*b*).

HITTING SET generalizes VERTEX COVER in the sense that VERTEX COVER is the same as 2-HITTING SET. The presented search tree algorithm for *d*-HITTING SET for constant *d* appears in Niedermeier and Rossmanith (2003*a*). Improved upper bounds can be found in Fernau (2004). Observe that for unlimited subset size HITTING SET is *W*[*2*]-complete (Downey and Fellows, 1999); hence there is basically no hope of finding a depth-bounded search tree as in the above cases.

The search tree for CLOSEST STRING is from Gramm *et al.* (2003*b*). The *NP*-completeness of CLOSEST STRING was first derived by Frances and Litman (1997). The algorithm presented has proven its usefulness in practical implementations. The more general problems CLOSEST SUBSTRING (Fellows *et al.*, 2002; Marx, 2005) and DISTINGUISHING SUBSTRING SELECTION (Gramm *et al.*, 2003) appear to be fixed-parameter intractable, though. By way of contrast, CLOSEST STRING and its generalizations *all* possess polynomial-time approximation schemes (Li *et al.*, 2002*a*; Li *et al.*, 2002*b*; Deng *et al.*, 2003).

The size-$O(8^k)$ search tree for DOMINATING SET IN PLANAR GRAPHS is proven in Alber *et al.* (2005*b*). It corrects an earlier flawed bound $O(11^k)$ appearing in Downey and Fellows (1999). The result generalizes to graphs of bounded genus (Ellis *et al.*, 2004).

The interleaving technique as described in Section 8.7 is presented along the lines of Niedermeier and Rossmanith (2000*a*). It is also applicable to MAXIMUM SATISFIABILITY (Chen and Kanj, 2004), 3-HITTING SET (Niedermeier and Rossmanith, 2003*b*), and many other problems solved by depth-bounded search trees and problem kernelization.

Concerning the automated generation and analysis of search trees we followed Gramm *et al.* (2004). Further work in this direction has been undertaken in Fedin and Kulikov (2004) and Nikolenko and Sirotkin (2003). In a somewhat different and more specialized setting, Robson (2001) applied computers to improve his older bounds from Robson (1986) on the search tree size for INDEPENDENT SET. Note that Robson (1986) contains a sophisticated search tree algorithm for INDEPENDENT SET where the combinatorial explosion is restricted with respect to the number of graph vertices but not with respect to the number of vertices in a maximum independent set for which one seeks.

There are numerous other depth-bounded search tree algorithms for various kinds of problem; a very small selection is contained in the papers Alber *et al.* (2004), Chen and Kanj (2003), Fernau and Niedermeier (2001), and Kaplan *et al.* (1999). In Chen *et al.* (2004*a*) the use of nondeterminism as an elegant way to perform and analyze search strategies is advocated. Finally, note that a very recent line of research takes the direction of investigating how to achieve better bounds, sometimes even with simplified algorithms, by improved mathematical analysis. Three examples here are Chen *et al.* (2002), Fernau (2004), and Fomin *et al.* (2005). To this end, multivariate recurrences often appear in the analysis of search tree sizes—Eppstein (2004) provides the necessary tools.

9

DYNAMIC PROGRAMMING

Avoiding time-consuming recomputations of the solutions of subproblems by storing intermediate results and doing table look-ups is the core idea behind dynamic programming. It is a fairly general problem-solving technique which applies to problems whose solution can be computed from solutions to subproblems. While the basic idea is very natural, it usually takes some time to verify the correctness of a given dynamic programming solution because, as a rule, some more or less complicated access to information on subproblems stored in tables has to be understood. That is why some people say that it might be easier to think about a dynamic programming solution for oneself than trying to understand a given one. In any case, however, dynamic programming is best learned by examples and by gaining personal experience by going through several applications.

One well-known example of dynamic programming is given by the computation of *binomial coefficients*. Pascal's formula states that, for all $n \geq 1$ and $1 \leq k \leq n$,

$$\binom{n}{k} = \binom{n-1}{k-1} + \binom{n-1}{k}.$$

This immediately leads to a recursive algorithm for computing $\binom{n}{k}$, knowing that $\binom{m}{m} = \binom{m}{0} = 1$ for all $m \geq 1$. Indeed, the table filled by computing $\binom{n}{k}$ is known as Pascal's triangle, which has $O(n \cdot k)$ entries. Hence filling this table in a "bottom-up" process yields a non-recursive polynomial-time algorithm for computing binomial coefficients. Note that a straightforward recursive implementation would always have branched into two recursive calls, yielding an exponential running time. The reason for this is that the same binomial coefficient is recomputed again and again in many different recursive calls.

Another standard dynamic programming algorithm from theoretical computer science is the cubic-time algorithm of Cocke, Younger, and Kasami that solves the word problem for context-free languages. Again the straightforward recursive implementation without using tables here would have ended up in exponential running time.

Both of the above examples are polynomial-time solvable problems and there are many others in this direction. In this chapter, however, we demonstrate the particular use of dynamic programming in the context of exponential-time fixed-parameter algorithms. To illustrate the various flavors in which dynamic programming may occur, we have chosen the STEINER PROBLEM IN GRAPHS, a flow problem in tree networks, and a tree-structured variant of the in general fixed-parameter "intractable" SET COVER problem. Finally, we discuss a combination of dynamic programming with depth-bounded search trees in order to save the exploration of some search tree parts, thus reducing the combinatorial explosion. To begin with, as a warm-up, we provide some notation and few classical examples concerning dynamic programming to solve hard combinatorial optimization problems.

9.1 Basic definitions and facts

In a sense, dynamic programming makes exhaustive search—which guarantees finding optimal solutions—more efficient by avoiding the computation of solutions of subproblems more than once. Dynamic programming is a promising solution strategy when a problem exhibits the property of having *optimal substructure*; that is, an optimal solution to the problem contains within it optimal solutions to subproblems. Moreover, dynamic programming uses this property in a *bottom-up* fashion. For instance, in our introductory example concerning binomial coefficients we can start with the entries $\binom{m}{m}$ and $\binom{m}{0}$ for all $1 \leq m \leq n$; then we may compute $\binom{2}{1} = \binom{1}{0} + \binom{1}{1} = 2$, by this $\binom{3}{1}$ and $\binom{3}{2}$, and so on. Two properties usually make dynamic programming particularly feasible:

- *independence*, and
- *overlapping subproblems.*

By independence of subproblems we mean that the solution of one subproblem does not affect the solution of another subproblem of the same problem. By overlapping subproblems we mean that the same problem occurs as a subproblem of different problems, hence giving an advantage over recursive algorithms where these solutions of subproblems would be computed more than once.

Dynamic programming applies to every problem that observes the *principle of optimality*. In a nutshell, this means that solutions to subproblems can be optimally extended only with regard to the current "state" of this solution—it is not necessary to know exactly which sequence of operations has been performed to date. In other words, this means that at a certain point one may also "forget" information because only the actual solution to a subproblem—and not the way to its construction—matters.

Finally, a brief word about a variation of dynamic programming: *memoization*. The basic difference from dynamic programming in the classical sense here is that the flow of control changes; the dynamic programming table is filled in a *top-down* manner instead of a bottom-up one. In fact, the idea is to *memorize* the natural but inefficient recursive solving algorithm. That is, again we maintain a table with subproblem solutions, but we fill these tables with the results from successful recursive calls. Section 9.6 describes a strategy to shrink depth-bounded search trees which can be interpreted as memoization. Note, however, that bottom-up dynamic programming is often more efficient because there is no administrative overhead for recursion and because sometimes regular access patterns to dynamic programming tables can be exploited. By way of contrast, the advantage of memoization is that it is guaranteed that only solutions to those subproblems that are actually needed are computed.

We conclude this section with a well-known exponential-time dynamic program which gives the to date best exact algorithm for the TRAVELING SALESPERSON PROBLEM (TSP):

> **Input:** A set $\{1, 2, \ldots, n\}$ of "cities" with pairwise nonnegative distances $d(i, j)$, $1 \leq i, j \leq n$.
> **Task:** Find an order of the cities such that by following this order each city is visited exactly once and the total distance traveled is minimized.

A trivial algorithm to solve TSP simply enumerates all possible $O(n!)$ tours. In 1962, the up to now best exact algorithm solving TSP was published. It brings the combinatorial explosion down to $O(2^n)$ by using a natural dynamic programming approach which we sketch in the following.

Clearly, without loss of generality, the tour we search for may start in city 1. For every non-empty subset $S \subseteq \{2, \ldots, n\}$ and for every city $i \in S$, denote by $Opt(S, i)$ the length of the shortest path that starts in city 1, then visits all cities in $S \setminus \{i\}$ in arbitrary order, and finally stops in city i. Obviously,

$$Opt(\{i\}, i) = d(1, i),$$

giving the "bottom entries" of the dynamic programming table. The decisive point fulfilling the principle of optimality now is based on the following recurrence:

$$Opt(S, i) = \min_{j \in S \setminus \{i\}} \{Opt(S \setminus \{i\}, j) + d(j, i)\}.$$

Then the optimal solution is given by

$$\min_{2 \leq j \leq n} \{Opt(\{2, \ldots, n\}, j) + d(j, 1)\}.$$

The combinatorial explosion now is basically determined by the number of subsets of $\{2, \ldots, n\}$ which is bounded by $O(2^n)$. Indeed, it is not difficult to see that the overall running time is bounded from above by $O(2^n \cdot n^2)$, giving a big improvement over enumerating all $O(n!)$ possibilities.

The above dynamic programming algorithm is not a parameterized one in the classical sense. In Section 5.4, however, we briefly discussed the special case of TSP restricted to the two-dimensional Euclidean plane and interpreting cities as points in the plane. Then, introducing the number of inner points with respect to the convex hull of the point set as a parameter called k, a fixed-parameter algorithm running in $O(2^k \cdot k^2 \cdot n)$ time was stated. For this special case, this will usually beat the above algorithm. See also Section 15.6.1 for further consideration of this problem.

9.2 Knapsack

The BINARY KNAPSACK problem is a classical *NP*-complete problem with many applications. It is particularly related to resource allocation scenarios. There are many variations of knapsack-like problems that could even fill a book on their own. Here, we focus on the following problem BINARY KNAPSACK:

> **Input:** A set of n items, each with a positive integer value v_i and a positive integer weight w_i, $1 \leq i \leq n$, and a positive integer bound W.
> **Task:** Find a subset of items such that their total value is maximized under the condition that their total weight does not exceed W.

Let I be an instance of BINARY KNAPSACK. By trying all possible 2^n subsets, one trivially finds an optimal solution in $2^n \cdot |I|^{O(1)}$ time. Using preprocessing methods, this can be improved to $2^{n/2} \cdot |I|^{O(1)}$ time using $2^{n/2} \cdot |I|^{O(1)}$ space. The focus in this section, however, is a dynamic programming algorithm related to the parameter W.

To begin with, let us clarify an important fact concerning BINARY KNAPSACK and similar problems. When describing an algorithm, we always measure its time and space complexity relative to the input size. In the case of BINARY KNAPSACK, however, the input instance is specified by $2n + 1$ integer numbers. These numbers are encoded in binary. In particular, this means that if we talk about number W we assume that the length of its encoding as an input is $\lceil \log W \rceil$. Hence one has to take care of what it means when we say that we have an algorithm with running time $O(W \cdot n)$ as we do in the following. This is an exponential running time, since $W = 2^{\log W}$. That is why we consider such an algorithm as a special case of a fixed-parameter algorithm with respect to parameter bound W. Before continuing this discussion, now let us see how to derive an $O(W \cdot n)$ algorithm for BINARY KNAPSACK.

In the case of "relatively small" weights, BINARY KNAPSACK is efficiently solvable: we employ dynamic programming with a table of size $O(W \cdot n)$. Obviously, we can throw away all elements whose weight exceeds W. Hence, without loss of generality, we assume that $w_i \leq W$ for all $1 \leq i \leq n$. Let $S \subseteq \{1, 2, \ldots, n\}$. Define the Boolean variable $R[X, S]$ for positive integer X: $R[X, S]$ is true iff there exists a subset of S whose total weight is exactly X. In other words, the weight X is "*realizable*" using only items from S. Step by step, we consider the subsets

$$S = \emptyset, S = \{1\}, S = \{1, 2\}, \ldots, S = \{1, 2, \ldots, n\},$$

and compute, for growing S, each of the values $R[X, S]$. Clearly, $R[X, \emptyset]$ is false for all $1 \leq X \leq W$. Let $S = \{1, 2, \ldots, i\}$. The value of $R[X, S \cup \{i + 1\}]$ is determined by

$$R[X, S \cup \{i + 1\}] = R[X, S] \vee R[X - w_{i+1}, S]. \qquad (9.1)$$

This is the central observation and its correctness is based on the fact that item $i + 1$ is either used for "realizing" weight X or it is not. Clearly, filling the corresponding size-$(W \times n)$ table can be done in $O(W \cdot n)$ time and, after all, we know all the realizable weights together with subsets realizing these weights.

So far we have neglected the value v_i of each item i, $1 \leq i \leq n$. To incorporate it, we replace the Boolean variables $R[X, S]$ by integer variables which express the *maximum* possible value that can be achieved by a subset S that realizes X. Thus, instruction (9.1) is then replaced by computing a maximum of two integer values instead of determining the Boolean "or" of two truth values. Recording here which option led to the maximum—if both did, we record an arbitrary one—we can by following the thus determined path through the dynamic programming table find out which subset of the n items actually gives a maximum value solution of BINARY KNAPSACK. Taken together, this proves the following.

Theorem 9.1 BINARY KNAPSACK *can be solved in* $O(W \cdot n)$ *time.*

Theorem 9.1 provides an efficient algorithm whenever the given total weight allowed is not too big. In fact, if the input instance of BINARY KNAPSACK were encoded in unary, that is, using a one-element alphabet, then BINARY KNAPSACK could be solved in polynomial time with respect to the input size. In the literature the above algorithm is called a *pseudo-polynomial-time algorithm* because it is polynomial with respect to the largest occurring value of an integer in the input. By way of contrast, note that VERTEX COVER remains NP-complete even when the size k of the vertex cover is encoded in unary. The distinction between problems that remain NP-complete even for input instances with unary encoding of integers and those that become polynomial-time solvable as BINARY KNAPSACK does is made rigorous through the concept of *strong NP-completeness.* Refer to the cited literature to learn more about that.

9.3 Steiner Problem in Graphs

Long before parameterized complexity theory was developed, in 1972 S. E. Dreyfus and R. A. Wagner gave an early example of a fixed-parameter algorithm for an important "connectivity problem" in graphs. The NP-complete STEINER PROBLEM IN GRAPHS is defined as follows.

> **Input:** An edge-weighted graph $G = (V, E)$ and a set of *terminal vertices* $S \subseteq V$ with $|S| =: k$.
> **Task:** Find a subgraph G' of G that connects all vertices in S and whose total edge weight is minimum.

The original STEINER PROBLEM is an old problem in geometry and requires finding the set of lines of minimum total length which connect a given set of points S in the Euclidean plane. Note that the lines may intersect not only in points from S but also on other points in the plane in order to minimize the total length. When studying the STEINER PROBLEM IN GRAPHS, observe that, opposite to the geometric problem as specified above, the triangle inequality must no longer hold making the problem more elusive. Still, it is easy to observe that the connecting subgraph G' must be a *tree*: assume that G' contains a cycle. Then there are two different paths available for connecting some vertex in S to the rest. Clearly, one can delete the heaviest edge on these paths without destroying the connectivity but reducing the total edge weight of G'. Hence we subsequently speak of minimum (weight) Steiner trees we are searching for.

Let us consider two simple special cases of STEINER PROBLEM IN GRAPHS.

- If $S = V$, then the problem reduces to computing a minimum weight spanning tree of $G = (V, E)$ and clearly can be solved in polynomial time.
- If $|S| = 2$, then the problem reduces to computing a shortest path between two vertices in G. Again, this clearly is polynomial-time solvable.

Considering the parameterization by the size of the terminal set S, we now show that STEINER PROBLEM IN GRAPHS is fixed-parameter tractable with respect to parameter $k := |S|$.

The STEINER PROBLEM IN GRAPHS carries a nice *decomposition property* that can be made use of to solve the problem by a recursive approach. To make things more efficient, that is, to avoid repeated calls solving one and the same subproblem, dynamic programming is employed. The fundamental idea is to compute the weight of a minimum Steiner tree for a given terminal set by considering the weights of the minimum Steiner trees of all proper subsets of this set. Starting the process with two-element subsets (where the Steiner tree can be determined by shortest path computations) one finally ends up with the k-element terminal set S.

The fixed-parameter algorithm follows easily from the subsequent key lemma. To this end, we have to introduce some notation. Let $X \subseteq S$ and $v \in V \setminus X$.

- $s(X \cup \{v\})$ denotes the weight of a minimum Steiner tree connecting all vertices from $X \cup \{v\}$ in the given graph G;
- $p(u, v)$ denotes the total weight of the shortest (lightest) path between vertices u and v.

Lemma 9.2 *Let $X \neq \emptyset$, $X \subseteq S$, and $v \in V \setminus X$. Then,*

$$s(X \cup \{v\}) = \min\{\min_{u \in X}\{s(X) + p(u, v)\}, \min_{u \in V \setminus X}\{s_u(X \cup \{u\}) + p(u, v)\}\}, \quad (9.2)$$

where

$$s_u(X \cup \{u\}) := \min_{X' \neq \emptyset, X' \subsetneq X}\{s(X' \cup \{u\}) + s((X \setminus X') \cup \{u\})\}. \quad (9.3)$$

Proof To prove equation 9.2, assume that T is a *minimum* Steiner tree for $X \cup \{v\}$. If v is a leaf in T, then define P_v as the longest (heaviest) path in T starting in v and in which all interior points have degree two in T. Distinguishing three cases and minimizing over all of them then gives equation (9.2):

1. If v is no leaf and has degree at least two in T, then it is easy to observe that there is a partition of X into two sets X' and X'' such that $T[X' \cup \{v\}]$ and $T[X'' \cup \{v\}]$ are the minimum Steiner trees for $X' \cup \{v\}$ and $X'' \cup \{v\}$, respectively. Moreover, vertex v is a leaf in both these Steiner trees. This case is clearly covered by the special case $u = v$ in (9.3).

2. If v is a leaf in T and P_v ends at a vertex $u \in X$, then T consists of a minimum Steiner tree for X and a shortest (lightest) path from u to v. Hence,

$$s(X \cup \{v\}) = s(X) + p(u, v),$$

which is covered by the first term in (9.2).
3. If v is a leaf in T and P_v ends at a vertex $u \in V \setminus X$, then u has at least three neighbor vertices in T. This implies that T consists of a minimum Steiner tree for $X \cup \{u\}$ in which u has degree at least two and a shortest (lightest) path from u to v. Observe that the minimum Steiner tree for $X \cup \{u\}$ exactly corresponds to the first case discussed above. Therefore, this case is covered by the second term in (9.2).

□

With the help of Lemma 9.2, the fixed-parameter algorithm solving the STEINER PROBLEM IN GRAPHS now easily derives.

Theorem 9.3 *The* STEINER PROBLEM IN GRAPHS *can be solved in* $O(3^k \cdot n + 2^k \cdot n^2 + n^2 \cdot \log n + n \cdot m)$ *time, where* n *denotes the number of vertices and* m *denotes the number of edges in the given graph.*

Proof We employ the following algorithm which is basically a one-to-one translation of the recursions from Lemma 9.2. The solution is given by the value of $s(S)$.

In an initialization step, compute $p(u, v)$ for all $u, v \in V$, and for all $x, y \in S$ initialize

$$s(\{x, y\}) := p(x, y).$$

Then perform recursive calls as prescribed by Lemma 9.2, assuming that the values on the right-hand sides are stored in a dynamic programming table.

With regard to the running time, observe that the initialization can be done in $O(n^2 \cdot \log n + n \cdot m)$ time using n times Dijkstra's shortest path algorithm. The number of recursive calls corresponding to (9.3) in Lemma 9.2 can be bounded from above by 3^k: for all $X \neq \emptyset$, $X \subseteq S$ we have to consider all $X' \neq \emptyset$, $X' \subseteq X$, and all $v \in V \setminus X$. The number of combinations can be upper-bounded by

$$\sum_{i=1}^{k} \binom{k}{i} \cdot \sum_{j=1}^{i-1} \binom{i}{j} \cdot n \leq n \cdot \sum_{i=1}^{k} \binom{k}{i} \cdot 2^{i-1} \leq n \cdot 3^k.$$

Since each combination leads to two table look-ups (recursive calls corresponding to $s(X' \cup \{v\})$ and $s((X \setminus X') \cup \{v\})$) and since we have to perform only constantly many operations for a fixed combination of X, X', and v, we obtain the upper bound $O(3^k \cdot n)$ for the running time here. As to equation (9.2), observe that by similar considerations we can conclude that in this case $O(2^k \cdot n)$ of pairs X and v are possible. For each fixed pair X and v, however, due to the consideration of

$u \in X$ and $u \in V \setminus X$, we get an additional factor of $O(n)$. Altogether, we have an upper bound of $O(2^k \cdot n^2)$ here, implying the claimed overall running time.

We have only described an algorithm for determining the weight of a minimum Steiner tree here—it is straightforward to extend it in order to construct the corresponding tree explicitly. □

9.4 Multicommodity Demand Flow in Trees

As a rule, most hard graph problems become easy when restricted to trees. For instance, VERTEX COVER restricted to trees is solvable in linear time by a straightforward algorithm that works bottom-up from the leaves to the (arbitrarily chosen) root. By way of contrast, the MULTICOMMODITY DEMAND FLOW IN TREES problem, MDFT for short, gives an example where *NP*-completeness even holds in the case of trees. Fortunately, for trees there is a fixed-parameter algorithm based on dynamic programming. It is the subject of this section.

MDFT is defined as follows.

> **Input:** A "capacitated tree" $T = (V, E, c)$, where each edge capacity $c(e)$ is an integer, and a collection F of *flows* which is encoded as a list of pairs of vertices of T,
>
> $$F = \{f_i \mid f_i := (u_i, v_i), u_i \in V, v_i \in V, u_i \neq v_i, 1 \leq i \leq m\}.$$
>
> Each flow $f \in F$ has associated an integer demand value $d(f) > 0$ and a real valued profit $p(f) > 0$.
>
> **Task:** Find a routable subset $F' \subseteq F$ which maximizes $p(F') := \sum_{f \in F'} p(f)$. A subset $F' \subseteq F$ is *routable* in T if the flows in F' can be simultaneously routed without violating any edge capacity of the tree. That is, for any edge e the sum of the demand values of the flows routed through e does not exceed $c(e)$.

The tree T is termed the *supply tree*. The vertices u, v are called the two *endpoints* of the flow $f = (u, v)$. Note that, for a tree, the path between two distinct vertices is uniquely determined and can be found in linear time. Thus, we can assume that each flow $f = (u, v) \in F$ is given as a path between u and v. We use F_v to denote the set of demand flows passing through vertex v. A demand flow *passes through* vertex v if it has v as one of its endpoints or v lies on the flow's path.

We will show that MDFT is fixed-parameter tractable with respect to the parameter $k :=$ "maximum number of demand flows passing through any vertex of the tree". It is important to note here that a parameterization by the "maximum number of demand flows passing through any *edge* of the tree" is of no "parameterized interest": it can be shown that MDFT is *NP*-complete even when there are at most six demand flows passing through any edge. Hence there is no hope for fixed-parameter tractability with respect to this parameter.

Before we present the dynamic programming algorithm for the general MDFT problem, as a "warm-up" we briefly discuss an easy special case with a restriction

$s := s_1 s_2 \dots s_{k_v}$	$D_e(s)$
$00\dots00$	
$00\dots01$	
$\vdots$	
$11\dots11$	

FIG. 9.1. Table D_e for edge $e = \{u, v\}$ with $F_v := \{f_1^v, f_2^v, \dots, f_{k_v}^v\}$ with $k_v \leq k$.

to paths instead of trees. Besides the restriction to paths we further require that only unit flow demands and unit profits are allowed. Then the following simple strategy works. The algorithm starts at the left end of the supply path. While not reaching the right end of the path, it does the following.

1. Check whether the capacity of the current edge $e = \{u, v\}$ suffices, where u is the left endpoint of e and v is the right endpoint.
2. If not, remove "longest" demand flows which start in u until the capacity of e suffices.
3. Replace each of the flows $f = (u, z)$ with $z \neq v$ by a flow between v and z.

The solution is simply given by the set of non-removed demand flows.

Proposition 9.4 *MDFT restricted to paths as supply tree, unit flow demands, and unit profits can be solved in $O(m + n)$ time, where n denotes the number of tree vertices and m denotes the number of flows.*

Proof The above algorithm is obviously correct. After simple preprocessing, we know the length of each demand flow. Then we can proceed from left to right in $O(n)$ main steps. All steps together need to remove at most $O(m)$ demand flows, yielding $O(m + n)$ running time. □

The central idea behind the dynamic programming algorithm for the general MDFT problem is to distinguish three main cases. This is what we describe next. We begin with some agreements and basic observations. We assume that we deal with arbitrarily rooted trees. Thus an edge $e = \{u, v\}$ reflects that u is the parent vertex of v. We use $T[v]$ to denote the subtree of the input tree rooted at vertex v. In particular, using the rooted tree structure, we will solve MDFT in a bottom-up fashion by dynamic programming from the leaves to the root.

The core idea of the dynamic programming is based on the following definition of tables which are used throughout the algorithm. For each edge $e = \{u, v\}$ with

$$F_v := \{f_1^v, f_2^v, \dots, f_{k_v}^v\} \subseteq F,$$

we construct a table D_e as illustrated in Figure 9.1. Thus, F_v denotes the set of flows from F running through v, and $k_v \leq k$ is the number of these flows. Table D_e has 2^{k_v} rows which correspond to all possible vectors s with k_v entries over $\{0, 1\}$—k_v-*vectors* for short. Each of these vectors represents a *route schedule* for the flows in F_v. The ith component s_i of s corresponds to flow f_i^v, and $s_i = 0$

means that we do not route flow f_i^v and $s_i = 1$ means that we do route f_i^v. Furthermore, to refer to the set of routed flows, we define

$$r(s) := \{i \mid s_i = 1, 1 \le i \le k_v\}.$$

Table entry $D_e(s)$ stores the maximum profit which we can achieve according to the route schedule encoded by s with the flows which have at least one of their endpoints in $T[v]$.

The algorithm works bottom-up from the leaves to the root. Having computed all tables D_e for the edges e connected to the root vertex, MDFT will be solved easily, as pointed out later on. The algorithm starts with "leaf edges" which are initialized as follows. For an edge $e = \{u, v\}$ connecting a leaf v with its parent vertex u, the table entries for the at most 2^k rows s are determined by

$$D_e(s) := \begin{cases} 0, \text{ if } c(e) < \sum_{i \in r(s)} d(f_i^v); \\ \sum_{i \in r(s)} p(f_i^v), \text{ otherwise.} \end{cases}$$

Then the main algorithm consists of distinguishing three cases.

Case 1: Consider an edge $e = \{u, v\}$ connecting two inner vertices u and v where v has only one child w connected to v by edge $e' = \{v, w\}$.

The sets of flows passing through u, v, and w, respectively, are denoted by $F_u := \{f_1^u, \ldots, f_{k_u}^u\}$, $F_v := \{f_1^v, \ldots, f_{k_v}^v\}$, and $F_w := \{f_1^w, \ldots, f_{k_w}^w\}$. Moreover, we use F_e and $F_{e'}$ to denote the sets of flows passing through e and e' and we have $F_e = F_u \cap F_v$ and $F_{e'} = F_v \cap F_w$.

First, if $F_e \cap F_{e'} = \emptyset$, then the given instance can be divided into two smaller instances, one consisting of subtree $T[v]$ and the flows inside it and the other consisting of the original tree without the vertices below v and the flows therein. The optimal solution for the original instance is then the sum of the optimal solutions of the two smaller instances. An optimal solution of the first smaller instance is already computed and is obtained from a maximum entry of table $D_{e'}$. In order to compute an optimal solution of the second smaller instance, we can treat v as a leaf and proceed as for leaf edges above.

Second, if $F_e \cap F_{e'} \neq \emptyset$, then there are some flows passing through both e and e'. Recall that entry $D_e(s)$ for a k_v-vector s will store the maximum profit with respect to the route schedule encoded by s that can be achieved by the flows with at least one of their endpoints in $T[v]$. We partition F_v into two sets, $F_v \cap F_w$ and $F_v \setminus F_w$. The value of $D_e(s)$ is thus the sum of the maximum of the entries of $D_{e'}$ which have the same route schedule for the flows in $F_v \cap F_w$ as encoded in s, and the profit achieved by the flows in $F_v \setminus F_w$ obeying the route schedule encoded by s. Let

$$B^v := \{i \mid f_i^v \in (F_v \cap F_w)\}$$

and

$$B^w := \{i \mid f_i^w \in (F_v \cap F_w)\},$$

and $j := |B^v| = |B^w|$. Clearly, B^v and B^w refer to the same sets of demand flows. Note, however, that they may differ due to different "naming" of the same flow in the two tables D_e and $D_{e'}$. To more easily obtain the maximum of the entries of $D_{e'}$ which have the same route schedule for the flows in $F_v \cap F_w$, we condense table $D_{e'}$ with respect to B^w. The *condensation of $D_{e'}$ with respect to B^w* is to keep only the components of the k_w-vector s' of $D_{e'}$ which correspond to the demand flows in $F_v \cap F_w$. More precisely, for a route schedule $\overline{s'}$ condensed with respect to B^w, we obtain

$$D_{e'}(\overline{s'}) := \max\{D_{e'}(s') \mid \overline{s'} = \pi_{B^w}(s')\}.$$

Herein, $\pi_{B^w}(s')$ returns the projection of s' onto the j components of s' that correspond to B_w. Then, using $A := \{i \mid f_i^v \in F_e\}$ to refer to the set of the flows in F_v passing through edge e, the entries of D_e are computed as follows.

$$D_e(s) := \begin{cases} 0, \text{ if } c(e) < \sum_{i \in A} d(f_i^v); \\ D_{e'}(\pi_{B^v}(s)) + \sum_{i \in (r(s) \setminus B^v)} p(f_i^v), \text{ otherwise.} \end{cases}$$

Obeying the route schedule encoded by s, $D_{e'}(\pi_{B^v}(s))$ and $\sum_{i \in (r(s) \setminus B^v)} p(f_i^v)$ denote the profits achieved by the flows in $F_v \cap F_w$ and in $F_v \setminus F_w$, respectively.

Case 2: Consider an edge $e = \{u, v\}$ connecting two non-leaf vertices u and v where v has two children w_1 and w_2 connected to v by edges $e' = (v, w_1)$ and $e'' = (v, w_2)$.

We use $F_u := \{f_1^u, \ldots, f_{k_u}^u\}$, $F_v := \{f_1^v, \ldots, f_{k_v}^v\}$, $F_{w_1} := \{f_1^{w_1}, \ldots, f_{k_{w_1}}^{w_1}\}$, and $F_{w_2} := \{f_1^{w_2}, \ldots, f_{k_{w_2}}^{w_2}\}$ to denote the sets of flows passing through vertices u, v, w_1, and w_2. As in Case 1, $F_e = F_u \cap F_v$, $F_{e'} = F_v \cap F_{w_1}$, and $F_{e''} = F_v \cap F_{w_2}$. With the same argument as in Case 1, if one of $F_e \cap F_{e'}$ and $F_e \cap F_{e''}$ is empty, we can divide the given instance into two smaller instances and solve them separately. Thus we assume that they are not empty. Similar to Case 1, we "partition" F_v into the three sets $F_v \cap F_{w_1}$, $F_v \cap F_{w_2}$, and $(F_v \setminus F_{w_1}) \setminus F_{w_2}$. For a k_v-vector s, one might simply set $D_e(s)$ equal to the sum of the maximum of the entries of $D_{e'}$ which have the same route schedule as encoded in s for the flows in $F_v \cap F_{w_1}$, the maximum of the entries of $D_{e''}$ which have the same route schedule as encoded in s for the flows in $F_v \cap F_{w_2}$, and the profit achieved by the flows in $(F_v \setminus F_{w_1}) \setminus F_{w_2}$ obeying the route schedule encoded by s. However, if $(F_v \cap F_{w_1}) \cap (F_v \cap F_{w_2}) \neq \emptyset$, that is, due to the tree structure, $F_{w_1} \cap F_{w_2} \neq \emptyset$, then the edges e' and e'' have some common flows. Then, for each flow between $T[w_1]$ and $T[w_2]$ scheduled to be routed in both subtrees, we must subtract its profit once from the sum to avoid double counting. Let

$$\begin{array}{ll} B_1^v := \{i \mid f_i^v \in (F_v \cap F_{w_1})\}; & B_1^{w_1} := \{i \mid f_i^{w_1} \in (F_v \cap F_{w_1})\}; \\ B_2^v := \{i \mid f_i^v \in (F_v \cap F_{w_2})\}; & B_2^{w_2} := \{i \mid f_i^{w_2} \in (F_v \cap F_{w_2})\}; \\ B_3^{w_1} := \{i \mid f_i^{w_1} \in (F_{w_1} \cap F_{w_2})\}; & B_3^{w_2} := \{i \mid f_i^{w_2} \in (F_{w_1} \cap F_{w_2})\}. \end{array}$$

Note that $|B_1^v| = |B_1^{w_1}|$, $|B_2^v| = |B_2^{w_2}|$, $|B_3^{w_1}| = |B_3^{w_2}|$, $B_3^{w_1} \subseteq B_1^{w_1}$, and $B_3^{w_2} \subseteq B_2^{w_2}$. As in Case 1, we condense $D_{e'}$ and $D_{e''}$. More specifically, we condense $D_{e'}$ with respect to $B_1^{w_1}$ and $D_{e''}$ with respect to $B_2^{w_2}$:

$$D_{e'}(\overline{s'}) := \max\{D_{e'}(s') \mid \overline{s'} = \pi_{B_1^{w_1}}(s')\};$$
$$D_{e''}(\overline{s''}) := \max\{D_{e''}(s'') \mid \overline{s''} = \pi_{B_2^{w_2}}(s'')\}.$$

Then, using $A := \{i \mid f_i^v \in F_e\}$, the entries of D_e are computed as follows.

$$D_e(s) := \begin{cases} 0, & \text{if } c(e) < \sum_{i \in A} d(f_i^v); \\ \alpha + \beta, & \text{otherwise.} \end{cases}$$

Herein,

$$\alpha := D_{e'}(\pi_{B_1^v}(s)) + D_{e''}(\pi_{B_2^v})(s) - \gamma,$$
$$\beta := \sum_{i \in (r(s) \setminus (B_1^v \cup B_2^v))} p(f_i^v),$$
$$\gamma := \sum_{i \in B_3^{w_1}, \pi_{\{i\}}(s)=1} p(f_i^{w_1}).$$

In order to avoid double counting of common flows of edges e' and e'', γ has been subtracted in the above determination of D_e.

Case 3: Consider an edge $e = \{u, v\}$ connecting two non-leaf vertices u and v where v has $l > 2$ children $w_1, w_2, \ldots, w_l$.

We add $l - 2$ new vertices $v_1, v_2, \ldots, v_{l-2}$ to T and we transform T into a binary tree. Each of these new vertices has exactly two children. Vertex v_1 is the parent vertex of w_1 and w_2, v_2 is the parent vertex of v_1 and w_3, and so on. Thus, v becomes the parent vertex of v_{l-2} and w_l. The edge between w_i and its new parent vertex is assigned the same capacity as the edge between w_i and v in the original tree. The edges $\{v, v_{l-2}\}, \{v_{l-2}, v_{l-1}\}, \ldots, \{v_2, v_1\}$ between the new vertices obtain unbounded capacity. The flows have the same endpoints as in the original instance. It is easy to see that the solutions for the new and the old tree are the same, and thus Case 1 and Case 2 suffice for handling the new binary tree. This concludes the description of the dynamic programming algorithm. It implies the following result.

Theorem 9.5 *MDFT can be solved in $O(2^k \cdot m \cdot n)$ time, where k denotes the maximum number of demand flows passing through any vertex of the given supply tree, n denotes the number of tree vertices and m denotes the number of flows.*

Proof The correctness of the algorithm follows directly from its description. To this end, however, note that when the D-tables of all edges of the supply tree are computed, we may easily determine the final optimum by comparing the tables without loss of generality of at most two edges leading to the root. Moreover, by means of a top-down traversal from the root to the leaves we can easily determine the actual subset of routable flows that led to the overall maximum profit. We omit the straightforward details here.

With regard to the algorithm's running time, observe that table D_e for edge $e = \{u, v\}$ has at most $O(2^k)$ entries. For a vertex v with more than two children, as described in Case 3 above, we add some new vertices to construct a binary tree. The resulting tree at most doubles in size. Moreover, since $F_{v'} \subseteq F_v$ for each of the new vertices v', the D-tables of the new edges have at most $O(2^k)$ entries. Assuming that all basic set operations such as union and set minus between two sets having at most m elements can be done in $O(m)$ time, the computation of a new table from its at most two child tables runs in $O(2^k \cdot m)$ time. Altogether, the algorithm then takes $O(2^k \cdot m \cdot n)$ time. □

The algorithm presented is a competitive or even superior alternative to heuristic or approximation algorithms for not too large values of parameter k, a realistic assumption for several application scenarios. Its space consumption grows with k as the running time does, though. Thus it is conceivable that its practical usefulness will be more limited by its memory usage than it is by its running time—a fairly frequent observation for dynamic programming in relation to tables of exponential size.

9.5 Tree-structured variants of Set Cover

In this section we present the technically most challenging example of dynamic programming. We deal with one of the most important problems of combinatorial optimization—SET COVER. This classical *NP*-complete problem is notoriously hard with respect to polynomial-time approximation—no constant-factor approximation is known—as well as from the viewpoint of parameterized complexity studies. That is why we introduce a tree-structured *NP*-complete variant of SET COVER which is motivated by practical applications. Thus, besides further presenting a highly non-trivial example of dynamic programming, the intentions behind this section are at least two-fold:

- To give an example of the creative process of discovering and exploiting practically relevant problem parameterizations, and
- to emphasize the importance of a "tree-likeness" respectively "acyclicity" property in dynamic programming; this will play a central role when investigating the use of tree decompositions of graphs in Chapter 10.

We divide this slightly longer section into three subsections. The first formally introduces the SET COVER problem and the tree-structured variants we study. Moreover, it summarizes some simple observations and the hardness of tree-like special cases to be studied. In the next subsection, we deal with the easier to grasp, polynomial-time solvable "very special" case where the tree structure actually boils down to only a path. This prepares the *NP*-complete case with arbitrary tree structures studied in the third and final subsection. We end by discussing a close relation to the MULTICUT IN TREES problem.

9.5.1 *Basic definitions and facts*

The basic SET COVER problem (optimization version) is defined as follows:

Input: A base set $S = \{s_1, s_2, \ldots, s_n\}$ and a collection C of subsets of S, $C = \{c_1, c_2, \ldots, c_m\}$, $c_i \subseteq S$ for $1 \leq i \leq m$, and $\bigcup_{1 \leq i \leq m} c_i = S$.
Task: Find a subset C' of C of minimum cardinality which covers all elements in S, that is, $\bigcup_{c \in C'} c = S$.

By assigning weights to the subsets and minimizing the total weight of the collection C' instead of its cardinality, one naturally obtains the WEIGHTED SET COVER problem. We call C' the *minimum (weight) set cover* of S. Define the *occurrence* of an element $s \in S$ in C as the number of the subsets in C which contain s. An element with occurrence of one is called *unique*. SET COVER remains NP-complete even if the occurrence of each element is bounded by two.

Definition 9.6. (Tree-like subset collection) *Given a base set $S = \{s_1, s_2, \ldots, s_n\}$ and a collection C of subsets of S, $C = \{c_1, c_2, \ldots, c_m\}$, $c_i \subseteq S$ for $1 \leq i \leq m$, we say that C is a* tree-like subset collection *of S if we can organize the subsets in C in an unrooted tree T such that every subset one-to-one corresponds to a node of T and, for each element $s_j \in S$, $1 \leq j \leq n$, the nodes in T corresponding to the subsets containing s_j induce a subtree of T.*

We call T the underlying *subset tree* and the property of T that, for each $s \in S$, the nodes containing s induce a subtree of T, is called the *consistency property* of T. Observe that the consistency property also will be of central importance in the notion of tree decompositions of graphs to be explored in Chapter 10—the dynamic programming techniques to be described heavily rely on this property. It is known from the literature that one can test whether a subset collection is a tree-like subset collection and, if yes, one can construct a subset tree for it in linear time. For this reason, in the following we always assume that the subset collection is given in form of a subset tree. For convenience, we denote the nodes of the subset tree by their corresponding subsets.

Example 9.7 For $S = \{s_1, s_2, s_3\}$, the subset collection $C = \{c_1, c_2, c_3\}$ where $c_1 = \{s_1, s_2\}$, $c_2 = \{s_2, s_3\}$, and $c_3 = \{s_1, s_3\}$ is not a tree-like subset collection. These three subsets can only be organized in a triangle. By way of contrast, if $c_1 = \{s_1, s_2, s_3\}$ instead, then we can construct a subset tree (actually a path) with these three nodes and two edges, one between c_1 and c_2 and one between c_1 and c_3.

We now define the tree-structured variant of SET COVER to be studied in the remainder of this section—TREE-LIKE WEIGHTED SET COVER (TWSC).

TREE-LIKE WEIGHTED SET COVER (TWSC):
Input: A base set $S = \{s_1, s_2, \ldots, s_n\}$ and a tree-like collection C of subsets of S, $C = \{c_1, c_2, \ldots, c_m\}$, $c_i \subseteq S$ for $1 \leq i \leq m$, and $\bigcup_{1 \leq i \leq m} c_i = S$. Each subset in C has a positive real weight $w(c_i) > 0$ for $1 \leq i \leq m$. The weight of a subset collection is the sum of the weights of all subsets in it.
Task: Find a subset C' of C with minimum weight which covers all elements in S, that is, $\bigcup_{c \in C'} c = S$.

The following simple observation will be helpful in developing our main results.

Lemma 9.8 *Given a tree-like subset collection C of S together with its underlying subset tree T, then each leaf of T is either a subset of its parent node or it has a unique element.*

Proof Consider a leaf $c = \{s_1, s_2, \ldots, s_r\}$ of the subset tree. Let c' be its parent node. We show that, if c is not a subset of c', that is, $c \setminus c' \neq \emptyset$, then c contains a unique element. Assume that c contains no unique element. Thus, for each $s_i \in (c \setminus c')$, $1 \leq i \leq k$, there is a subset c'' different from c which contains s_i. According to the consistency property of the subset tree, all subsets on the path from c to c'' must also contain s_i. Since the uniquely determined path from c to c'' must pass through c', s_i has to be in c'. This contradicts $s_i \in (c \setminus c')$. □

Unlike the general unweighted SET COVER which is *NP*-complete, TREE-LIKE UNWEIGHTED SET COVER can be solved by a simple polynomial-time algorithm:

Process the subset tree T in a bottom-up manner, that is, begin with the leaves. By Lemma 9.8, each leaf of T is either a subset of its parent node or it contains a unique element from S. If a leaf c contains a unique element, the only choice to cover this element is to put c into the set cover. Then we delete c from T and c's elements from other subsets. If c is contained in its parent node, it is never better to take c into the set cover than to take its parent node. Hence we can safely delete it from the subset tree. After processing its children, an internal node becomes a leaf and we can iterate the described process. Thus we obtain the following result.

Proposition 9.9 TREE-LIKE UNWEIGHTED SET COVER *can be solved in $O(m \cdot n)$ time.*

Proof The correctness of the above algorithm is clear from its description. Furthermore, it is easy to observe that the algorithm goes through each node of the tree only once. At each node, we compare it with its parent node and delete some elements from it. All these operations can be done in $O(n)$ time. Altogether, the whole subset tree can be processed in $O(m \cdot n)$ time. □

The key idea of the above algorithm is that one never puts a set into the desired subset collection C' which is a subset of other subsets in the collection C. If we associate each subset with an arbitrary positive weight and ask for the set cover with minimum weight, however, this strategy is no longer valid: a simple reduction of the *NP*-complete unweighted SET COVER problem to TWSC shows that the decision version of TWSC is *NP*-complete. Indeed, this reduction also implies that TWSC is $W[2]$-hard with respect to the parameter "total weight of the solution" of the set cover. This excludes fixed-parameter tractability for this parameterization. Similarly, one can show that TWSC remains *NP*-complete when the occurrence of each element is limited to at most three. Thus we can also conclude that parameterization by "occurrence number" is not useful concerning fixed-parameter tractability studies for TWSC. We therefore propose parameterization by maximum subset size. To this end, as a warm-up to get acquainted with the employed dynamic programming strategy, we first study the simpler special case PATH-LIKE WEIGHTED SET COVER. Here, however, the parameter "maximum subset size" may be still completely unbounded.

9.5.2 *Algorithm for Path-like Weighted Set Cover*

We begin with studying the polynomial-time solvable special case of TREE-LIKE WEIGHTED SET COVER where the underlying subset tree is only a path. The techniques introduced here will be further refined and developed when presenting a fixed-parameter algorithm for the general, *NP*-complete case. Replacing trees with paths in Definition 9.6, we naturally end up with the notion of PATH-LIKE WEIGHTED SET COVER. Note that PATH-LIKE WEIGHTED SET COVER is also known by the name SET COVER WITH CONSECUTIVE ONES PROPERTY. The dynamic programming algorithm to solve this problem works as follows.

Given $S = \{s_1, s_2, \ldots, s_n\}$ together with a path-like subset collection $C = \{c_1, c_2, \ldots, c_m\}$ with positive real weights $w(c_i) > 0$ for $1 \leq i \leq m$, assume that these subsets are ordered on a line, that is, from the left end to the right end, $c_1, c_2, \ldots, c_m$. Every two consecutive subsets are connected by an edge. Additionally, we define for every subset c_i two functions $A(c_i)$ and $B(c_i)$:

$$A(c_i) := c_1 \cup c_2 \cup \cdots \cup c_i,$$

and

$$B(c_i) := \text{ minimum weight to cover } A(c_i) \text{ by using only subsets from } \{c_1, \ldots, c_i\}.$$

Clearly, $B(c_m)$ stores the minimum weight to cover all elements, which basically solves PATH-LIKE WEIGHTED SET COVER. To compute $B(c_m)$ based on the values of $B(c_1), B(c_2), \ldots, B(c_{m-1})$, the dynamic programming algorithm processes the subsets from left to right and, afterwards, by way of traceback from right to left constructs a minimum set cover with weight $B(c_m)$.

Initialization. We can assume that every element occurs in at least two subsets; otherwise, by preprocessing, we add the subsets into the set cover that contain at least one such unique element and delete the elements already covered by these subsets from the remaining subsets. For ease of presentation, we additionally introduce $c_0 := \emptyset$ with $w(c_0) := 0$, $A(c_0) := \emptyset$, and $B(c_0) := 0$. The values of $A(c_1)$ and $B(c_1)$ are trivial to compute, that is, $A(c_1) := c_1$ and $B(c_1) := w(c_1)$.

Main algorithm. We compute the value of $A(c_i)$ and $B(c_i)$ for an arbitrary i, $2 \leq i \leq m$. Clearly, $A(c_i) := A(c_{i-1}) \cup c_i$. To compute $B(c_i)$, we distinguish two cases.

Case 1: If $c_i \nsubseteq c_{i-1}$, then c_i contains an element which is not in $A(c_{i-1})$. To cover $A(c_i)$ by only using $c_1, \ldots, c_i$, we have to take c_i. Hence,

$$B(c_i) := w(c_i) + \min_{0 \leq l \leq i-1} \{\, B(c_l) \mid (A(c_i) \setminus c_i) \subseteq A(c_l) \,\}.$$

Case 2: If $c_i \subseteq c_{i-1}$, then $A(c_i) = A(c_{i-1})$. We compare the two alternatives to cover $A(c_i)$—either to take c_i or not—and we take the minimum of the two. If we do not take c_i, the minimum weight to cover $A(c_i)$ is stored in $B(c_{i-1})$; thus,

$$B(c_i) := \min\{B(c_{i-1}),\, w(c_i) + \min_{0 \leq l \leq i-1}\{\, B(c_l) \mid (A(c_i) \setminus c_i) \subseteq A(c_l)\,\}\,\}.$$

After computing $B(c_m)$, we know the minimum weight to cover all elements. Based on whether the minimum weight was achieved by taking c_m or not, we can construct the minimum set cover by a traceback.

Theorem 9.10 PATH-LIKE WEIGHTED SET COVER *can be solved in* $O(m^2 \cdot n)$ *time.*

Proof The correctness of the above algorithm follows directly from its description.

The preprocessing needs $O(m \cdot n)$ time. With regard to the running time of the dynamic programming, we go through each subset from left to right once in order to compute the minimum weight. For each subset c_i, the most time-consuming work is to find out the subset c_l with minimum $B(c_l)$ among those subsets satisfying $(A(c_i) \setminus c_i) \subseteq A(c_l)$. Clearly, the subset c_j, $j < i$, with minimum index satisfying $(A(c_i) \setminus c_i) \subseteq A(c_j)$ can be found in $O(m \cdot n)$ time. Note that there can be two subsets $c_{j'}, c_{j''}$, $j \leq j' < j'' < i$, with $A(c_{j'}) \subseteq A(c_{j''})$ and $B(c_{j'}) > B(c_{j''})$. Hence, in order to find the subset c_l with the minimum $B(c_l)$ among the subsets between c_j and c_i, we go through all these subsets. This takes $O(m)$ time. Since all employed set operations such as set minus and union between two sets with a maximum size of n can be done in $O(n)$ time, we need $O(m \cdot n)$ time at each node. The traceback is clearly doable in $O(m)$ time. Thus, in summary, the algorithm takes $O(m^2 \cdot n)$ time. □

9.5.3 *Algorithm for Tree-like Weighted Set Cover*

By extending the polynomial-time dynamic programming algorithm for the path-like case given in Section 9.5.2, we will here present a fixed-parameter algorithm for TWSC with respect to the parameter maximum subset size k, that is,

$$k := \max_{c \in C}\{\, |c|\,\}.$$

To facilitate the presentation of the algorithm, we will only describe how to solve the problem for binary subset trees, an also *NP*-complete special case of TWSC. It is not very hard to extend the method presented here in order to solve the problem for arbitrarily structured subset trees (see Exercises).

By analogy with the dynamic programming algorithm from Section 9.5.2, which processes the subset collection in a manner from left to right, we process the underlying subset tree bottom-up, that is, first the leaves, then the nodes having leaves as their children, and finally the root. Hence, for a given tree-like subset collection C with its underlying subset tree T, we define for each node c_i of T the function $A(c_i)$. However, taking into account the tree structure, $A(c_i)$ contains all elements occurring in the nodes of the subtree with c_i at the root:

$$A(c_i) := \bigcup_{c \in T[c_i]} c,$$

where $T[c_i]$ denotes the node set of the subtree of T rooted at c_i.

During the bottom-up process, consider an internal node c_i. By analogy with the dynamic programming in Section 9.5.2, we try to cover $A(c_i)$ by using only the subsets in $T[c_i]$. If c_i contains an element which does not occur in any node of $T[c_i]$ but c_i, we have no other choice to cover $A(c_i)$ than to take c_i. Otherwise, we have to consider two possibilities: one is to use c_i to cover the elements in c_i, and the other is to use some subsets in $T[c_i]$ other than c_i to cover the elements in c_i. Unlike in PATH-LIKE WEIGHTED SET COVER, we now have a subtree instead of a line. In the path-like subset collection, c_i has only one preceding (child) node c_{i-1}, where we can find, in $B(c_{i-1})$, the best alternative to cover the elements in c_i by using some subsets to the left of c_i. In contrast, a node in the subset tree can have two children, each of these children having a subtree rooted at it. Then, if an element in c_i occurs in both of these child subsets, it is possible to cover it by using one or several subsets from any of the subtrees. In order to find the best alternative to cover the elements in c_i by using subsets in these subtrees, a naive approach would have to try all the possibilities to combine the subsets in $T[c_i]$. It is clear that there can be exponentially many (in the size of $T[c_i]$) of these subset combinations.

To get a fixed-parameter algorithm with respect to k, we thus have to bound the number of subset combinations by a function depending only on k. To this end, we associate with each node c of T a table D_c. Table D_c has three columns, the first two corresponding to the two children of c and the third to c. The rows of the table correspond to the elements of the power set of c, that is, there are $2^{k'}$ rows if $c = \{s_1, s_2, \ldots, s_{k'}\}$, $k' \leq k$. Figure 9.2 illustrates the structure of table D_c for a node c having two children c' and c''. Table D_c has $3 \cdot 2^{k'} = O(2^k)$ entries. Entry $D_c(x, y)$ of the row corresponding to $x \subseteq c$ and of the column corresponding to $y \in \{c, c', c''\}$ is the minimum weight to cover the elements in $x \cup (A(y) \setminus c)$ by using the subsets in the subtree $T[y]$. At each node, we have to fill out such a table. Since we process the subset tree in a bottom-up manner, the entries of the columns corresponding to c' and c'' can be directly retrieved from $D_{c'}$ and $D_{c''}$, which have been already computed before we arrive at node c. Note that these two columns are only used to simplify the description of the computation of the last column. For the purpose of implementation, the table D_c needs only the column corresponding to c. Using the values from the first two columns, we can compute the entries in the column of c. After D_r for the root r of the subset tree is computed, we can find the minimum weight to cover all elements in S in the entry $D_r(r, r)$. In the following, we describe the

D_c	c'	c''	c
$\emptyset$			
$\{s_1\}$			
$\{s_2\}$			
$\vdots$			
$c := \{s_1, s_2, \ldots, s_{k'}\}$			

FIG. 9.2. Table D_c for node $c := \{s_1, s_2, \ldots, s_{k'}\}$ with $k' \leq k$ having two children c' and c''.

subtle details of how to fill out the table for a node in the tree. We distinguish three cases:

Case 1: Node $c := \{s_1, s_2, \ldots, s_{k'}\}$ is a leaf:

Since c has no child, columns c' and c'' are empty. We can easily compute the third column:

$$D_c(x, c) := \begin{cases} 0, & \text{if } x = \emptyset; \\ w(c), & \text{otherwise.} \end{cases}$$

Case 2: Node $c := \{s_1, s_2, \ldots, s_{k'}\}$ has only one child c':

The column c'' of D_c is empty. The first step to fill out the table is to get the values of the first column from the table $D_{c'}$. According to the definition of the entries of D_c, entry $D_c(x, c')$ stores the minimum weight to cover the elements of $x \cup (A(c') \setminus c)$ by using the subsets in $T[c']$. If there is one element s_j, $1 \leq j \leq k'$, in set x which does not occur in $T[c']$, that is, $x \not\subseteq A(c')$, then it is impossible to cover $x \cup (A(c') \setminus c)$ by using only the subsets in $T[c']$. The entry $D_c(x, c')$ is then set to ∞. Otherwise, that is, $x \subseteq A(c')$, in order to get the value of $D_c(x, c')$, we have to find the uniquely determined row in table $D_{c'}$ which corresponds to the subset x' of c' satisfying $x' \cup (A(c') \setminus c') = x \cup (A(c') \setminus c)$. Due to the consistency property of tree-like subset collections, each element in c also occurring in $T[c']$ is an element of c'. Hence we get

$$x \cup (A(c') \setminus c) = x \cup (c' \setminus c) \cup (A(c') \setminus c').$$

We set $x' := x \cup (c' \setminus c)$. Since $x \subseteq A(c')$ and $x \subseteq c$, it follows that $x \subseteq c'$. Therefore, also $x' \subseteq c'$ and there is a row in $D_{c'}$ corresponding to x'. Thus, $D_c(x, c')$ is set equal to $D_{c'}(x', c')$. Altogether, we have:

$$D_c(x, c') := \begin{cases} \infty, & \text{if } x \not\subseteq c'; \\ D_{c'}(x \cup (c' \setminus c), c'), & \text{if } x \subseteq c'. \end{cases}$$

The second step is to compute the last column of D_c using the values from the column for c'. For each row corresponding to a subset x of c, we have to compare the two possibilities to cover the elements of $x \cup (A(c) \setminus c)$, either using c to cover elements in x and using some subsets in $T[c']$ to cover the remaining elements or using solely subsets in $T[c']$ to cover all elements:

$$D_c(x, c) := \min\{w(c) + D_c(\emptyset, c'),\ D_c(x, c')\}.$$

Case 3: Node $c := \{s_1, s_2, \ldots, s_{k'}\}$ has two children c' and c'':

In this case, the first step can be done in the same way as in Case 2, that is, retrieving the values of the columns c' and c'' of D_c from tables $D_{c'}$ and $D_{c''}$.

In order to compute the value of $D_c(x, c)$, for a row x corresponding to a subset of c, we also compare the two possibilities to cover $x \cup (A(c) \setminus c)$, either using c to cover x or not. In this case, however, we have two subtrees $T[c']$ and $T[c'']$, and hence we have more than one alternative to cover $x \cup (A(c) \setminus c)$ by only using subsets in $T[c']$ and $T[c'']$. As a simple example consider a subset $x \subseteq c$ that has only two elements, that is, $x = \{s'_1, s'_2\}$. We can cover it by using only subsets in $T[c']$, only subsets in $T[c'']$, or a subset in $T[c']$ to cover $\{s'_1\}$ and a subset in $T[c'']$ to cover $\{s'_2\}$ or vice versa. Therefore, for $x := \{s'_1, s'_2, \ldots, s'_{k''}\} \subseteq c$ with $k'' \leq k'$,

$$D_c(x, c) := \min \left\{ \begin{array}{l} w(c) + D_c(\emptyset, c') + D_c(\emptyset, c''), \\ D_c(\emptyset, c') + D_c(x, c''), \\ D_c(\{s'_1\}, c') + D_c(x \setminus \{s'_1\}, c''), \\ D_c(\{s'_2\}, c') + D_c(x \setminus \{s'_2\}, c''), \\ \vdots \\ D_c(x \setminus \{s'_2\}, c') + D_c(\{s'_2\}, c''), \\ D_c(x \setminus \{s'_1\}, c') + D_c(\{s'_1\}, c''), \\ D_c(x, c') + D_c(\emptyset, c'') \end{array} \right\}.$$

With these three cases, we can fill out all tables. The entry $D_r(r, r)$ stores the minimum weight to cover all elements where r denotes the root of the subset tree. In order to construct the minimum set cover, we can, using table D_r, find out whether the computed minimum weight is achieved by taking r into the minimum set cover or not. Then, doing a traceback, we can recursively, from the root to the leaves, determine the subsets in the minimum weight set cover. Note that, if one only wants to know the minimum weight, we can discard the tables $D_{c'}$ and $D_{c''}$ after filling out D_c, for each internal node c with children c' and c'', to reduce the required memory space from $O(2^k \cdot m)$ to $O(2^k)$.

Theorem 9.11 TREE-LIKE WEIGHTED SET COVER *with an underlying binary subset tree can be solved in* $O(3^k \cdot m \cdot n)$ *time, where* k *denotes the maximum subset size, that is,* $k := \max_{c \in C} |c|$.

Proof The correctness of the above algorithm directly follows from its description.

With regard to the running time of the algorithm, the size of table D_c is upper-bounded by $2^k \cdot 3$ for each node c, since $|c| \leq k$. Using a proper data structure, such as a hash table, the retrieval of a value from one of the tables corresponding to the children can be done in constant time. Thus the two columns of D_c corresponding to the two children c' and c'' can be filled out in $O(2^k)$ time.

To compute an entry in the column c, which corresponds to a subset x of c, the algorithm compares all possibilities to cover some elements of x by some subsets in $T[c']$. There can be only $2^{|x|}$ such possibilities. Hence it needs $O(2^{|x|})$ steps to compute $D_c(x, c)$ for each subset x of c. Since all set operations needed between two sets with maximum size of n can be done in $O(n)$ time, the running time for computing D_c is

$$n \cdot \left(\sum_{j=1}^{|c|} \binom{|c|}{j} \cdot O(2^j) \right) + O(2^{|c|}) = O(3^{|c|} \cdot n).$$

Therefore the computation of the tables of all nodes can be done in $O(3^k \cdot m \cdot n)$. During the traceback, we visit, from the root to leaves, each node only once and, at each node, can in constant time find out whether or not to put this node into the set cover and with which entries in the tables of the children to continue the traceback. Thus the traceback works in $O(m)$ time. □

Observe that the path-like subset collection is a special case of the tree-like subset collection with a binary subset tree. We can also use the fixed-parameter algorithm to solve PATH-LIKE WEIGHTED SET COVER. For instances where the maximum subset size is $o(\log m)$, the fixed-parameter algorithm is faster than the dynamic programming algorithm from Section 9.5.2. However, it needs more space to store the tables.

In a sense, by reorganizing an arbitrary tree into a binary tree, it is possible to show the following consequence of Theorem 9.11. We omit the details.

Corollary 9.12 TREE-LIKE WEIGHTED SET COVER *can be solved in* $O(3^k \cdot m \cdot n)$ *time, where* k *denotes the maximum subset size, that is,* $k := \max_{c \in C} |c|$.

Recall that we have already discussed the *NP*-complete (unweighted) MULTICUT IN TREES problem in Section 1.3. WEIGHTED MULTICUT IN TREES is defined as follows.

> **Input:** An undirected tree $T = (V, E)$, $n := |V|$, and a collection H of m pairs of nodes in V, $H = \{(v_i, u_i) \mid v_i, u_i \in V, v_i \neq u_i, 1 \leq i \leq m\}$. Each edge $e \in E$ has a positive real weight $w(e) > 0$.
> **Task:** Find a subset E' of E with minimum total weight whose removal separates each pair of nodes in H.

It is possible to show that WEIGHTED MULTICUT IN TREES can be reduced to TWSC in polynomial time in a "parameter-preserving" way. Hence, again omitting the details, by Theorem 9.11 and Corollary 9.12 we end up with the following fixed-parameter tractability result for WEIGHTED MULTICUT IN TREES.

Corollary 9.13 WEIGHTED MULTICUT IN TREES *can be solved in* $O(3^k \cdot m \cdot n)$ *time, where* k *denotes the maximum number of paths passing through a node or an edge,* m *denotes the number of node pairs, and* n *denotes the number of tree nodes.*

In conclusion, we note that the original motivation for studying TWSC is drawn from efforts to reduce the memory requirement of dynamic programming in connection with tree decompositions of graphs and the solution of corresponding problems such as VERTEX COVER or DOMINATING SET. We will further discuss this issue in Chapter 10. Interestingly, the problem appears also in a completely different context in computational biology, appearing there in phylogenetic studies; refer to the bibliographical remarks.

9.6 Shrinking search trees

Depth-bounded search trees are of fundamental importance in fixed-parameter algorithmics (Chapter 8). In Section 8.7 we have already seen how search trees can be cleverly combined with data reduction and problem kernels in order to achieve significant polynomial-time speed-ups. In what follows, we describe how to combine depth-bounded search trees with dynamic programming in order to obtain even (small) improvements in the exponential running time terms; that is, we shrink the size of the search trees. Note, however, that this does not come for free. It goes along with the use of an exponential amount of space, something that is usually not the case for ordinary depth-bounded search trees. In other words, the approach trades space for time. Hence it remains open to clarify the practical usefulness of the subsequently described approach invented by John M. Robson in a 1986 paper containing an exact algorithm for the INDEPENDENT SET problem. Other authors used the term *memoization* for this approach. The key idea—not surprisingly—is that if the same subproblem appears many times, then it may be helpful to store its solution instead of recomputing it again and again.

Depth-bounded search tree algorithms, as described in Chapter 8 and, in particular, those for VERTEX COVER incur exponential running times but only use a polynomial amount of space. This is true because working through a search tree in a depth-first manner only requires storing the data related to a path of bounded length. This can be improved by dynamic programming: we focus our considerations on the VERTEX COVER problem. Choose a size s (that is, number of vertices) and store all induced subgraphs of the input graph G of size s in a database D. Solve all instances in D and store an optimal solution for each of them. Then apply a "regular" search tree algorithm for VERTEX COVER, given graph G. Such a regular algorithm finds an optimal solution for G by recursively computing optimal solutions for induced subgraphs of G. "Regular search tree algorithm" here means that the only operation allowed to reduce the size of a graph is the deletion of vertices. Normally, the size of the graph is then recursively reduced in the branches of the search tree until the sizes of the graphs—note that each node of the search tree corresponds to an induced subgraph of G—in the leaves come down to 0. Having the database D at its disposal, the search tree algorithm can stop earlier: as soon as the size of the graphs in the nodes of the search tree are as small as s, a dictionary lookup replaces the remaining part of the search tree. In this way, one may save time to (re-)compute optimal vertex covers for the same (small) induced subgraphs again and again. Clearly, this cuts down the size of the search tree at the cost of storing optimal solutions for all induced subgraphs of G consisting of up to s vertices.

Before we continue our description of how to transfer the described technique into the fixed-parameter context we point out a subtle issue concerning the applicability of the whole scenario. There is a fixed-parameter algorithm for VERTEX COVER that uses only polynomial space and takes time $O(1.286^k + kn)$, where k denotes the size of the desired vertex cover set and n denotes the number of graph vertices. Unfortunately, it does not seem possible to apply dynamic programming to this algorithm since this algorithm uses the so-called folding technique that contracts edges (see Section 7.4.4). This leads to graphs that are *not* induced subgraphs of the original instance. Hence the memoization technique only applies to search tree algorithms for VERTEX COVER that do not apply the folding technique. For instance, there exists an algorithm which fulfills these requirements and which employs a search tree of size $O(1.292^k)$.

Instead of pruning the search tree and looking up the optimal vertex cover of the remaining graph in a database when the *size* of the graph drops below some predetermined size, in the parameterized setting as applies here we prune the search tree when the *parameter* reaches or drops below some predetermined value.

In this way, making use of a $2k$-vertices problem kernel (Section 7.4), for VERTEX COVER at the expense of exponential space usage, the search tree size 1.292^k could be lowered to 1.275^k. Meanwhile, by using a further refined memoization technique for VERTEX COVER, running times of $O(1.2759^k k^{1.5} + kn)$ and $O(1.2745^k k^4 + kn)$ could be proven (see bibliographical remarks).

One might argue that these improvements are of purely theoretical interest. Note, however, that the dynamic programming method presented shrinks the search tree by a larger amount, the bigger the original search tree and the smaller the (linear) problem kernel is. For instance, if the best known search tree for VERTEX COVER only were of size 2^k then we could reduce the search tree size to approximately 1.89^k. By way of contrast, the known linear-size problem kernel for DOMINATING SET IN PLANAR GRAPHS (see Section 7.6) is too big to significantly improve the corresponding search tree of size 8^k (see Section 8.6). If we had a $2k$-vertices problem kernel for DOMINATING SET IN PLANAR GRAPHS (which might be possible but is far from what we have now) the search tree size could be improved from 8^k to approximately 4.67^k. Hence, the practical relevance seems realistic.

9.7 Summary and concluding remarks

Dynamic programming, a core method of algorithm design as a whole, also plays an important role in the context of fixed-parameter algorithms. With BINARY KNAPSACK (Section 9.2) we see how the concept of pseudo-polynomial-time algorithms naturally fits into the concept of fixed-parameter algorithms as a special case. The dynamic programming solution for the STEINER PROBLEM IN GRAPHS (Section 9.3) is one of the earliest examples of a successful parameterized complexity analysis. A very hard flow problem even *NP*-complete in trees—MULTICOMMODITY DEMAND FLOW IN TREES—can be solved by means of dynamic programming incorporating the "thickness of flow" as a parameter (Section 9.4). The TREE-LIKE WEIGHTED SET COVER problem requires a fairly complex dynamic programming solution, showing that a decisive point is the right definition of the table entries (Section 9.5). Moreover, TREE-LIKE WEIGHTED SET COVER in a sense prepares the dynamic programming used in the context of tree decompositions of graphs. Here and there the so-called consistency property plays a decisive role. Finally, dynamic programming also combines nicely with other techniques of parameterized algorithm design. The usage for shrinking depth-bounded search trees (Section 9.6) gives a perhaps more surprising example of this sort.

In summary, however, there is one point worth particular notice. Except for the last example (Section 9.6) all preceding applications of dynamic programming in the fixed-parameter context are related to some form of "*structural parameterization*" and not to parameterization by the size or quality of the desired output:

- For BINARY KNAPSACK the parameter was the maximum weight W allowed and not the total value to be maximized.
- For the STEINER PROBLEM IN GRAPHS the parameter is the number of vertices to be connected and not the weight of the connecting tree network to be minimized.
- For MULTICOMMODITY DEMAND FLOW IN TREES the parameter is the maximum thickness of flow and not the profit to be maximized.
- For TREE-LIKE WEIGHTED SET COVER the parameter is the maximum subset size and not the weight of the set cover that is to be minimized.

The seeming closeness of dynamic programming to structural parameterization will find its culmination when using it in the solution of many hard problems on graphs of bounded treewidth—the most important general graph parameter in this book.

9.8 Exercises

1. The LONGEST COMMON SUBSEQUENCE problem (for two strings) is, given two strings s_1 and s_2 over some fixed alphabet, find a maximum length subsequence[6] of s_1 and s_2. Show how to solve LONGEST COMMON SUBSEQUENCE in $O(|s_1| \cdot |s_2|)$ time by dynamic programming.
2. Show that BINARY KNAPSACK can be solved in $2^{n/2} \cdot |I|^{O(1)}$ time.
3. Develop a dynamic programming algorithm for BINARY KNAPSACK with respect to the parameter $V := \max\{v_i \mid 1 \leq i \leq n\}$.
4. Generalize (the proof of) Theorem 9.3 in order to explicitly construct a minimum Steiner tree.

[6]Note that, for instance given $s_1 = ababab$ and $s_2 = ababb$, then $s = aabb$ is a maximum-length common subsequence but s is not a common substring.

5. Try to bring the combinatorial explosion in the algorithm for STEINER TREE IN GRAPHS below 3^k. Warning: this is a hard exercise.
6. Describe the top-down traversal needed for the algorithm solving MULTICOMMODITY DEMAND FLOW IN TREES (omitted in the proof of Theorem 9.5).
7. Using a concrete example, explain why the simple approach for TREE-LIKE UNWEIGHTED SET COVER (Proposition 9.9) no longer works in the weighted case.
8. Generalize the algorithm for TREE-LIKE WEIGHTED SET COVER to arbitrarily structured subset trees.
9. Prove Corollary 9.13.

9.9 Bibliographical remarks

Dynamic programming is a classical algorithm design technique, see, for instance, Cormen *et al.* (2001) and Skiena (1998). The CYK algorithm of Cocke, Younger, and Kasami for recognizing context-free languages is a standard part of introductions to formal language theory such as Hopcroft *et al.* (2001).

Held and Karp (1962) and independently Bellman (1962) presented the "best" algorithm to date for TSP. The algorithm dealing with the TSP with few inner points has been proposed in Deĭneko *et al.* (2004). This one is also based on dynamic programming and essentially generalizes the above algorithm.

There is a book dealing exclusively with KNAPSACK problems: Kellerer *et al.* (2004). The mentioned $2^{n/2} \cdot |I|^{O(1)}$ time solution for BINARY KNAPSACK is due to Horowitz and Sahni (1974). Its space consumption was lowered by Schroeppel and Shamir (1981), which is the state of the art today. The presentation of the dynamic programming algorithm here is inspired by the one in Papadimitriou (1994). The connection between fixed-parameter algorithms and pseudo-polynomial time algorithms is emphasized by Hromkovič (2002, Section 3.3). The concept of strong *NP*-completeness is treated in Garey and Johnson (1979) and Papadimitriou (1994).

The fixed-parameter algorithm for the STEINER PROBLEM IN GRAPHS first appeared in Dreyfus and Wagner (1972). We followed the presentation given in Prömel and Steger (2002) which is exclusively dedicated to STEINER TREE problems. Very recently, an improvement of the exponential factor 3^k to almost 2^k has been stated in Mölle *et al.* (2005).

The dynamic programming solution for MULTICOMMODITY DEMAND FLOW IN TREES appears in Guo and Niedermeier (2006). Its *NP*-completeness (and *MaxSNP*-hardness) is shown in (Garg *et al.*, 1997). Moreover, see Chekuri *et al.* (2003) with respect to its constant-factor approximability within polynomial time.

The TREE-LIKE WEIGHTED SET COVER problem was introduced and studied in Guo and Niedermeier (2005*a*). We followed the presentation there concerning the dynamic programming solving algorithm. PATH-LIKE WEIGHTED SET COVER or, equivalently, SET COVER WITH CONSECUTIVE ONES PROPERTY is already studied in Veinott and Wagner (1962) and Nemhauser and Wolsey (1988). A practical motivation together with corresponding empirical results can be found in Betzler *et al.* (2004).

The idea to use dynamic programming for shrinking search trees is originally due to Robson (1986). The application to a fixed-parameter algorithm for VERTEX COVER was observed in Niedermeier and Rossmanith (2003*b*) and is further refined in Chandran and Grandoni (2005).

10

TREE DECOMPOSITIONS OF GRAPHS

Many hard graph problems become easy when restricted to trees. For instance, VERTEX COVER and DOMINATING SET can be solved in linear time when restricted to trees: start at the leaves and perform a straightforward bottom-up approach. Based on this fact, one would naturally like to find out what makes trees such an algorithmically nice class of graphs—for the moment ignoring the "exceptions" of *NP*-complete problems such as MULTICUT IN TREES (Section 1.3) and MULTICOMMODITY DEMAND FLOW IN TREES (Section 9.4)—and whether and how these properties can be extended to more general classes of graphs. These considerations lead to the following central question.

How "tree-like" is a given graph?

This question is the cradle of the concept of tree decompositions of graphs, introduced by Neil Robertson and Paul D. Seymour about twenty years ago. Tree decompositions nowadays play a central role in algorithmic graph theory. In this chapter, which is far from giving an exhaustive presentation, we will survey several important aspects of tree decompositions of graphs and their algorithmic use with respect to fixed-parameter tractability.

The notion of *tree decomposition* and the related *treewidth* measure—the smaller the treewidth number the more tree-like the graph is—are introduced as a kind of compromise between the generality of graphs and the algorithmic feasibility of trees. In a nutshell, tree decompositions of small width demonstrate algorithmic tractability—in our sense, often fixed-parameter tractability—for many problems on graphs that are "almost" trees. Needless to say, there are several real-world applications of tree decompositions of graphs, ranging from compiler optimization and natural language processing over expert systems and probabilistic inference to telecommunication network design and frequency assignments, to name just a few.

In the following, we begin by introducing basic definitions and facts about tree decompositions. After that, we discuss how to construct a tree decomposition for a given graph, paying special attention to the case of planar graphs. We continue by illustrating the algorithmic use of tree decompositions once found; dynamic programming is again the key technique here and we exhibit in detailed examples concerning VERTEX COVER and DOMINATING SET. In addition, we present monadic second-order logic as a classification tool for deciding on fixed-parameter tractability of graph problems with respect to the parameter treewidth. Finally, we briefly discuss other width metrics for graphs such as pathwidth, local treewidth, and branchwidth, and we conclude by summarizing the contributions of this chapter. It must be emphasized here, however, that tree decompositions and related concepts constitute a very deep and far-reaching element of algorithmic graph theory that deserves treatment in a book on its own. Thus this chapter only scratches the surface of this important and prospective field.

10.1 Basic definitions and facts

This section is based on the following, at first sight somewhat technical, definition.

Definition 10.1 *Let $G = (V, E)$ be a graph. A* tree decomposition *of G is a pair $\langle \{X_i \mid i \in I\}, T\rangle$ where each X_i is a subset of V, called a* bag, *and T is a tree with the elements of I as nodes. The following three properties must hold:*

1. $\bigcup_{i \in I} X_i = V$;
2. *for every edge $\{u, v\} \in E$, there is an $i \in I$ such that $\{u, v\} \subseteq X_i$; and*
3. *for all $i, j, k \in I$, if j lies on the path between i and k in T then $X_i \cap X_k \subseteq X_j$.*

The width *of $\langle \{X_i \mid i \in I\}, T\rangle$ equals* $\max\{|X_i| \mid i \in I\} - 1$. *The* treewidth *of G is the minimum k such that G has a tree decomposition of width k.*

A clique of n vertices has treewidth $n-1$. The corresponding tree decomposition trivially consists of one bag containing all graph vertices. In fact, no tree decomposition with smaller width is attainable. More generally, it is known that every complete subgraph of a graph G is completely "contained" in a bag of G's tree decomposition. By way of contrast, a tree has treewidth 1 and the bags of the corresponding tree decomposition are simply the two-element vertex sets formed by the edges of the tree. Note that the third, algorithmically most important, requirement in Definition 10.1 is fulfilled in this way. Due to its importance in dynamic programming this third condition is also called *consistency property.* An equivalent formulation of this property is to demand that for every graph vertex v, all bags containing v form a connected subtree. Observe that we have already encountered the consistency property—phrased somewhat differently—in Definition 9.6 (Section 9.5) when introducing tree-like subset collections. The subsets there correspond one-to-one with the bags here. Here and there it is essential for making dynamic programming techniques applicable. Finally, there are several equivalent notions for tree decompositions; amongst others, graphs of treewidth at most k are known as *partial k-trees.* Figure 10.1 shows a graph together with an optimal tree decomposition of width two.

Importantly, an $\ell \times \ell$-grid graph has treewidth ℓ. The upper bound can be easily shown by going—in a row-by-row manner—through the grid, always taking one more vertex from the next row and deleting one from the previous row when constructing the bags of the tree decomposition. Actually, the underlying decomposition tree is only a path, and it can be shown that this tree decomposition is optimal, that is, it yields minimum width. The important consequence of this is, however, that graphs that "contain" large grids have large treewidth.

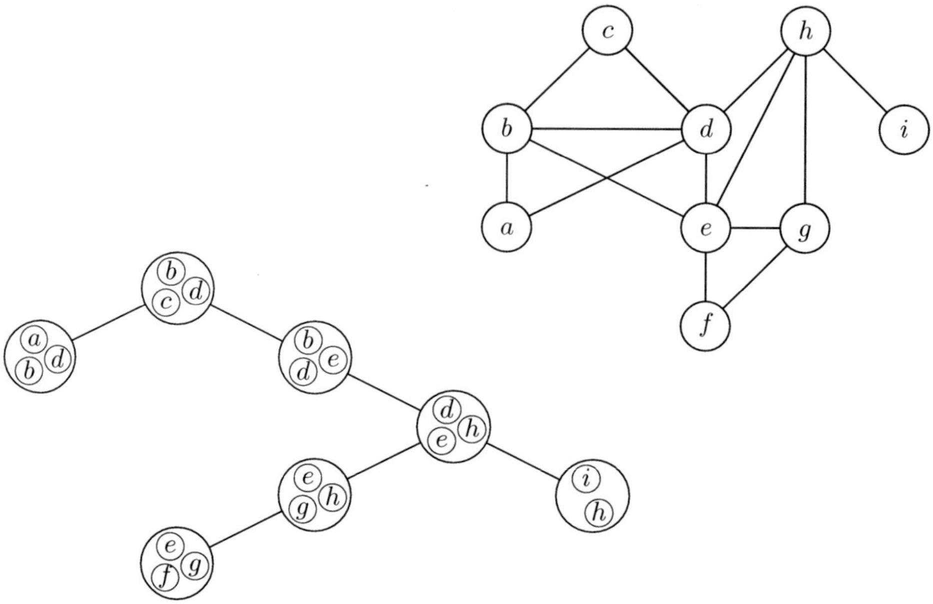

FIG. 10.1. A graph together with a tree decomposition of width 2. Observe that—as demanded by the consistency property—each graph vertex induces a subtree in the decomposition tree.

In particular, because grid graphs are planar it follows that planar graphs in general will not have small treewidth.

There is a very helpful and intuitively appealing characterization of tree decompositions in terms of a game. Consider the following *robber–cop game.* The robber stands on a graph vertex and, at any time, he can run at arbitrary speed to any other vertex of a graph as long as there is a path connecting both. He is not permitted to run through a cop, though. A cop, at any time, either stands at a graph vertex or is in a helicopter (that is, he is above the game board). The goal is to land a helicopter on the vertex occupied by the robber. Note that, due to the use of helicopters, cops are not constrained to move along graph edges. The robber can see a helicopter approaching its landing vertex and he may run to a new vertex before the helicopter actually lands. Thus, for a set of cops the goal is to occupy all vertices adjacent to the robber's vertex and to land one more remaining helicopter on the robber's place. The treewidth of the graph is then simply the minimum number of cops needed to catch a robber minus one.

A tree decomposition with a particularly simple structure is given by the following. Its usefulness will be exhibited when solving problems by dynamic programming on tree decompositions, as shown in Section 10.4.

Definition 10.2 *A tree decomposition* $\langle\{X_i \mid i \in I\}, T\rangle$ *is called a* nice tree decomposition *if the following conditions are satisfied:*

1. *Every node of the tree T has at most two children.*
2. *If a node i has two children j and k, then $X_i = X_j = X_k$; in this case, i is called a* JOIN NODE.
3. *If a node i has one child j, then one of the following situations must hold*
   (a) $|X_i| = |X_j| + 1$ *and* $X_j \subset X_i$; *in this case, i is called an* INTRODUCE NODE, *or*
   (b) $|X_i| = |X_j| - 1$ *and* $X_i \subset X_j$; *in this case, i is called a* FORGET NODE.

It is not hard to transform a given tree decomposition into a nice tree decomposition. More precisely, without stating the proof, the following result holds.

Lemma 10.3 *Given a width-k and n-nodes tree decomposition of a graph G, one can find a width-k and $O(n)$-nodes nice tree decomposition of G in $O(n)$ time.*

Tree decompositions of graphs are connected to another central concept in algorithmic graph theory: *graph separators* are vertex sets whose removal from the graph separates the graph into two or more connected components.

Definition 10.4 *Let $G = (V, E)$ be a connected graph. A subset $S \subseteq V$ is called a* separator *of G if the subgraph $G[V \setminus S]$ is disconnected.*

Actually, each bag of a tree decomposition forms a separator of the corresponding graph. Here, however, we are more interested in the reverse direction, that is, constructing tree decompositions from graph separators. The fundamental idea is to find small separators of the graph and to merge tree decompositions of the resulting subgraphs using the separator sets as "interfaces".

For any given separator splitting a graph into different components, we obtain a simple upper bound for the treewidth of this graph which depends on the size of the separator and the treewidth of the resulting components.

Proposition 10.5 *If a connected graph can be decomposed into components of treewidth of at most t by means of a separator of size s, then the whole graph has treewidth of at most $t + s$.*

Proof The separator splits the graph into different components. Suppose that we are given the tree decompositions of these components of width at most t. The goal is to construct a tree decomposition for the original graph. This can be achieved by firstly adding the separator to every bag in each of these given tree decompositions. In a second step, add some arbitrary connections preserving acyclicity between the trees corresponding to the components. It is straightforward to check that this forms a tree decomposition of the whole graph of width at most $t + s$. □

10.2 On the construction of tree decompositions

Constructing a tree decomposition of minimum width for a given graph is a difficult task. In fact, the subproblem to determine, given a graph G and an integer k, whether the treewidth of G is at most k is NP-complete. For several special graph classes (such as bipartite graphs or graphs of maximum degree nine) the problem remains NP-complete, whereas for others (such as chordal graphs or permutation graphs) it is polynomial-time solvable. As to the question of whether this problem is fixed-parameter tractable with respect to the width parameter k, the answer is positive. Hans L. Bodlaender in 1996 published a "linear-time-FPT" algorithm—it runs in linear time when the parameter k is constant. Unfortunately, the hidden constant factor is huge and seems too large for practical purposes. Thus it is a major open question whether there is a significantly better fixed-parameter algorithm. Another longstanding open problem in this context refers to the class of planar graphs. Whereas we learned that the determination of treewidth is NP-complete for general graphs, it is open whether this also holds for planar graphs or whether it is polynomial-time solvable.

To apply the concept of tree decompositions in practice, as a rule heuristic approaches are employed for their construction. Although these methods do not guarantee optimal tree decompositions, their output is often good enough for the desired application. For instance, there is a relatively simple and efficient ratio-4 approximation algorithm for constructing tree decompositions—the obtained treewidth is at most a factor of 4 from the optimum value. This approximation algorithm is a fixed-parameter algorithm in the sense that its running time is exponential in the treewidth. It is a longstanding open problem whether there exists a polynomial-time approximation algorithm for treewidth with a constant approximation factor. To describe general tree decomposition construction methods in detail is beyond the scope of this book. Roughly speaking, many algorithms and, in particular, heuristics for tree decomposition finding are based upon the same principle:

1. Try to find a linear ordering of the vertices (which allows a one-to-one assignment of numbers to them) such that the higher numbered neighbors of every vertex form a clique—in other words, the graph is chordal. Then an optimal tree decomposition can be found using this so-called *perfect elimination scheme*. In this case, the graph can be interpreted as the "intersection graph" of subtrees of trees, naturally leading to a tree decomposition.

2. In general, the ordering found is not a perfect elimination scheme as described before. Hence one has to run some "fill-in" procedure (making the graph chordal, which means that there is no induced cycle of length at least four without a chord) to "triangulate" the graph by adding edges between non-adjacent higher-numbered neighbors of every vertex. After this task is accomplished, again the triangulation is turned into a tree decomposition as indicated in the first step, also giving a tree decomposition for the original graph.

It is important to note, however, that although one usually obtains non-optimal tree decompositions in this way, using these tree decompositions in a

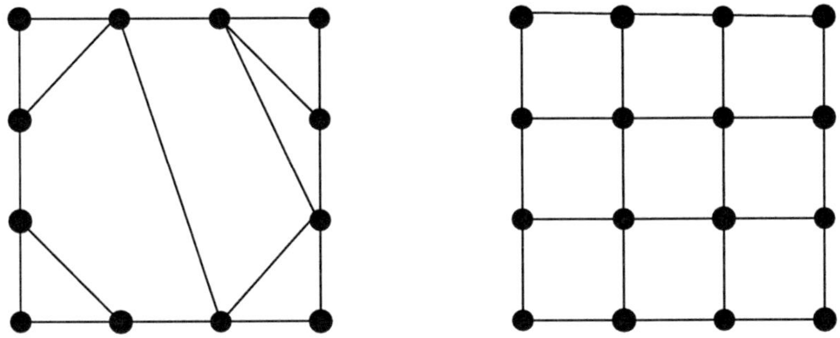

FIG. 10.2. An outerplanar graph (left) and a 2-outerplanar graph (right).

subsequent dynamic programming phase one can nevertheless obtain optimal solutions for the underlying graph problem, such as VERTEX COVER or DOMINATING SET, to be solved. The "only" price one has to pay for non-optimal tree decompositions is that the dynamic programming will consume more time and memory space because both resource requirements usually grow exponentially with respect to the width of the given tree decomposition.

In the context of fixed-parameter tractability with respect to solution size of the underlying graph problem, most algorithms based on tree decompositions deal with planar graphs or slight generalizations thereof. In these cases, efficient construction algorithms are known. That is why we devote the next section solely to planar graphs and how to find "problem-specific tree decompositions" for them. In a few words, by problem-specific we mean that if we know that a planar graph has a vertex cover or dominating set of size k, then we can make use of that to efficiently find a tree decomposition of width only depending on k.

10.3 Planar graphs

It is known that every n-vertex planar graph has treewidth $O(\sqrt{n})$ which, in a certain sense, can also be interpreted as the famous Planar Separator Theorem in disguise. Graph separators play an important role in the construction of tree decompositions. Moreover, in the case of planar graphs, there is a constructive way towards small separators. This is partially based on the "layer view" of planar graphs, expressed by the notion of r-outerplanarity.

Definition 10.6 *A plane graph G is called* outerplanar *if each vertex lies on the boundary of the outer face. A graph G is called* outerplanar *if it admits an outerplanar embedding in the plane.*

See the left-hand side of Figure 10.2 for an example outerplanar graph. The following generalization of the notion of outerplanarity is very helpful in our context.

Definition 10.7 1. *A plane embedding of a graph G is called* r-outerplanar *if, for $r = 1$, the embedding is outerplanar, and, for $r > 1$, inductively,*

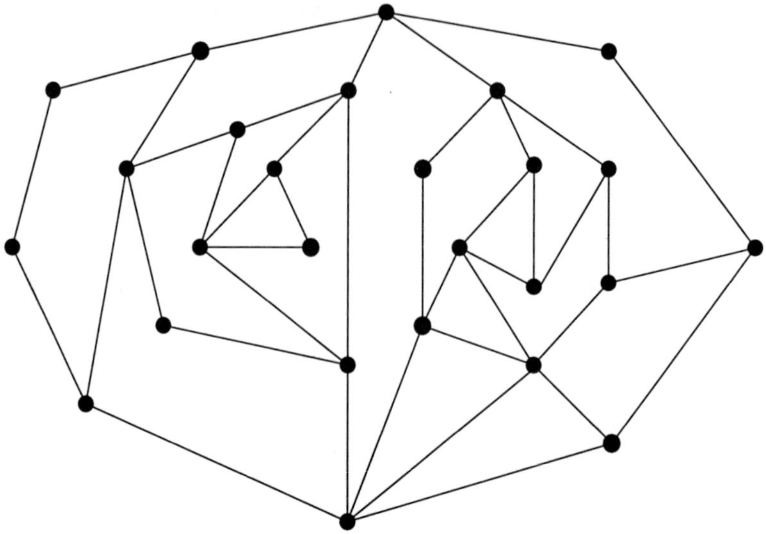

FIG. 10.3. A 3-outerplanar embedding of a graph.

when removing all vertices on the boundary of the outer face and their incident edges the embedding of the remaining subgraph is $(r-1)$*-outerplanar.*

2. *A graph* G *is called* r-outerplanar *if it admits an* r*-outerplanar embedding.*
3. *The smallest number* r *such that* G *is* r*-outerplanar is called the* outerplanarity number.

See the right-hand side of Figure 10.2 for an example 2-outerplanar graph, namely a 4×4-grid graph.

In this way, we may speak of the layers $L_1, \ldots, L_r$ of an embedding of an r-outerplanar graph.

Definition 10.8 *For a given* r*-outerplanar embedding of a graph* $G = (V, E)$*, we define the* ith layer L_i *inductively as follows. Layer* L_1 *consists of the vertices on the boundary of the outer face, and, for* $i > 1$*, the layer* L_i *is the set of vertices that lie on the boundary of the outer face in the embedding of the subgraph* $G[V \setminus (L_1 \cup \ldots \cup L_{i-1})]$.

Figure 10.3 shows a plane graph with three layers. Notably, by peeling off the outer layer one would obtain *two* connected components, both being 2-outerplanar.

Now, using the layer decomposition of plane graphs, there is an iterated version of Proposition 10.5.

Proposition 10.9 *Let* G *be a plane graph with layers* L_i*,* $i = 1, \ldots, r$*. For* $i = 1, \ldots, \ell$*, let* $\mathcal{L}_i$ *be a set of consecutive layers, that is,*

$$\mathcal{L}_i = \{L_{j_i}, L_{j_i+1}, \ldots, L_{j_i+n_i}\},$$

such that $\mathcal{L}_i \cap \mathcal{L}_{i'} = \emptyset$ *for all* $i \neq i'$. *Moreover, suppose that* G *can be decomposed into components, each of treewidth of at most* t, *by means of separators* $S_1, \ldots, S_\ell$, *where* $S_i \subseteq \bigcup_{L \in \mathcal{L}_i} L$ *for all* $i = 1, \ldots, \ell$.

Then G *has treewidth of at most* $t + 2s$, *where* $s = \max_{i=1,\ldots,\ell}\{|S_i|\}$.

Proof The proof uses the merging technique illustrated in Proposition 10.5: suppose that, without loss of generality, the sets $\mathcal{L}_i$ appear in successive order, that is, $j_i < j_{i+1}$. For each $i = 0, \ldots, \ell$, consider the component G_i of treewidth at most t which is cut out by the separators S_i and S_{i+1}—by default, we set $S_0 = S_{\ell+1} = \emptyset$. We add S_i and S_{i+1} to every bag in a given tree decomposition of G_i. In order to obtain a tree decomposition of G, we successively add an arbitrary connection between the trees T_i and T_{i+1} of the so-modified tree decompositions that correspond to the subgraphs G_i and G_{i+1}. □

Thus, for plane graphs the goal now can be set as follows: decompose the given graph into various "small" components by using "small" separators, construct a tree decomposition for each component, and get an overall tree decomposition by applying Proposition 10.9. It remains to show how to find these small separators and how to get the tree decompositions for the graph components. This is explained next.

As already mentioned, we aim at *problem-specific* tree decompositions. The reason for this is that in this way we can relate the solution size of the graph problem we want to solve with the size of the separators we can find in the underlying planar graph. In what follows, we proceed in two steps. To this end, the example graph problems we consider are VERTEX COVER and DOMINATING SET. Analogous considerations hold for many other graph problems.

1. In the first step we demonstrate that the vertex cover or domination number k and the treewidth of a planar graph are linearly related. We show this without explicit construction of graph separators.
2. In the second step, we show that the linear bound $O(k)$ from Step 1 can be improved to $O(\sqrt{k})$. Here, me make explicit use of graph separators.

With regard to the first step, one easily observes the following central relation between the vertex cover number and the domination number on the one side and the outerplanarity number of a planar graph on the other side.

Proposition 10.10 1. *If a planar graph* $G = (V, E)$ *has a size-*k *vertex cover then all plane embeddings of* G *can be at most* k*-outerplanar.*

2. *If a planar graph* $G = (V, E)$ *has a size-*k *dominating set then all plane embeddings of* G *can be at most* $3k$*-outerplanar.*

Proof We show only the result for DOMINATING SET. For a given crossing-free embedding of G in the plane, each vertex in the dominating set can dominate vertices from the previous, the next, or its own layer only. Hence each vertex in the dominating set can contribute to at most three new layers. □

To understand the techniques also used in the second step, it is helpful to consider the concept of a layer decomposition of an r-outerplanar embedding of graph G. A *layer decomposition* of an r-outerplanar embedding of G is a forest of height $r-1$: the trees of the forest correspond to different connected components of G. The nodes correspond to the various layers.

One more result needed is the following relation between r-outerplanarity and treewidth. We skip the proof.

Theorem 10.11 *An r-outerplanar graph has treewidth of at most $3r-1$.*

The proof of Theorem 10.11 can be made constructive, so a tree decomposition of width at most $3r-1$ can be computed in polynomial time.

Proposition 10.10 and Theorem 10.11 immediately imply the following relationship between the domination number and the treewidth of a planar graph.

Corollary 10.12 *If a planar graph has a k-dominating set then its treewidth is at most $9k-1$.*

With regard to the second step mentioned above, the basic idea of constructing a tree decomposition of small width is the following. If the given graph has few layers, then use Theorem 10.11 directly. If not,

1. find small graph separators that decompose the graph into chunks of small outerplanarity,
2. apply Theorem 10.11 to these graph chunks, and
3. finally combine the tree decompositions of the various chunks into a big one for the overall graph using Proposition 10.9.

Next, we describe how to find these small graph separators.

It is clear that in a layer decomposition $(L_1, \ldots, L_r)$ of a given planar graph of outerplanarity r each layer L_i, $1 \leq i \leq r$, forms a separator. What makes the problem mathematically demanding is that the sizes of the layers L_i might be too large. Thus it remains to be shown that, nevertheless, small separators can be found in a layerwise fashion. To this end, one makes use of the special properties of the underlying parameterized graph problem. We focus on VERTEX COVER and DOMINATING SET to see how this works. Both problems possess problem kernels with a number of vertices linearly depending on the size of the solution set; see Sections 7.4 and 7.6. Hence in both cases we trivially have that $|\bigcup_{i=1}^{r} L_i| = O(k)$.

Theorem 10.13 *If a planar graph has a k-vertex cover or a k-dominating set, then its treewidth is bounded from above by $O(\sqrt{k})$.*

Proof Using the graph layers L_i as separators, go through the sequence of layers $L_1, L_2, L_3, \ldots$ and look for separators of size $s(k) := O(\sqrt{k})$. Due to $|\bigcup_{i=1}^{r} L_i| = O(k)$ such separators of size at most $s(k)$ must appear within each $n(k) := O(\sqrt{k})$ layers in the sequence. In this manner, we obtain a set of disjoint separators of size at most $s(k)$ each, such that any two consecutive separators from this set are at most $O(\sqrt{k})$ layers apart. Clearly, the separators chosen in this way fulfill the requirements in Proposition 10.9.

The components cut out in this way each have $O(\sqrt{k})$ layers, and hence their treewidth is bounded by $O(\sqrt{k})$ due to Theorem 10.11. Using Proposition 10.9, we can upperbound the treewidth of the planar graph by $O(\sqrt{k})$. □

The tree structure of the tree decomposition obtained in the above proof corresponds to the structure of the layer decomposition forest.

Remark 1 *Up to constant factors, the relation exhibited in Theorem 10.13 is optimal. This can be seen, for example, by considering a grid graph G_ℓ of size $\ell \times \ell$, that is, with ℓ^2 vertices and $2(l^2 - l)$ edges: it is known that the treewidth of G_ℓ is exactly ℓ and that a minimum vertex cover as well as a minimum dominating set for G_ℓ both consist of $\Theta(l^2)$ vertices.*

A mathematically more refined analysis than the above one—not making use of linear problem kernels as we did here but employing a more direct way of constructing the separators layerwisely—gives upper bounds $O(\sqrt{k})$ on the treewidths with somewhat smaller constants hidden in the O-notation. Still, however, these worst-case factors are fairly large (in the tens but not astronomical).

Summarizing, we join together the preceding considerations into an algorithm that constructs tree decompositions of width $O(\sqrt{k})$ in the case that we are given a planar graph that possesses a vertex cover or a dominating set of size at most k. It is important to note here that the tree decompositions are only constructed for reduced graphs that are obtained by the reduction to a problem kernel for the underlying parameterized problem (VERTEX COVER or DOMINATING SET in this context). The algorithm proceeds in the following steps.

1. Perform a reduction to a problem kernel that yields a reduced planar graph whose number of vertices is $O(k)$.
2. Embed the reduced planar graph $G = (V, E)$ crossing-free into the plane. Linear-time algorithms are known for that. Determine all layers $L_1, \ldots, L_r$ and the outerplanarity number r of this embedding. By default, we set $L_i = \emptyset$ for all $i \leq 0$ and $i > r$.
3. VERTEX COVER: If $r > k$, then exit (there exists no size-k vertex cover). This is justified by Proposition 10.10.

   DOMINATING SET: If $r > 3k$ then exit (there exists no size-k dominating set). This is justified by Proposition 10.10.
4. Find separators of size $O(\sqrt{k})$ according to the proof of Theorem 10.13.
5. Decompose the graph into subgraphs by removing all the graph separators found in the preceding step. Note that each of these subgraphs has outerplanarity $O(\sqrt{k})$.
6. Construct tree decompositions for all the subgraphs using Theorem 10.11. In this way, all subgraphs obtain tree decompositions of width $O(\sqrt{k})$.
7. Merge the tree decompositions of all subgraphs into a tree decomposition of the overall graph. To do so, use the tree decompositions of the subgraphs and the separators that generated these subgraphs (see fifth step above)

and apply the "separator merging technique" described in the proof of Proposition 10.9.

The above algorithm outline constructively shows how to obtain tree decompositions of width $O(\sqrt{k})$ for problems such as VERTEX COVER or DOMINATING SET in planar graphs parameterized by k. Clearly, to demonstrate the definite usefulness of constructing tree decompositions of graphs it remains to be shown how the underlying graph problem can be efficiently solved using dynamic programming on tree decompositions. This will be the topic of Sections 10.4 and 10.5.

What about the constant factors hidden in the O-notation used throughout the previous considerations? Note that the given presentation traded comprehensibility for exact determination of the considered running time and treewidth values. A thorough mathematical analysis can be found in the literature. As already indicated, the proven upper bounds involve quite high constant factors. For instance, based on the above approach and several more technical details the running times $O(2^{4\sqrt{3k}} \cdot n)$ and $O(2^{12\sqrt{34k}} \cdot n)$ were proven for VERTEX COVER and DOMINATING SET in planar graphs, respectively. The bounds are worst-case, however. Recent work has lowered the exponential bound for DOMINATING SET in planar graphs.

10.4 Dynamic programming for Vertex Cover

To understand the practical usefulness of tree decompositions, one has to learn how algorithmic techniques used for graph problems restricted to trees can be generalized to also work for graphs of bounded treewidth. The standard approach here is dynamic programming. Notably, making this dynamic programming as efficient and practical as possible is not only a question of running time but also of memory consumption. Finally, observe that our dynamic programming approaches provide optimal solutions for the problems considered. This also underpins the relevance of approximative or heuristic algorithms for constructing tree decompositions as long as the delivered treewidths remain small enough.

Typically, tree decomposition based algorithms proceed according to the following scheme in two stages:

1. Find a tree decomposition of bounded width for the input graph; and
2. solve the problem by dynamic programming on the tree decomposition.

So far in this chapter on tree decompositions, we have dealt with the first stage. Now, we describe how the second stage works. In fact, what we show is a fixed-parameter algorithm with respect to the parameter treewidth. Notably, the parameter is not the size of the vertex cover set but it is the width of the tree decomposition. This implies that we are able to solve VERTEX COVER for almost arbitrarily large graphs as long as their treewidth is small enough.

Theorem 10.14 *For a graph G with given tree decomposition $\langle \{X_i \mid i \in I\}, T \rangle$ an optimal vertex cover can be computed in $O(2^{\omega} \cdot \omega \cdot |I|)$ time. Here, ω denotes the width of the tree decomposition.*

Proof The basic idea is to check for all of the $|I|$ many bags each time for all of the at most $2^{|X_i|}$ possibilities to obtain a vertex cover for the subgraph $G[X_i]$ of G induced by the vertices from X_i. This information is stored in tables A_i, $i \in I$. Adjacent tables will be updated in a bottom-up process starting at the leaves of the decomposition tree. Each bag of the tree decomposition thus has a table associated with it. During this updating process it is guaranteed that the "local" solutions for each subgraph associated with a bag of the tree decomposition are combined into a "globally optimal" solution for the overall graph G.

The algorithmic details are as follows.

Step 0: For each bag $X_i = \{x_{i_1}, \ldots, x_{i_{n_i}}\}$, $|X_i| = n_i$, compute a table

$$A_i \;=\; \begin{array}{cccccc||c} x_{i_1} & x_{i_2} & \cdots & x_{i_{n_i-1}} & x_{i_{n_i}} & & m_i() \\ \hline 0 & 0 & \cdots & 0 & 0 & & \\ 0 & 0 & \cdots & 0 & 1 & & \\ 0 & 0 & \cdots & 1 & 0 & & \\ 0 & 0 & \cdots & 1 & 1 & & \\ & & \vdots & & & & \\ 1 & 1 & \ldots & 1 & 0 & & \\ 1 & 1 & \ldots & 1 & 1 & & \end{array} \left.\vphantom{\begin{array}{c}0\\0\\0\\0\\0\\0\\0\end{array}}\right\} 2^{n_i}$$

The table consists of 2^{n_i} rows and $|n_i| + 1$ columns. Each row represents a so-called "coloring" of subgraph $G[X_i]$. By this we mean a 0-1 sequence of length n_i that determines which of the respective bag vertices from X_i should be put into the current vertex cover—this corresponds to value 1—and which should not—this corresponds to value 0. Formally, a coloring is a mapping

$$C_i \,:\, X_i = \{x_{i_1}, \ldots, x_{i_{n_i}}\} \to \{0, 1\}.$$

For each of the 2^{n_i} different possibilities for a coloring, the table has one further entry. This last column stores, for each specific coloring C_i, the number $m_i(C_i)$ of vertices of a minimal vertex cover that contains those vertices from X_i selected by the coloring C_i. More precisely, this means that it stores the value

$$\begin{aligned} m_i(C_i) := \min\{|V'| : V' \subseteq V & \text{ is a vertex cover for } G, \\ & \text{such that } v \in V' \text{ for all } v \in (C_i)^{-1}(1) \\ & \text{and } v \notin V' \text{ for all } v \in (C_i)^{-1}(0)\}. \end{aligned}$$

Note that $(C_i)^{-1}(1)$ denotes the set of all vertices in X_i that are colored 1, and $(C_i)^{-1}(0)$ denotes all vertices in X_i that are colored 0 under coloring C_i. This value is determined by dynamic programming as described in Step 2 to follow.

Of course, not every possible coloring may lead to a vertex cover. Such a coloring is called *invalid*. To check whether or not a coloring C_i is *valid*, for each edge $\{u, v\}$ of the subgraph $G[X_i]$ induced by X_i, consider $C_i(u)$ and $C_i(v)$. If there is at least one edge where $C_i(u) = C_i(v) = 0$ then the coloring is invalid; otherwise, it is valid.

Step 1: Table initialization.
For all tables X_i and each coloring $C_i : X_i \to \{0, 1\}$ set

$$m_i(C_i) := \begin{cases} \left|(C_i)^{-1}(1)\right|, & \text{if } C_i \text{ is valid;} \\ +\infty, & \text{otherwise.} \end{cases}$$

Clearly, once when value $+\infty$ appears the corresponding "computation path" can be immediately stopped because no vertex cover of the whole graph can be computed in this way.

Step 2: Dynamic programming.
We now go through the decomposition tree T from the leaves to the root and compare the corresponding tables against each other.
Let $i \in I$ be the parent node of $j \in I$. We show how the table for X_i with m_i can be updated by the table for X_j with m_j. To this end, assume that

$$X_i = \{z_1, \ldots, z_s, v_1, \ldots, v_{t_i}\} \text{ and}$$
$$X_j = \{z_1, \ldots, z_s, u_1, \ldots, u_{t_j}\},$$

that is, $X_i \cap X_j = \{z_1, \ldots, z_s\}$.
Subsequently, by an *extension of a coloring* $C : W \to \{0, 1\}$ (where $W \subseteq V$) we mean a coloring $\tilde{C} : \tilde{W} \to \{0, 1\}$ with $\tilde{W} \supseteq W$ and $\tilde{C}$ restricted to ground set W yields C, in signs, $\tilde{C}|_W = C$. Then, for each possible coloring

$$C : \{z_1, \ldots, z_s\} \to \{0, 1\}$$

and each extension $C_i : X_i \to \{0, 1\}$ of C we set

$$\begin{aligned} m_i(C_i) := m_i(C_i) & \\ & + \min\{ m_j(C_j) \mid C_j : X_j \to \{0, 1\} \text{ is an extension of } C\} \\ & - \left|C^{-1}(1)\right|. \end{aligned}$$

This is the central instruction of the algorithm. In other words, what happens here is that the value of $m_i(C_i)$ grows by the minimum value for the vertex cover of the subgraph induced by all the vertices contained in the subtree rooted at j. Here, however, one has to take care of double counting with respect to vertices that are also contained in X_i, which is why $|C^{-1}(1)|$ is subtracted. In fact, we record some more (pointer) information in order to be able to construct a solution set efficiently in a subsequent traceback phase.
If a node $i \in V_T$ has several children $j_1, \ldots, j_l \in V_T$ then table A_i is successively updated against all tables $A_{j_1}, \ldots, A_{j_l}$ in the same way. All this is repeated until the root node is finally updated.

Step 3: Construction of a minimum vertex cover V'.
The size of V' is derived from the minimum entry of the last column of the root node table A_r. The coloring of the corresponding row shows which of the vertices of the "root bag" X_r are contained in V'. By recording during Step 2 how the respective minimum of each bag was determined by its "child values", one can easily compute all vertices of an optimal vertex cover by following the pointers mentioned in Step 2.

This concludes the description of the dynamic programming algorithm. It remains to show its correctness and its running time.

1. *Correctness of the algorithm.*
   a) The first condition in Definition 10.1, that is, $V = \bigcup_{i \in I} X_i$, makes sure that every graph vertex is taken into account during the computation.
   b) The second condition in Definition 10.1, that is, $\forall e \in E\ \exists i_0 \in I : e \in X_{i_0}$, makes sure that after the treatment of invalid colorings right after the initialization in Step 0, during the dynamic programming process only actual vertex covers are dealt with.
   c) The third condition in Definition 10.1 guarantees the consistency of the dynamic programming. If a vertex $v \in V$ occurs in two different bags X_{i_1} and X_{i_2} then it is guaranteed that for the computed minimum vertex cover this vertex cannot receive different colors in the two respective rows in the tables A_{i_1} and A_{i_2}. This is handled in the bag of the least common ancestor i_0 of i_1 and i_2 in T.

2. *Running time of the algorithm.*
By keeping the tables sorted adequately, the comparison of a table A_j against a table A_i can be done in time proportional to the maximum table size, that is, $O(2^\omega \cdot \omega)$. For each edge $e \in E_T$ in tree T such a comparison has to be done, that is, the overall running time of the algorithm is $O(2^\omega \cdot \omega \cdot |I|)$. □

Combining Theorem 10.14 with Theorem 10.13 and the corresponding algorithm that constructs a tree decomposition (see Section 10.3) results in a fixed-parameter algorithm for VERTEX COVER IN PLANAR GRAPHS that provides an exponential speedup when compared with the depth-bounded search tree fixed-parameter algorithms for VERTEX COVER on general graphs in Section 8.3. Note, however, that now the constants in the O-term are significantly larger.

Corollary 10.15 VERTEX COVER IN PLANAR GRAPHS *can be solved in* $2^{O(\sqrt{k})} \cdot n$ *time, where* k *denotes the size of the vertex cover and* n *is the number of graph vertices.*

Doing a more refined, deep mathematical analysis, according to the literature the exponential factor in the statement of Corollary 10.15 can be upper-bounded by $2^{4.5\sqrt{k}}$.

10.5 Dynamic programming for Dominating Set

The basic technique for solving DOMINATING SET on a "tree-decomposed" graph is the same as for VERTEX COVER. We have already experienced in earlier parts of this book, however, that from a combinatorial point of view DOMINATING SET is a problem that is more elusive than VERTEX COVER. This also reflects in the larger overhead needed to solve DOMINATING SET efficiently via dynamic programming on tree decompositions. The very first observation is that we need three colors for the dynamic programming tables instead of only two as we had for VERTEX COVER: suppose that $G = (V, E)$ and $V = \{x_1, \ldots, x_n\}$. Assume that the vertices in the bags are given in increasing order when used as indices of the dynamic programming tables, that is, $X_i = \{x_{i_1}, \ldots, x_{i_{n_i}}\}$ with $i_1 \leq \ldots \leq i_{n_i}$, $1 \leq i \leq |I|$. We use *three* different "colors" that will be assigned to the vertices in the bag:

- "black": represented by 1, meaning that the vertex belongs to the dominating set;
- "white": represented by 0, meaning that the vertex is already dominated at the current stage of the algorithm; and
- "grey": represented by $\hat{0}$, meaning that, at the current stage of the algorithm, one is still asking for a domination of this vertex.

Again, mapping

$$C_i : \{x_{i_1}, \ldots, x_{i_{n_i}}\} \to \{0, \hat{0}, 1\}$$

will be called a *coloring* for the bag $X_i = \{x_{i_1}, \ldots, x_{i_{n_i}}\}$, and the *color* assigned to vertex x_{i_t} by C_i is given by $C_i(x_{i_t})$. Hence, a coloring can be represented as a vector $(C(x_{i_1}), \ldots, C(x_{i_{n_i}}))$.

For each bag X_i with $|X_i| = n_i$, in the same spirit as for VERTEX COVER we use a mapping

$$m_i : \{0, \hat{0}, 1\}^{n_i} \longrightarrow \mathbb{N} \cup \{+\infty\}.$$

For a coloring C_i, the value $m_i(C_i)$ stores how many vertices are needed for a minimum dominating set of the graph visited up to the current stage of the algorithm under the restriction that the color assigned to vertex x_{i_t} is $C_i(x_{i_t})$, $t = 1, \ldots, n_i$. We end up with tables of size 3^{n_i}. Now, by performing a table updating process analogous to the case for VERTEX COVER described in Section 10.4, it is not difficult to finally come up with an algorithm that solves DOMINATING SET on graphs with given tree decomposition of $|I|$ nodes and width ω in $O(9^\omega \cdot \omega \cdot |I|)$ time. The significant increase from base value 2 to 9 is due to the more complicated "dependence structure" in the combinatorics of DOMINATING SET when implemented in a basically straightforward way. Thus, comparing two tables A_i and A_j now takes $O(3^{|X_i|} \cdot 3^{|X_j|} \cdot \max\{|X_i|, |X_j|\}) = O(9^\omega \cdot \omega)$ time.

There is room for improvement, however. This needs some further definitions: to simplify matters, we identify colorings simply by their naturally corresponding vectors $\{0, \hat{0}, 1\}^{n_i}$. On the color set $\{0, \hat{0}, 1\}$, let $\prec$ be the partial ordering given by $\hat{0} \prec 0$ and $d \prec d$ for all $d \in \{0, \hat{0}, 1\}$. This ordering naturally extends to colorings: for $c = (c_1, \ldots, c_m), c' = (c'_1, \ldots, c'_m) \in \{0, \hat{0}, 1\}^m$, let $c \prec c'$ iff $c_t \prec c'_t$ for all $t = 1, \ldots, m$. It is essential for the improved dynamic programming that the mappings m_i are *monotonic* functions from $(\{0, \hat{0}, 1\}, \prec)$ to $(\mathbb{N} \cup \{+\infty\}, \leq)$; that is, for $c, c' \in \{0, \hat{0}, 1\}^{n_i}$, $c \prec c'$ implies that $m_i(c) \leq m_i(c')$. A coloring $c \in \{0, \hat{0}, 1\}^{n_i}$ is *locally invalid* for a bag X_i if

$$\left(\exists s \in \{1, \ldots, n_i\} : c_s = 0\right) \wedge \left(\nexists t \in \{1, \ldots, n_i\} : ((x_{i_t} \in N(x_{i_s})) \wedge (c_t = 1))\right).$$

In other words, a coloring is locally invalid if there is some vertex in the bag that is colored white but this color is not "justified" within the bag, that is, this vertex is not dominated by a vertex colored 1 within the bag. Note that a locally invalid coloring may still be a correct coloring if the white vertex whose color is not justified *within* the bag is dominated by a vertex from bags that have been considered earlier. Also, for a coloring $c = (c_1, \ldots, c_m) \in \{0, \hat{0}, 1\}^m$ and a color $d \in \{0, \hat{0}, 1\}$, we use the notation

$$\#_d(c) := |\{t \in \{1, \ldots, m\} \mid c_t = d\}|.$$

Now we are ready to describe the various steps of the improved dynamic programming. To make things easier, we subsequently assume that we work with a *nice* tree decomposition (see Definition 10.2 and Lemma 10.3 in Section 10.1).

Step 1: Table initialization.

For all tables X_i and each coloring $c \in \{0, \hat{0}, 1\}^{n_i}$ set

$$m_i(c) := \begin{cases} +\infty, & \text{if } c \text{ is locally invalid for } X_i, \\ \#_1(c), & \text{otherwise} \end{cases} \tag{10.1}$$

With this initialization step we make sure that only colorings are taken into consideration where an assignment of color 0 is justified.

Since the check for local invalidity takes $O(n_i)$ time, this step can be carried out in $O(3^{n_i} \cdot n_i)$ time.

Trivially, the mappings m_i of leaf bags are monotonic.

Step 2: Dynamic programming.

After the initialization, we visit the bags of our tree decomposition from the leaves to the root, evaluating the corresponding mappings in each step according to the following rules.

FORGET NODES: Suppose i is a FORGET NODE with child node j and suppose that $X_i = \{x_{i_1}, \ldots, x_{i_{n_i}}\}$. Without loss of generality—possibly after rearranging the vertices in j's bag X_j and the entries of m_j accordingly—we may assume that $X_j = \{x_{i_1}, \ldots, x_{i_{n_i}}, x\}$ for some graph vertex x not in X_i. Evaluate the mapping m_i of X_i as follows: For all colorings $c \subset \{0, \hat{0}, 1\}^{n_i}$ set

$$m_i(c) := \min_{d \in \{0,1\}} \{m_j(c \times \{d\})\}. \tag{10.2}$$

Note that a coloring $c \times \{\hat{0}\}$ for X_j means that the vertex x is assigned color $\hat{0}$, that is, x is not yet dominated by a graph vertex. Since, by the consistency property of tree decompositions, the vertex x will never appear in a bag for the rest of the algorithm, a coloring $c \times \{\hat{0}\}$ will not lead to a dominating set because vertex x is not dominated. That is why the minimum in the assignment (10.2) is taken over colors 1 and 0 only.

Clearly, the evaluations can be carried out in $O(3^{n_i} \cdot n_i)$ time. It is trivial to observe that the monotonicity of the mapping m_j implies that m_i also is monotonic.

INSERT NODES: Suppose that i is an INSERT NODE with child node j and suppose that $X_j = \{x_{j_1}, \ldots, x_{j_{n_j}}\}$. Without loss of generality—possibly after rearranging the vertices in X_i and the entries of A_i accordingly—we may assume that $X_i = \{x_{j_1}, \ldots, x_{j_{n_j}}, x\}$. Let

$$N(x) \cap X_j = \{x_{j_{p_1}}, \ldots, x_{j_{p_s}}\}$$

be the neighbors of the "introduced" vertex x which are contained in the bag X_i. We now define a function

$$\phi : \{0, \hat{0}, 1\}^{n_j} \to \{0, \hat{0}, 1\}^{n_j}$$

on the set of colorings of X_j. For $c = (c_1, \ldots, c_{n_j}) \in \{0, \hat{0}, 1\}^{n_j}$, we define $\phi(c) := (c'_1, \ldots, c'_{n_j})$ such that

$$c'_t = \begin{cases} \hat{0}, & \text{if } t \in \{p_1, \ldots, p_s\} \text{ and } c_t = 0, \\ c_t, & \text{otherwise.} \end{cases}$$

Then, evaluate the mapping m_i of X_i as follows: for all colorings $c = (c_1, \ldots, c_{n_j}) \in \{0, \hat{0}, 1\}^{n_j}$ set

$$m_i(c \times \{0\}) := m_j(c) \text{ if } x \text{ has a neighbor } x_{j_q} \in X_i \text{ with } c_q = 1; \tag{10.3}$$

$$m_i(c \times \{1\}) := m_j(\phi(c)) + 1; \tag{10.4}$$

$$m_i(c \times \{\hat{0}\}) := m_j(c). \tag{10.5}$$

Concerning the correctness of the assignments (10.3) and (10.4) we remark the following: it is clear that, if we assign color 0 to vertex x (assignment (10.3)), we again—as already done in the initializing assignment (10.1)—have to check whether this color can be justified at the current stage of the algorithm. Such a justification is given if and only if the coloring under examination already assigns a 1 to some neighbor of x in X_j resp. X_i. This is true, since the consistency property of tree decompositions implies that at the current stage of the dynamic programming process, x can only be dominated by a vertex in X_j (as checked in assignment (10.3)).

If we assign color 1 to vertex x (assignment (10.4)), we thereby dominate all vertices $\{x_{j_{p_1}}, \ldots, x_{j_{p_s}}\}$. Suppose now that we want to evaluate $m_i(c \times \{1\})$ and suppose that some of these vertices are assigned color 0 by c, say $c_{p'_1} = \ldots = c_{p'_q} = 0$, where $\{p'_1, \ldots, p'_q\} \subseteq \{p_1, \ldots, p_s\}$. Since the "1-assignment" of x already justifies the "0-values" of $c_{p'_1}, \ldots, c_{p'_q}$, and since our mapping m_j is monotonic, we obtain $m_i(c \times \{1\})$ by taking entry $m_j(c')$, where $c'_{p'_1} = \ldots = c'_{p'_q} = \hat{0}$, that is, $c' = \phi(c)$.

Since we need time $O(n_i)$ in order to check whether a coloring is locally invalid, the evaluation of m_i can be carried out in $O(3^{n_i} \cdot n_i) \subseteq O(4^{\omega})$ time.

Considering assignments (10.4) and (10.5), it is easy to see that m_i is monotonic if m_j is monotonic.

JOIN NODES: Suppose i is a JOIN NODE with children j and k and suppose that $X_i = X_j = X_k = (x_{i_1}, \ldots, x_{i_{n_i}})$.

Let $c = (c_1, \ldots, c_{n_i}) \in \{0, \hat{0}, 1\}^{n_i}$ be a coloring for X_i. We say that $c' = (c'_1, \ldots, c'_{n_i})$, $c'' = (c''_1, \ldots, c''_{n_i}) \in \{0, \hat{0}, 1\}^{n_i}$ *divide* c if

1. $(c_t \in \{\hat{0}, 1\} \Rightarrow c'_t = c''_t = c_t)$, and
2. $\big(c_t = 0 \Rightarrow \big[(c'_t, c''_t \in \{0, \hat{0}\}) \wedge (c'_t = 0 \vee c''_t = 0)\big]\big)$.

Then, evaluate the mapping m_i of X_i as follows:

For all colorings $c \in \{0, \hat{0}, 1\}^{n_i}$ set

$$m_i(c) := \min\{m_j(c') + m_k(c'') - \#_1(c) \mid c' \text{ and } c'' \text{ divide } c\}. \quad (10.6)$$

In other words, in order to determine the value $m_i(c)$ we look up the corresponding values for coloring c in m_j (corresponding to the left subtree) and in m_k (corresponding to the right subtree), add the corresponding values, and subtract the number of "1-assignments" in c, since they would be counted twice otherwise.

Clearly, if coloring c of node i assigns the colors 1 or $\hat{0}$ to a vertex x from X_i, we have to make sure that we use colorings c' and c'' of the children j and k which assign the same color to x. However, if c assigns color 0 to x, it is sufficient to justify this color by *only* one of the colorings c' or c''. Observe that, from the monotonicity of m_j and m_k we obtain the same "min" in assignment (10.6) if we replace condition 2 in the definition of "divide" by:

2′. $\big(c_t = 0 \Rightarrow [(c'_t, c''_t \in \{0, \hat{0}\}) \wedge (c'_t \neq c''_t)]\big)$.

Note that, for given $c \in \{0, \hat{0}, 1\}^{n_i}$, with $z := \#_0(c)$, we have 2^z many pairs (c', c'') that divide c if we use condition (2') instead of (2). Since there are $2^{n_i - z}\binom{n_i}{z}$ many colorings c with $\#_0(c) = z$, we obtain that

$$\{(c', c'') : c \in \{0, \hat{0}, 1\}^{n_i},\ c' \text{ and } c'' \text{ divide } c\ \}$$

has size

$$\sum_{z=0}^{n_i} 2^{n_i - z} \binom{n_i}{z} \cdot 2^z = 4^{n_i}.$$

This shows that evaluating m_i can be done in $O(4^{n_i} \cdot n_i)$ time. Again, it is not hard to see that m_i is monotonic if m_j and m_k are monotonic. This basically follows by the definition of "divide".

Step 3: Let r denote the root of T. The domination number is given by

$$\min\{m_r(c) \mid c \in \{0,1\}^{n_r}\}. \tag{10.7}$$

The minimum in (10.7) is taken only over colorings containing only colors 0 and 1 because a color $\hat{0}$ would mean that the corresponding vertex still needs to be dominated.

This concludes the description of the dynamic programming algorithm and leads to the following result.

Theorem 10.16 *For a graph G with given tree decomposition $\langle\{X_i \mid i \in I\}, T\rangle$ an optimal dominating set can be computed in $O(4^\omega \cdot \omega \cdot |I|)$ time. Here, ω denotes the width of the tree decomposition.*

Proof Obviously, the total running time of the algorithm is $O(4^\omega \cdot \omega \cdot |I|)$. For the correctness of the algorithm, we observe the following. In initialization Step 1, as well as in the updating process for INSERT NODES and JOIN NODES of Step 2, we made sure that the assignment of color 0 to a vertex x always guarantees that at the current stage of the algorithm x is already dominated by a vertex from previous bags. Since, by definition of tree decompositions, any pair of neighbors appears in at least one bag, the validity of the colorings was checked for each such pair of neighbors. Finally, the consistency property of tree decompositions together with the comments given in Step 2 of the algorithm imply that the updating of each mapping is done consistently with all mappings that have been visited earlier in the algorithm.

When bookkeeping how the minima in assignments (10.2), (10.6), and (10.7) of Step 2 and Step 3 were obtained, this method also constructs a dominating set D delivering the domination number. □

In conclusion, the most important point in dynamic programming on tree decompositions is the sizes of the tables involved. The table sizes are usually bounded by c^ω, where ω denotes the width of the underlying tree decomposition and c usually depends on the underlying combinatorial problem. Hence two optimization goals are immediate:

1. keep the width of the tree decomposition as small as possible; and
2. closely investigate the combinatorics of the underlying graph problem in order to keep the base c as small as possible.

DOMINATING SET provides a striking example of the second goal, as the constant could be improved from the naturally given 9 to 4. To illustrate the significance of

Runtime	$\omega = 5$	$\omega = 10$	$\omega = 15$	$\omega = 20$
$9^\omega \cdot \omega \cdot \lvert I \rvert$	0.25 sec	10 hours	100 years	$8 \cdot 10^6$ years
$4^\omega \cdot \omega \cdot \lvert I \rvert$	0.005 sec	10 sec	4.5 hours	260 days

Table 10.1 *Comparing the* $O(4^\omega \cdot \omega \cdot |I|)$ *algorithm for* DOMINATING SET *with the* $O(9^\omega \cdot \omega \cdot |I|)$ *algorithm for* $|I| = 1000$*; we assume a computer executing* 10^9 *instructions per second and we neglect the constants hidden in the O-terms (which are comparable in both cases).*

such a result, Table 10.1 compares hypothetical running times of the $O(9^\omega \cdot \omega \cdot |I|)$ algorithm to the $O(4^\omega \cdot \omega \cdot |I|)$ algorithm for some realistic values of ω and $|I| = 1000$. Observe that this gives a realistic comparison of the relative performances since both algorithms incur comparable, small constant factors hidden in the O-notation. The $O(4^\omega \cdot \omega \cdot |I|)$ time algorithm has been successfully implemented. As a non-surprising result, this tells us that improving exponential terms often is a "big issue" for fixed-parameter algorithms.

Finally, it must be emphasized that besides (exponential) running time (exponential) memory usage is also an important issue in making tree decomposition-based algorithms useful in practice. The exponential table sizes lie at the heart of the exponential running times as well as of the exponential memory usage. In order to avoid the "memory boundedness" of dynamic programming on tree decompositions, all tricks and techniques should be tried—another promising research challenge connected with the development of efficient fixed-parameter algorithms.

In complete analogy to the case of VERTEX COVER, Theorem 10.16 yields the following:

Corollary 10.17 DOMINATING SET IN PLANAR GRAPHS *is solvable in* $2^{O(\sqrt{k})} \cdot n$ *time, where* k *denotes the size of the dominating set and* n *is the number of graph vertices.*

10.6 Monadic second-order logic (MSO)

In the preceding two sections we have explicitly seen two efficient fixed-parameter algorithms with respect to the parameter treewidth to determine optimal vertex covers and dominating sets for graphs given together with their tree decompositions. It might not always be easy to see whether a problem is fixed-parameter tractable with respect to the parameter treewidth in this way. In fact, dynamic programming can become an elusive matter here. There are also results, however, that state that large *classes* of problems can be solved in linear time when a tree decomposition with constant treewidth is known; in other words, these problems are in "linear-time *FPT*". The main work in this direction was done by Bruno Courcelle, providing a powerful classification tool for such problems. It must be emphasized, however, that the now described methodology is of purely theoretical interest because the associated running times suffer from huge constant factors and combinatorial explosions with respect to the parameter treewidth. Still, it provides an excellent tool for quickly deciding whether a problem is fixed-parameter tractable on graphs parameterized by treewidth. After establishing fixed-parameter tractability in this way, as a second step one should then head for a concrete, problem-specific algorithm with improved efficiency. Because the underlying theory is much beyond the scope of this book, we subsequently focus on describing central concepts and results and how to use them in a profitable way.

Our tool is called *monadic-second order logic* (*MSO*). Roughly speaking, it is an extension of first-order logic that also allows quantification over sets. The point here is that whenever a graph problem is expressible using the formalism of this logic, we can infer that the problem is fixed-parameter tractable with respect to the parameter treewidth. We start by describing the language of MSO.

What do MSO-formulae expressing graph properties look like, that is, what is their syntax? First of all, we have an infinite supply of *"individual" variables*, denoted by small letters x, y, z, etc. Moreover, there is an infinite supply of *set variables*, denoted by capital letters X, Y, Z, etc. To specify graphs and their properties, we use the particular vocabulary $\{E, V, I\}$, where V and E are unary relation symbols interpreted as the vertex and the edge set of a graph, and I is a binary relation symbol interpreted as the incidence relation between vertices and edges. Continuing with MSO syntax for graphs, using mainly postfix notation, we introduce *atomic* MSO-formulae as formulae of the forms $x = y$, Vx, Ex, Ixy, and Xx, where X is a set variable and x, y are individual variables. Based on atomic formulae, more complicated MSO-formulae can be built as follows, thus specifying the whole class of MSO-formulae.

- If ϕ is an MSO-formula, then $\neg\phi$ is one as well.
- If ϕ and ψ are MSO-formulae, then $\phi \wedge \psi$, $\phi \vee \psi$, and $\phi \rightarrow \psi$ are as well.
- If ϕ is an MSO-formula and x is an individual variable and X is a set variable, then $\exists x\phi$, $\forall x\phi$, $\exists X\phi$, and $\forall X\phi$ are as well.

Here, "$\rightarrow$" denotes the Boolean implication. This concludes the description of the syntax. Let us next come to the semantics of MSO-formulae. To simplify notation, for a graph $G = (V, E)$ let $U := V \cup E$. An *assignment* α for an MSO-formula ϕ maps each individual variable of ϕ to an element of U and every set variable to a subset of U. Thus we can inductively define the concept of an assignment α *satisfying* an MSO-formula ϕ, written as $(G, \alpha) \models \phi$ for a given graph G.

- $(G, \alpha) \models x = y$ iff $\alpha(x) = \alpha(y)$;
- $(G, \alpha) \models Vx$ iff $\alpha(x) \in V$;
- $(G, \alpha) \models Ex$ iff $\alpha(x) \in E$;
- $(G, \alpha) \models Ixy$ iff $\alpha(x) \in V$, $\alpha(y) \in E$, and vertex $\alpha(x)$ is endpoint of edge $\alpha(y)$;
- $(G, \alpha) \models Xx$ iff $\alpha(x) \in \alpha(X)$;

- $(G, \alpha) \models \neg\phi$ iff $(G, \alpha) \not\models \phi$;
- $(G, \alpha) \models \phi \wedge \psi$ iff $(G, \alpha) \models \phi$ and $(G, \alpha) \models \psi$, and analogously for the logical or "$\vee$" and the logical implication "$\rightarrow$";
- $(G, \alpha) \models \exists x \phi$ iff there exists an $a \in U$ such that $(G, \alpha\frac{a}{x}) \models \phi$, where $\alpha\frac{a}{x}$ denotes the assignment with $\alpha\frac{a}{x}(x) = a$ and $\alpha\frac{a}{x}(v) = \alpha(v)$ for all $v \neq x$;
- $(G, \alpha) \models \forall x \phi$ iff for all $a \in U$ we have $(G, \alpha\frac{a}{x}) \models \phi$;
- $(G, \alpha) \models \exists X \phi$ iff there exists an $A \subseteq U$ such that $(G, \alpha\frac{A}{X}) \models \phi$, and similarly for $\forall X$ meaning "for all $A \subseteq U$".

The relation $(G, \alpha) \models \phi$ depends only on the values of α at the *free variables* of ϕ, that is, those variables v not occurring in the scope of a quantifier $\exists v$ or $\forall v$, where v may denote an individual as well as a set variable. One writes

$$\phi(x_1, \ldots, x_i, X_1, \ldots, X_j)$$

to indicate that the free individual variables of ϕ are $x_1, \ldots, x_k$ and the free set variables are $X_1, \ldots, X_j$. Then for a graph G and $a_1, \ldots, a_i \in U$, $A_1, \ldots, A_j \subseteq U$ one writes

$$G \models \phi(a_1, \ldots, a_i, A_1, \ldots, A_j)$$

if for every assignment α with

$$\alpha(x_1) = a_1, \ldots, \alpha(x_i) = a_i$$

and

$$\alpha(X_1) = A_1, \ldots, \alpha(X_j) = A_j$$

it holds that $(G, \alpha) \models \phi$. A *sentence* is a formula without free variables.

For example, to express that a graph G is bipartite we write $G \models \phi$ with ϕ being the following formula

$$\begin{aligned} &\exists X \exists Y (\forall x (Vx \rightarrow (Xx \vee Yx)) \wedge \\ \forall x \forall y (((x \neq y) \wedge \exists z (Ixz \wedge Iyz)) &\rightarrow \neg((Xx \wedge Xy) \vee (Yx \wedge Yy)))). \end{aligned}$$

The core result in this field, due to Bruno Courcelle, now reads as follows.

Theorem 10.18 *Let $\omega \geq 1$ and let*

$$\phi(x_1, \ldots, x_i, X_1, \ldots, X_j)$$

be an MSO-formula. Then there is a linear-time algorithm, given a graph G with a tree decomposition of width at most ω and $a_1, \ldots, a_i \in U$, $A_1, \ldots, A_j \subseteq U$, decides whether

$$G \models \phi(a_1, \ldots, a_i, A_1, \ldots, A_j).$$

So far, we have seen how MSO-formulae can be used to express decision problems. There are *extensions of MSO* allowing us to deal with optimization problems without giving up linear-time solvability as stated in Theorem 10.18.

For instance, the following formula expresses the (optimization version of the) VERTEX COVER problem:

$$\min X\ \forall y \exists x (Xx \wedge Ey \wedge Ixy).$$

Similarly, the optimization version of DOMINATING SET can be expressed as follows:

$$\min X\ \forall y \exists x \exists z (Xx \wedge Vy \wedge Ez \wedge Ixz \wedge Iyz).$$

In the purely parameterized (decision) version searching for a vertex cover of size k, we would write

$$\exists x_1 \exists x_2 \ldots \exists x_k \forall y (Vx_1 \wedge \ldots \wedge Vx_k \wedge (Ey \rightarrow (Ix_1y \vee \ldots \vee Ix_ky)))$$

instead. The parameterized version of DOMINATING SET can be dealt with similarly. Observe that, of course, the huge constant factor hidden in the O-notation in Theorem 10.18 depends on the formula size and its complexity.

Summarizing, monadic second-order logic offers a *descriptive* way of the classifying the computational complexity of problems restricted to graphs of bounded treewidth. It is a very elegant and powerful tool for quickly deciding about fixed-parameter tractability, but it is far from any efficient implementations. Algorithmic analysis of the concrete problem at hand always seems beneficial in order to come up with more practical results once we know that fixed-parameter tractability is achievable due to a description in the language of MSO.

10.7 Related graph width parameters

Besides treewidth there are several *graph width* parameters of algorithmic use. We briefly mention three of them here.

Pathwidth. This is a simple special case of treewidth. It is obtained by restriction of the underlying trees in the definition of tree decompositions to paths. Pathwidth closely resembles a form of "narrowness" of graphs. Graphs with small pathwidth typically appear in natural language processing. An important problem in VLSI layout theory, the so-called GATE MATRIX LAYOUT problem, is equivalent to the pathwidth problem. Trivially, the treewidth of a graph always gives a lower bound for the pathwidth of a graph. Moreover, trees may have arbitrarily large pathwidth. As for tree decompositions, there are many different equivalent characterizations for graphs of bounded pathwidth. Recall from Section 10.1 that the optimal tree decomposition of a grid graph is actually a path decomposition. Concerning the determination of optimal path decompositions of graphs, again it is known that the corresponding decision problem is NP-complete. As to fixed-parameter tractability with respect to the parameter pathwidth for the problem to construct path decompositions, analogous statements as for the treewidth case apply—the algorithms are far from practical. That is why here also approximation and heuristic algorithms are important.

With regard to the algorithmic use of path decompositions, we note only that it is obvious that dynamic programming is easier for path-like than for tree-like structures. Actually, in Section 9.5 we experienced with the WEIGHTED SET COVER problem polynomial-time solvability for arbitrary "pathwidth" values, whereas it turned out *NP*-complete for tree-structured cases in general. For the latter case, however, we could show fixed-parameter tractability with respect to the parameter "width of the corresponding tree structure". By way of contrast, path decompositions can lead to larger width values than tree decompositions do—simply consider trees which have treewidth one—and so the use of tree decompositions may become favorable when we encounter a combinatorial explosion with respect to the corresponding width.

Local treewidth. In Section 10.3 we saw that graphs which have a size-k vertex cover or a size-k dominating set possess tree decompositions of width $O(\sqrt{k})$. In Section 10.6 we learned that for graphs of bounded treewidth all graph properties (such as having a size-k vertex cover) that are expressible in monadic second-order logic are decidable in linear time. The notion of local treewidth arose in an attempt to extend and generalize these sorts of results.

To this end, we need to define the concept of *r-neighborhood.* Let $G = (V, E)$ be a graph and $v \in V$ be a vertex in G. Then $N_r[v]$ is the set of all vertices with distance at most r to v, that is, $N_r[v]$ includes v and all vertices that are connected to v via a path of at most r edges in G. Let $tw(G)$ denote the treewidth of graph G.

Definition 10.19 *The* local treewidth *of a graph $G = (V, E)$ is*

$$ltw(G, r) := \max\{tw(G[N_r[v]]) \mid v \in V\}.$$

Then, G has bounded local treewidth *if there is a function $f : \mathbb{N} \longrightarrow \mathbb{N}$ such that $ltw(G, r) \leq f(r)$ for $r \in \mathbb{N}$.*

The definition of bounded local treewidth requires that the local treewidth depends only on r and is independent of the number of graph vertices etc. Clearly, if a graph has treewidth t then its local treewidth is bounded from above by t. However, whereas planar graphs have unbounded treewidth, their local treewidth is bounded, that is, it can be shown that $ltw(G, r) \leq 3r$ for an arbitrary planar graph G and $r \geq 1$. Without going into any details here, we only mention in passing that with the help of the concept of bounded local treewidth numerous fixed-parameter tractability results extending those for planar graphs could be achieved. These results, however, suffer from large hidden constant factors, and so far are of only theoretical interest.

Branchwidth. This is a concept similar to treewidth. In contrast to tree decompositions, however, it is known that optimal branch decompositions of planar graphs can be computed in polynomial time. For general graphs, branchwidth determination again becomes an *NP*-complete problem.

Definition 10.20 *Let $G = (V, E)$ be a graph. A* branch decomposition *of G is a pair $\langle \sigma, T = (I, F) \rangle$ where T is a tree with every node in T of degree one or three, and σ is a one-to-one mapping from E to the set of leaves in T.*

The order *of an edge $f \in F$ is the number of vertices $v \in V$ with incident edges $\{v,u\},\{v,w\} \in E$ such that the path in T from $\sigma(\{v,u\})$ to $\sigma(\{v,w\})$ uses f.*

The width *of branch decomposition $\langle \sigma, T = (I,F)\rangle$ is the maximum order over all edges $f \in F$. The* branchwidth *of G is the minimum k such that G has a branch decomposition of width k.*

There is a close relationship between treewidth and branchwidth. Let k be the treewidth of a graph and let k' be its branchwidth. Then it is known that

$$\max\{2, k'\} \leq k + 1 \leq \max\{2, \lfloor 3 \cdot k'/2 \rfloor\}.$$

As with tree decompositions, branch decompositions receive particular attention from an algorithmic point of view because they are amenable to dynamic programming techniques. The basic ideas are analogous to dynamic programming on tree decompositions and we omit the details. Note that the currently best fixed-parameter "$c^{\sqrt{k}}$-algorithm" with constant base c solving DOMINATING SET IN PLANAR GRAPHS is based on branch decompositions. More specifically, it runs in $O(36000^{\sqrt{k}} \cdot k + k^4 + n^4)$ time. Observe, however, that this refers to proven worst-case bounds concerning the constant c in the base. For the time being it is not clear what the best algorithms in practice are. Moreover, algorithmically the various tree decomposition and branch decomposition based algorithms are similar—the improvements in the worst-case time bounds are mainly due to refined and improved mathematical analysis.

10.8 Summary and concluding remarks

Treewidth is a sophisticated concept one has to get used to. To this end, its characterization by the robber–cop game (see Section 10.1) is extremely useful. It needs time, however, to become familiar with this methodology—only starting material is presented in this chapter. Already the construction of (semi-)optimal tree decompositions is a thing that needs much more attention than we could pay to it here. For the purpose of studying parameterized problems in planar graphs, however, Section 10.3 presents the fundamental ideas. Note that, so far, most known fixed-parameter complexity results using tree decompositions and related concepts are connected to planar and somewhat more general graphs. The algorithmic usefulness of tree decompositions is tightly connected to dynamic programming—here we presented two concrete examples for VERTEX COVER and DOMINATING SET. In the literature one can find a generalized description of how dynamic programming for graphs of bounded treewidth works. With monadic second-order logic a powerful classification tool is given in order to determine the solvability of graph problems for bounded treewidth in a "descriptive way". Intuitively speaking, in particular for graph problems that carry some "non-local" properties—that is, for instance, the choice of a vertex into a solution set may directly affect other vertices at arbitrary distance from that vertex—fixed-parameter tractability with respect to the parameter treewidth is often far from being clear. Monadic second-order logic may help to decide on that. Finally, it must be noted that the concept of tree decompositions has several relatives—local treewidth and branchwidth being two of them which have already played a role in fixed-parameter complexity studies.

Treewidth and related concepts offer a new view of parameterization—that is, *structural parameterization*. Whereas the most conventional way to choose problem parameters is based on the sizes of the desired solution sets, structural parameters are basically independent of these. Thus, structural parameters such as treewidth might also be considered as ways to define special problem instances—graphs of bounded treewidth also belong into the large field of studying special graph classes and their algorithmic properties. It is plausible to assume that future research will reveal more and more connections and mutual interaction between parameterized computational complexity and the vast field of special graph classes. Another point to note here is that it seems less appropriate to search in the case of structural parameterizations such as by treewidth for problem kernels: clearly, a graph of bounded treewidth k may have size completely independent of k, so there appears to be no point in asking for bounding the graph size by its treewidth as is inherent by the concept of reduction to a problem kernel (see Chapter 7).

Tree decomposition-based algorithms bring forward a so far somewhat neglected point in fixed-parameter algorithmics—efficient *usage of memory*. Combinatorial explosions with respect to memory consumption must be kept as small as possible—a system may run almost forever on hard problems, but it will immediately break down when running out of memory. Keeping this in mind is essential for successful applications of tree decomposition-based and related memory-intensive algorithms.

Tree decompositions of graphs, studied thoroughly, deserve a book on their own. The basic concepts and definitions are technically demanding. Nevertheless, it seems as though anybody doing advanced fixed-parameter algorithmics should gain some familiarity with this mathematically beautiful and algorithmically strong methodology. Discovering and exploiting "width parameters" in hard problems is a core issue that fixed-parameter algorithmics should be occupied with.

10.9 Exercises

1. Describe in detail what the width-ℓ tree decomposition of an $\ell \times \ell$-grid looks like.
2. Construct an optimal tree decomposition for a graph that is simply a cycle.
3. Prove Lemma 10.3.
4. Show that a graph that has a vertex cover of size k has pathwidth at most $k + 1$.
5. Show that a complete graph with n vertices has treewidth exactly $n - 1$.
6. Show that if a graph G has an induced complete subgraph K then every tree decomposition of G must contain a bag that contains all vertices from K.

7. Prove Proposition 10.10 for the case of VERTEX COVER.
8. The *NP*-complete 3-COLORING problem is to color (if possible) each vertex of a graph with one out of three colors such that no pair of neighboring vertices has the same color.
   Show how 3-COLORING can be solved by dynamic programming on graphs of bounded treewidth (with given tree decomposition).
9. Describe bounded treewidth dynamic programs for the following variations of DOMINATING SET:
   (a) INDEPENDENT DOMINATING SET: Here, the desired dominating set must additionally form an independent set.
   (b) TOTAL DOMINATING SET: Here, each vertex in the dominating set additionally must have a neighbor in the dominating set.
   (c) PERFECT DOMINATING SET: Here, every vertex not in the dominating set must have exactly one neighbor in the dominating set.
10. The *NP*-complete FEEDBACK VERTEX SET problem is defined as follows.
    Input: A graph $G = (V, E)$ and a nonnegative integer k.
    Task: Find a subset of vertices $V' \subseteq V$ with k or fewer vertices such that each cycle in G contains at least one vertex from V'. Thus, removing the vertices in V' from G results in a forest.
    Show how to express FEEDBACK VERTEX SET using monadic second-order logic.

10.10 Bibliographical remarks

Tree decompositions of graphs go back to Robertson and Seymour (1986). Survey papers (Bodlaender, 1993; Bodlaender, 1997; Bodlaender, 1998; Bodlaender, 2005) provide good overviews focussing on different aspects. See also the book Kloks (1994). The linear-time algorithm for bounded treewidth (and the construction of corresponding tree decompositions) is due to Bodlaender (1996). The practical heuristic mentioned for constructing tree decompositions is due to Reed (1993). See Bodlaender (2005) for an up-to-date survey of constructing tree decompositions. The description of planar graphs follows Alber *et al.* (2002). The current best bound for a $c^{\sqrt{k}}$-algorithm for DOMINATING SET IN PLANAR GRAPHS appears in Fomin and Thilikos (2003). See also Demaine *et al.* (2004*a*) and Demaine *et al.* (2004*b*) for several more related issues. The dynamic programming for DOMINATING SET is described in Alber *et al.* (2002) and Alber and Niedermeier (2002).

Related graph width parameters are discussed in Bodlaender (1993) and Bodlaender (1998). The polynomial-time algorithm for computing branchwidth in planar graphs is due to Seymour and Thomas (1994). Local treewidth has been introduced by Eppstein (2000). Important theoretical fixed-parameter tractability results in this direction are Frick and Grohe (2001) and Demaine *et al.* (2004*a*).

11

FURTHER ADVANCED TECHNIQUES

In the preceding chapters we presented in some depth—using several concrete problems and the fixed-parameter algorithms for solving them—the fundamental techniques of reduction to a problem kernel, depth-bounded search trees, dynamic programming, and tree decompositions of graphs. Now we describe a few more techniques which belong in the toolkit of every "parametric algorithm designer".

We begin with color-coding, a very elegant randomized method designed in the context of subgraph isomorphism problems. The method can be derandomized and it seems to apply to a whole list of problems. Our illustrative example is LONGEST PATH, which concerns the parameter path length. Future research must show its breadth and how it performs with respect to practical applications.

Our next technique is integer linear programming. The point here is that due to a ground-breaking result of Hendrik W. Lenstra it is known that bounding the number of variables in an integer linear program by a function depending only on the problem parameter yields fixed-parameter tractability with respect to this parameter. Because of the hidden constants and the large combinatorial explosion involved this method seems more suitable as a classification tool, though. Still, the corresponding integer linear programs can be implemented and handled using standard solvers. Our illustrative example is CLOSEST STRING with respect to the parameter number of input strings—no other way to show fixed-parameter tractability is known here.

We continue with a very new and promising method from the year 2004—iterative compression. The point here is to make use of size-$(k+1)$ solutions in order to find size-k solutions in an iterative manner. This technique is designed for minimization problems. It was used for the breakthrough result to show that GRAPH BIPARTIZATION is fixed-parameter tractable with respect to the number of vertices to be deleted. To illustrate the ideas, we employ the conceptually simpler cases VERTEX COVER and FEEDBACK VERTEX SET. We believe that future research might turn this technique into one that fills a whole chapter.

As with iterative compression, the technique called greedy localization is from 2004. It tries to make use of a greedily found solution in order to find a better or even optimal one. It works for maximization problems where small solution set sizes—which are the parameters—might be less frequent than for minimization problems. Still, the method definitely carries potential for more applications. We illustrate the method using the SET SPLITTING and 3-SET PACKING problems.

We conclude our series of advanced techniques with a brief glimpse at the celebrated graph minor theory—one of the deepest achievements of modern mathematics—and its particular use with respect to fixed-parameter tractability. This beautiful theory, due to its universality and enormous hidden factors in the running times of the corresponding algorithms, must at the current state of knowledge be considered as a classification tool only.

11.1 Color-coding

Many graph problems studied in this work are special versions of the SUBGRAPH ISOMORPHISM problem:

> **Input:** Two graphs $G = (V, E)$ and $G' = (V', E')$.
> **Task:** Determine whether there is a subgraph of G that is isomorphic to G'.

For instance, CLIQUE asks for a subset $U \subseteq V$ of size at least k such that in the induced subgraph $G[U]$ each pair of vertices is connected by an edge.

Color-coding is a method that can be used to derive (randomized) fixed-parameter algorithms for several subgraph isomorphism problems. We study this technique through an example application of the *NP*-complete LONGEST PATH problem:

> **Input:** A graph $G = (V, E)$ and a nonnegative integer k.
> **Task:** Find a *simple* path in G that contains $k - 1$ edges and k vertices.

Note that the restriction to simple paths where no vertex may appear more than once is crucial here—otherwise, computing the kth power of the adjacency matrix of G, one can easily find in polynomial time all pairs of vertices that are connected by a path of at most $k-1$ edges: assume that one matrix multiplication takes $t(n)$ time. Then using repeated squaring one obtains after $O(\log k)$ rounds all pairs of vertices connected by a path of length at most k. Altogether, this needs $O(\log k \cdot t(n))$ time.

The central idea behind color-coding is that to find the desired vertex set U with $|U| = k$ one randomly colors the whole graph with k colors and "hopes" that all vertices in U will obtain different colors. If so, the task of finding U is greatly simplified: color-coding makes use of *dynamic programming* for finding paths of vertices with pairwisely different colors. The randomized fixed-parameter algorithms derived in this way can be transformed into deterministic fixed-parameter algorithms through the use of hashing techniques at the cost of increased running time. On the negative side, color-coding does not lead to a fixed-parameter algorithm for the $W[1]$-complete problem CLIQUE.

The key to solving LONGEST PATH lies in the concept of *colorful* paths which simply means that each vertex of the path has another color. Each colorful path is clearly simple, and by coloring the graph vertices uniformly at random with k colors, a simple path will consist of k different colors with probability $k!/k^k$. Using Stirling's approximation, this probability is lower-bounded by e^{-k}.

For the time being, let us assume that there is a colorful simple path of k vertices in G. The following lemma shows that it can be found quickly by dynamic programming.

Lemma 11.1 *Let $G = (V, E)$ and let $C : V \to \{1, \ldots, k\}$ be a coloring. Then a colorful path of k vertices can be found (if it exists) in time $2^{O(k)} \cdot |E|$.*

Proof In what follows, we describe an algorithm that finds all colorful paths of k vertices starting at some fixed vertex s. This is not really a restriction because to solve the general problem, we may just add some extra vertex s' to V, color it with the new color 0, and connect it with each of the remaining vertices of V by an edge.

To find the described paths, we use dynamic programming. Assume that for all $v \in V$ already all possible *color sets* of colorful paths between s and v consisting of i vertices have been found. Clearly, the length-0 path starting and ending at s is assigned the color set $\{0\}$. For each v, there are at most $\binom{k}{i}$ of these sets. Let now F be such a color set belonging to v. We consider every F belonging to v and every edge $\{u, v\} \in E$: if $C(u) \notin F$ then build the new color set $F' := F \cup \{C(u)\}$—so F' becomes part of the set of color sets belonging to u. In this way, we obtain all color sets belonging to paths of length $i + 1$ and so on. Thus G contains a colorful path with respect to coloring C iff there exists a vertex $v \in V$ that has at least one color set that corresponds to a path of k vertices.

The described algorithm performs $O(\sum_{i=1}^{k} i \cdot \binom{k}{i} \cdot |E|)$ steps. Here the factor i refers to the test whether or not $C(u)$ is already contained in F. The factor $\binom{k}{i}$ refers to the number of possible sets F and the factor $|E|$ refers to the time to check all edges $\{u, v\} \in E$. The whole expression is upper-bounded by $O(k \cdot 2^k \cdot |E|)$. It is not hard to effectively construct a colorful path for a given color set. □

Observe that in the proof of Lemma 11.1 it was crucial not that the paths (that is, all vertices on them) were stored but only that their corresponding color sets were recorded. Thus, for a path of $i \leq k$ vertices at most $\binom{k}{i} = O(k^i)$ candidate colorings are possible. By way of contrast, there are $\binom{n}{i} = O(n^i)$ different vertex sets of size i. This difference exactly reflects the gap between fixed-parameter tractability (combinatorial explosion $f(k)$) and fixed-parameter intractability (combinatorial explosion n^k).

Now, using standard techniques for randomized algorithms, a randomized fixed-parameter algorithm for LONGEST PATH follows.

Theorem 11.2 LONGEST PATH *can be solved in expected running time $2^{O(k)} \cdot |E|$.*

Proof According to the above remarks a simple path of k vertices is colorful with probability at least e^{-k}. According to Lemma 11.1, such a colorful path can be found in time $2^{O(k)} \cdot |E|$; more precisely, all colorful paths of k vertices can be found.

We repeat the following $e^k = 2^{O(k)}$ times:

1. Randomly choose a coloring $C : V \to \{1, \ldots, k\}$.
2. Check using Lemma 11.1 whether or not there is a colorful path; if so then this is a simple path of k vertices.

In this way, the expected value of the number of colorful paths found—if any exist—after trying $2^{O(k)}$ random colorings is at least one. □

Theorem 11.2 is based on a randomized algorithm. Using *hashing*, it can be derandomized at the cost of somewhat increased running time. To this end, we need a list of colorings of the vertices in V such that for *each* subset $V' \subseteq V$ with $|V'| = k$ there is at least one coloring in the list that gives to each vertex in V' a unique color. This is formalized by the concept of a k-perfect family of hash functions from $\{1, 2, \ldots, |V|\}$ onto $\{1, 2, \ldots, k\}$.

Definition 11.3 *A* k-perfect family of hash functions *is a family* $\mathcal{H}$ *of functions from* $\{1, \ldots, n\}$ *onto* $\{1, \ldots, k\}$ *such that for each* $S \subseteq \{1, \ldots, n\}$ *with* $|S| = k$ *there exists an* $h \in \mathcal{H}$ *such that* h *is one-to-one when restricted to* S.

In the literature one can find the following statement.

Theorem 11.4 *It is possible to construct families of* k*-perfect hash functions from* $\{1, \ldots, n\}$ *onto* $\{1, \ldots, k\}$ *which consist of* $2^{O(k)} \cdot \log n$ *hash functions. For such a hash function* h *the value* $h(i)$, $1 \leq i \leq n$, *can be computed in linear time.*

Based on Theorem 11.4, we obtain deterministic fixed-parameter tractability for LONGEST PATH.

Theorem 11.5 LONGEST PATH *can be solved deterministically in* $2^{O(k)} \cdot |E| \cdot \log |V|$ *time.*

Proof Assume that a simple path with k vertices exists. Color the graph using all possible hash functions from the family given in Theorem 11.4. According to Definition 11.3 at least one of these colorings must lead to a colorful simple path. Such a colorful path then can again be found using Lemma 11.1.

Because the family from Theorem 11.4 consists of $2^{O(k)} \cdot \log n$ hash functions, the time complexity of the algorithm from Lemma 11.1 has to be multiplied by this factor. In total, we obtain the upper bound

$$2^{O(k)} \cdot \log |V| \cdot 2^{O(k)} \cdot |E| = 2^{O(k)} \cdot |E| \cdot \log |V|$$

for the overall running time. □

Although (randomized) color-coding appears to be an elegant and promising tool for designing fixed-parameter algorithms, we are not aware of any substantial practical experience with this method.

There are several other applications of color-coding to subgraph isomorphism problems to be found in the literature. Still, let us briefly discuss why $W[1]$-hard problems such as CLIQUE seem inaccessible to color-coding. Consider CLIQUE.

In Lemma 11.1, it was decisive that a path could be represented by its start vertex s, its end vertex v, and the color set corresponding to the path. To extend a path by one further vertex, it was sufficient to consider edges with endpoint v and to know the colors already used. By way of contrast, this would not be sufficient when constructing a clique in such a step-by-step manner. Here we would need to check the existence of edges of the new vertex to all the already selected vertices—the mere information about the colors of these vertices would not suffice. Then, however, it seems impossible to avoid the "$\binom{n}{i}$-behaviour" instead of the "$\binom{k}{i}$-behaviour" as discussed before.

We mention in passing that families of k-perfect hash functions also can be used directly to obtain fixed-parameter algorithms—similar to the above described derandomization—by systematically going through a search space testing all hash functions. To the author's knowledge, these approaches suffer from bad running times and still seem impractical.

11.2 Integer linear programming

In spite of the enormous significance that (integer) linear programming generally has for approximation algorithms and combinatorial optimization in general, so far it has only played a minor role in the context of fixed-parameter algorithms. Future research might bring significant changes here. One first link will be described now and it will be illustrated using the aforementioned CLOSEST STRING problem.

There is a deep result of Hendrik W. Lenstra that applies to fixed-parameter algorithms. It basically says that integer linear programs (ILP's for short) with a constant number of variables can be solved in linear time. More precisely, with improvements due to Ravi Kannan, we have the following theorem. It refers to the so-called INTEGER LINEAR PROGRAMMING FEASIBILITY problem where one has to decide on the existence of (not necessarily optimal) solutions fulfilling all the constraints given by linear inequalities over a set of integer-valued variables.

Theorem 11.6 INTEGER LINEAR PROGRAMMING FEASIBILITY *can be solved with $O(p^{9p/2}L)$ arithmetic operations in integers of $O(p^{2p}L)$ bits in size, where p is the number of ILP variables and L is the number of bits in the input.*

Note that this fixed-parameter result with respect to p also needs space exponential in the parameter p. In what follows, we will explain the relevance of this result when designing fixed-parameter algorithms. To give a detailed exposition of integer linear programming and, in particular, Lenstra's result, is beyond the scope of this work. Using Theorem 11.6, we show that CLOSEST STRING is fixed-parameter tractable with respect to the parameter "number of input strings". Using a depth-bounded search tree, in Section 8.5, we have shown that CLOSEST STRING is fixed-parameter tractable with respect to the parameter "maximum distance allowed" called d.

Recall the definition of CLOSEST STRING:

Input: A set of k strings $s_1, s_2, \ldots, s_k$ over alphabet Σ of length L each, and a nonnegative integer d.
Task: Find a string s such that $d_H(s, s_i) \leq d$ for all $i = 1, \ldots, k$.

We use the term "closest" string here for a string which has Hamming distance *at most* d to all given strings. As a warm-up in thinking about the problem, one might think about how to solve CLOSEST STRING for an input instance with $k = 2$ (which is easy) and for one with $k = 3$ (which is hard) by a direct combinatorial algorithm.

The goal is to give an ILP formulation for CLOSEST STRING such that the number of variables depends solely on the parameter value k, the number of input strings. The key to this lies in the notion of column types. Given a set of k strings of length L, we can think of these strings as a $k \times L$ character matrix. By *columns* of a CLOSEST STRING instance we refer to the columns of this matrix. In the following, we state that after reordering the columns of the CLOSEST STRING instance, we can easily obtain solutions for the original instance from solutions for the reordered instance. For reordering the columns, we introduce a permutation on strings as follows. Given a string $s = c_1 c_2 \ldots c_L$ of length L with $c_1, \ldots, c_L \in \Sigma$ and a permutation $\pi : \{1, \ldots, L\} \to \{1, \ldots, L\}$. Then, $\pi(s) = c_{\pi(1)} c_{\pi(2)} \ldots c_{\pi(L)}$. The following lemma is obvious.

Lemma 11.7 *Given a set of strings $S = \{s_1, s_2, \ldots, s_k\}$, each of length L, and a permutation $\pi : \{1, \ldots, L\} \to \{1, \ldots, L\}$. Then s is an optimal closest string for $\{s_1, s_2, \ldots, s_k\}$ iff $\pi(s)$ is an optimal closest string for $\{\pi(s_1), \pi(s_2), \ldots, \pi(s_k)\}$.*

Several columns can be identified due to *isomorphism*: we make use of the fact that the columns are independent of each other in the sense that the distance from the closest string is measured columnwise. For instance, consider the case of the two columns $(a, a, b)^T$ and $(b, b, a)^T$ when $k = 3$. Clearly, these two columns are isomorphic because they express the same structure except that the symbols play different roles. For the process of finding the optimal closest string, however, only the structure matters. Isomorphic columns form *column types,* that is, a column type is a set of columns isomorphic to each other.

This can be generalized as follows. Without loss of generality, let a always denote the letter that occurs most often in a column, let b always denote the letter that has the secondly most often occurrences, and so on. This property of being *normalized*, as we will refer to it in the following, can be easily achieved by a simple linear time preprocessing of the input instance. In addition, solving the normalized problem optimally, one can again compute the optimal solution of the original problem instance by simply reversing the above mapping done by the preprocessing. Hence:

Lemma 11.8 *To compute an optimal closest string it is sufficient to solve a normalized and reordered instance. From this, the solution of the original instance can be derived in linear time.*

In the following, we call two input instances *isomorphic* if there is a one-to-one correspondence between the columns of both instances such that each in this way determined pair of columns is isomorphic. The following lemma shows that it is sufficient to solve an instance with alphabet size $|\Sigma'| \leq k$.

Lemma 11.9 *A* Closest String *instance with arbitrary alphabet* Σ, $|\Sigma| > k$, *is isomorphic to a* Closest String *instance with alphabet* Σ', $|\Sigma'| = k$.

Proof Assume that there is an input instance with $|\Sigma| > k$. Clearly, in each column at most k different symbols from Σ appear. Since columns are independent from each other, to solve the underlying Closest String problem it suffices to represent the structure of a column by an isomorphic input instance. To do this, at most k symbols are always enough. □

Example 11.10 For $k = 3$, the set of all possible column types for a Closest String instance consists of

$$(a, a, a)^T, (a, a, b)^T, (a, b, a)^T, (b, a, a)^T, (a, b, c)^T.$$

Generally, the number of column types for k strings depends only on k—it is given by the so-called Bell number $B(k) \leq k!$. By using the column types, Closest String can be formulated as an ILP having only $B(k) \cdot k$ variables. Let the underlying alphabet be Σ. The ILP can be formulated as follows. It uses $B(k) \cdot k$ variables $x_{t,\varphi}$, where t denotes a column type which is represented by a number between 1 and $B(k)$, and $\varphi \in \Sigma$ where $|\Sigma| = k$. The value of $x_{t,\varphi}$ denotes the number of columns of column type t whose corresponding character in the desired solution string of Closest String is set to φ. Thus, the corresponding ILP seeks to minimize

$$\max_{1 \leq i \leq k} \sum_{1 \leq t \leq B(k)} \sum_{\varphi \in (\Sigma \setminus \{\varphi_{t,i}\})} x_{t,\varphi}, \tag{11.1}$$

where $\varphi_{t,i}$ denotes the alphabet symbol at the ith entry of column type t. The following two *constraints* must be fulfilled when minimizing the above function.

1. All variables $x_{t,\varphi}$ must be nonnegative integers.
2. Let $\#_t$ denote the number of columns of type t in the input instance (taking into account isomorphism as described before). Then,

$$\sum_{\varphi \in \Sigma} x_{t,\varphi} = \#_t$$

for every column type t. That is, for each column there must be a letter in the solution.

We must explain one more subtle point. Above, we formulated Closest String as an integer linear programming *optimization* problem looking for the smallest possible distance d, with a solution that fulfills the given constraints, that is, a feasible solution. The difficulty to overcome now is that Theorem 11.6 (Lenstra) refers to the integer linear programming feasibility and not the optimization problem. A parameterized CLOSEST STRING instance formulated as a decision problem gives the maximum distance d allowed. Thus we may obtain the following "feasibility formulation" where the above two constraints remain unchanged, but the objective function (11.1) that was to be minimized is replaced by a third constraint, namely:

$$\max_{1 \le i \le k} \sum_{1 \le t \le B(k)} \sum_{\varphi \in (\Sigma \setminus \{\varphi_{t,i}\})} x_{t,\varphi} \le d.$$

Taken together, we arrive at three types of constraint which can all be expressed by linear (in)equalities over nonnegative integer variables. In this way, we meet all the conditions to apply Theorem 11.6. Therefore, we obtain fixed-parameter tractability of CLOSEST STRING with respect to the parameter k because the parameter p (number of ILP variables) can be bounded from above by a function exclusively depending on k. Note, however, that the combinatorial explosion in k is huge and this approach appears to be impractical for $k > 4$ as some experimental investigations indicated. In this case, ILP heuristics such as branch-and-bound strategies might somewhat extend the range of practical applicability, still using the above ILP formulation.

The ILP approach described can serve as a tool to help classify whether a problem is fixed-parameter tractable. As to CLOSEST STRING, the ILP approach seems to be the only one that yields fixed-parameter tractability with respect to parameter k. Finally, note that there exists an alternative ILP formulation for CLOSEST STRING where the variables have only binary values but the number of variables is $|\Sigma| \cdot L$ (for alphabet Σ and string length L). Hence such an ILP formulation does *not* imply the fixed-parameter tractability of CLOSEST STRING with respect to parameter k. In conclusion, it remains to investigate further examples besides CLOSEST STRING where the described ILP approach turns out to be applicable. More generally, it would be interesting to discover more connections between fixed-parameter algorithms and (integer) linear programming.

11.3 Iterative compression

Iterative compression is a very new technique published in 2004. It has already led to significant breakthroughs in showing fixed-parameter tractability results. For instance, in this way the problem GRAPH BIPARTIZATION, that is, the task of finding a minimum set of vertices whose deletion transforms a graph into a bipartite graph, has been shown fixed-parameter tractable with respect to the number of the deleted vertices. For years this has been a central open problem in parameterized complexity. Moreover, ongoing research indicates that the corresponding algorithm, improved by performance tuning through algorithm engineering, is competitive in practice. In this section we present the fundamental, intuitively very appealing ideas behind iterative compression. We reveal the central technical hurdle that has to be overcome in order to apply iterative compression successfully to a problem. To this end, we present two illustrative examples—the VERTEX COVER problem and the FEEDBACK VERTEX SET problem.

In a nutshell, iterative compression can be described as follows. The technique serves for developing fixed-parameter algorithms for minimization problems parameterized by the size of the solution set. The basic idea is that it is sufficient to give a fixed-parameter algorithm which, given a size-$(k+1)$ solution, either constructs a size-k solution or proves that there is no size-k solution. For example, one can show that, given an n-vertex graph and a bipartization vertex set with $k+1$ vertices, one can find in $O(3^k \cdot k \cdot m)$ time a bipartization vertex set with k vertices, or decide that there is no such vertex bipartization. This is the *compression step*, the workhorse of the whole algorithm. Based on this compression step, the algorithm iteratively considers for $i = 1, 2, \ldots, n$ the induced subgraphs G_i of the input graph $G = (V, E)$ where $V := \{v_1, \ldots, v_n\}$ and $G_i = G[\{v_1, \ldots, v_i\}]$. Clearly, the optimal vertex bipartization for G_1 is the empty set. For $i \geq 1$, if G_i has no vertex bipartization of size k, then neither has G_{i+1}; otherwise, let X denote a vertex bipartization of G_i with $|X| \leq k$. Then, $X \cup \{v_{i+1}\}$ is a vertex bipartization of G_{i+1}. If the set $X \cup \{v_{i+1}\}$ has $k+1$ vertices, the algorithm attempts to "compress" it to a set of size k. The running time of the algorithm is clearly $O(3^k \cdot k \cdot m \cdot n)$. Although the overall algorithmic strategy is very intuitive, observe that a lot of technical work can lie in showing—if at all possible—that the compression step can be done in $O(f(k) \cdot n^{O(1)})$ time.

11.3.1 *Vertex Cover*

We already have seen simple and elaborate depth-bounded search tree algorithms for VERTEX COVER (see Chapter 8). Recall the definition.

> **Input:** A graph $G = (V, E)$ and a nonnegative integer k.
> **Task:** Find a subset $C \subseteq V$ with k or fewer vertices such that each edge in E has at least one of its endpoints in C.

Even the straightforward observation that one or both endpoints of an edge must be in a vertex cover set leads to a combinatorial explosion bounded from above by $O(2^k)$. The best known depth-bounded search tree algorithms can bring this below $O(1.28^k)$. Thus, the subsequently described, completely different approach to solve VERTEX COVER currently is of no real practical interest but serves only to give a simple, easy-to-overview introductory example of the iterative compression technique.

As mentioned above, the workhorse of iterative compression is some sort of compression step. In case of VERTEX COVER, this reads as follows.

Lemma 11.11 *Let $G = (V, E)$ be a graph with $V = \{v_1, \ldots, v_n\}$. Assume that we have a size-k vertex cover for the subgraph G_i of G induced by $\{v_1, \ldots, v_i\}$ where $1 \leq i \leq n$. Then we can check in $O(2^k \cdot |G|)$ time whether there exists a size-k vertex cover for G_{i+1}.*

Proof Let C be the size-k vertex cover for G_i and let $C' = C \cup \{v_{i+1}\}$. Clearly, C' is a size-$(k+1)$ vertex cover for $G_{i+1} = G[\{v_1, \ldots, v_{i+1}\}] = (V_{i+1}, E_{i+1})$.

With the following algorithm we can check whether or not there exists a size-k vertex cover for G_{i+1}:

Consider all possible partitions of C' into two sets C'_0 and C'_1. For each pair C'_0 and C'_1, where C'_0 contains the vertices from C' that will be discarded when constructing a size-k vertex cover whereas C'_1 will be chosen to be part of a size-k vertex cover, we perform the following steps. Hence the subsequent steps are repeated 2^{k+1} times.

1. Set $C'' := \emptyset$.
2. For every edge $\{u, v\} \in E_{i+1}$ with $u \notin C'_1$ and $v \notin C'_1$ distinguish three cases:
   (a) If $u \in C'_0$ and $v \in C'_0$, then there is no vertex cover without a vertex from C'_0 and the partition into C'_0 and C'_1 gives no solution candidate.
   (b) If $u \in C'_0$ and $v \in (V_{i+1} \setminus C')$, then set $C'' := C'' \cup \{v\}$.
   (c) If $v \in C'_0$ and $u \in (V_{i+1} \setminus C')$, then set $C'' := C'' \cup \{u\}$.
3. If $|C'_1 \cup C''| \leq k$, then return $C'_1 \cup C''$ as a solution.

If none of the 2^{k+1} repetitions led to the output of a solution in Step 3 then no size-k vertex cover for G_{i+1} exists. This can be seen as follows. First, note that the set C'_1 exactly represents the vertices from size-$(k+1)$ vertex cover C' for G_{i+1} which are tested as being part of a size-k vertex cover for G_{i+1}. Hence, trying all partitions of C' checks all possible situations. In Step 2 observe that it is essential that no edge can have both its endpoints in $V' \setminus C'$ because C' is a vertex cover for G_{i+1}. This is the decisive point for "compressing the search space". Here, we make sure that edges not covered by C'_1 will be covered by some other vertices. In Step 2.(a) this fails and we have to proceed with the next partition. Otherwise, one of the two cases 2.(b) or 2.(c) must apply and their correctness is obvious because C'_0 represents the vertices not being in the size-$(k+1)$ vertex cover. Evidently, in Step 3, $C'_1 \cup C''$ being small enough then yields a desired size-k (or smaller) vertex cover. Steps 1–3 can be performed in linear time, giving the overall time complexity of $O(2^k \cdot |G|)$. □

Obviously, for $G_1 = (\{v_1\}, \emptyset)$ we have the initial vertex cover $C = \emptyset$. Going through all graphs $G_1, \ldots, G_n = G$ in n rounds, always applying Lemma 11.11, implies that we can solve the VERTEX COVER problem for graph G in $O(2^k \cdot |G| \cdot n)$ time.

Of course, this algorithm is not competitive with the known search tree algorithms. The point here is to see a relatively simple compression lemma—as a rule, the proofs of compression lemmas, if at all possible, require significant efforts and in interesting cases are far less evident than in case of VERTEX COVER. The case of FEEDBACK VERTEX SET to be discussed next provides an example in this direction. Moreover, unlike for VERTEX COVER, here the application of the iterative compression technique really means an algorithmic step forward.

11.3.2 *Feedback Vertex Set*

In the mid-1980s, FEEDBACK VERTEX SET was considered to be one of the least understood of the classical *NP*-complete problems. It is defined as follows.

> **Input:** A graph $G = (V, E)$ and a nonnegative integer k.
> **Task:** Find a subset $V' \subseteq V$ with k or fewer vertices such that each cycle in G contains at least one vertex from V'. Thus, removing the vertices in V' from G results in a forest.

In this section we show that FEEDBACK VERTEX SET can be solved in $O(c^k \cdot n^{O(1)})$ time for a constant c by presenting an algorithm based on iterative compression. Observe that after 10 years of fixed-parameter algorithms for FEEDBACK VERTEX SET, this was the first one to have a combinatorial explosion of the form c^k with constant c. The following lemma provides the compression step.

Lemma 11.12 *Given a graph G and a size-$(k+1)$ feedback vertex set (fvs) X for G, we can decide in $O(c^k \cdot m)$ time whether there exists a size-k fvs X' for G, and if so, provide one.*

Proof Consider the smaller fvs X' as a modification of the larger fvs X. The smaller fvs retains vertices $Y \subseteq X$ and replaces the other vertices $S := X \setminus Y$ with at most $|S| - 1$ new vertices from $V \setminus X$. The idea is to try by brute force all $2^{|X|}$ partitions of X into such sets Y and S. In each case, we then have significant information about a possible smaller fvs X'—it contains Y, but not S—and it turns out that there is only a "small" set V' of candidate vertices to draw from in order to complete Y to X'. More precisely, we show later in Lemma 11.14 that the size of V' can be bounded from above by $14 \cdot |S|$ and that, given S, we can compute V' in $O(m)$ time. Since $|S| \leq k+1$, $|V'|$ thus depends only on the problem parameter k and not on the input size. We again use brute force and consider each of the at most $\binom{14 \cdot |S|}{|S|-1}$ possible choices of $|S| - 1$ vertices from V' that can be added to Y in order to form X'. The test of whether a choice of vertices from V' together with Y forms an fvs can be easily done in $O(m)$ time: delete the vertices in Y and the vertices chosen from V' together with their incident edges from G; this can be done in $O(m)$ time. In the resulting graph, one can detect a cycle with depth-first search in $O(m)$ time. We can now estimate the overall running time T, where the subsequent index i corresponds to a partition of X into Y and S with $|Y| = i$ and $|S| = |X| - i$:

$$T = O\left(\sum_{i=0}^{k} \binom{|X|}{i} \cdot \left(O(m) + \binom{14 \cdot (|X| - i)}{|X| - i - 1} \cdot O(m)\right)\right),$$

and with Stirling's inequality to evaluate the second binomial coefficient,

$$= O\left(2^k \cdot m + \sum_{i=0}^{k} \binom{|X|}{i} (36.7)^{k+1-i} \cdot m\right) = O((1 + 36.7)^k \cdot m),$$

which gives the lemma's claim with $c = 37.7$. The value of c can be improved by a more careful analysis in Lemma 11.14 to a value close to 10. □

Using Lemma 11.12, we can prove the central result:

Theorem 11.13 FEEDBACK VERTEX SET *can be solved in* $O(c^k \cdot m \cdot n)$ *time for a constant* c.

Proof Given a graph G with vertex set $\{v_1, v_2, \ldots, v_n\}$, we can apply iterative compression to solve FEEDBACK VERTEX SET for G by iteratively considering the graph G_i that is the subgraph of G induced by $\{v_1, \ldots, v_i\}$.

For $i = 1$, the optimal fvs is the empty set. For $i > 1$, assume that an optimal fvs X_i for G_i is known. We can compute an optimal fvs X_{i+1} for G_{i+1} as follows. We consider set $X_i \cup \{v_{i+1}\}$, which is obviously an fvs for G_{i+1}. Using Lemma 11.12, we can in $O(c^k \cdot m)$ time either determine that $X_i \cup \{v_{i+1}\}$ is an optimal fvs for G_{i+1}, or, if not, compute an optimal fvs of size $|X_i|$ for G_{i+1}.

When $i = n$, we will thus have computed an optimal fvs for G in $O(c^k \cdot m \cdot n)$ time. □

It remains to show the size bound of the "candidate vertices set" V' for fixed partition Y and S of a size-$(k + 1)$ fvs X. To this end, we make use of two simple data reduction rules: it is intuitively clear that degree-one and degree-two vertices are of no particular interest when trying to destroy cycles with a minimum number of vertex removals. Hence these rules accomplish getting rid of (most of) them. This plays a significant role in order to prove the subsequent central technical lemma.

Lemma 11.14 *Given a graph* $G = (V, E)$, *a size-*$(k + 1)$ *fvs* X *for* G, *and a partition of* X *into two sets* Y *and* S. *Let* X' *be a size-*k *fvs for* G *with* $X' \cap X = Y$ *and* $X' \cap S = \emptyset$. *In* $O(m)$ *time, we can either decide that no such* X' *exists or compute a subset* V' *of* $V \setminus X$ *with* $|V'| < 14 \cdot |S|$ *such that there exists an* X' *as desired consisting of* $|S| - 1$ *vertices from* V' *and all vertices from* Y.

Proof The idea of the proof is to use a well-known data reduction technique for FEEDBACK VERTEX SET to get rid of degree-one and degree-two vertices from $V \setminus X$ and to show that if the resulting instance is too large compared with the part S (whose vertices we are not allowed to add to X'), then there exists no X' as desired. The details are a little technical.

First, check that S does not induce a cycle; otherwise, no X' with $X' \cap S = \emptyset$ can be an fvs for G. Then, remove in G all vertices from Y as they are determined to be in X'. Finally, use the mentioned standard data reduction technique for FEEDBACK VERTEX SET to get rid of degree-one and degree-two vertices in $V \setminus X$, removing degree-one vertices and bypassing any degree-two vertices by a new edge between its neighbors (thereby removing the bypassed degree-two vertices) with two exceptions: one is that we do not bypass a degree-two vertex in $V \setminus X$ which has two neighbors in S, and the other is the way we deal with parallel edges. If we create two parallel edges between two vertices during the data reduction process—these two edges form a length-two cycle—then exactly one of the two

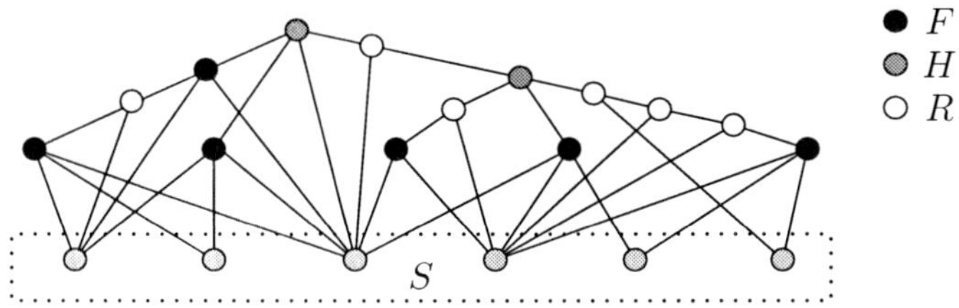

FIG. 11.1. Partition of the vertices in V' into three disjoint subsets F, H, and R. Observe that removing S yields a tree.

endpoints of these edges must be in S since S is an fvs of the subgraph $G[V \setminus Y]$ of G induced by $V \setminus Y$, and the induced subgraph $G[S]$ contains no cycle. Thus, we have to delete the other endpoint and add it to X' since we are not allowed to add vertices from S to X'. Given an appropriate graph data structure, all of the above steps can be accomplished in $O(m)$ time. Proofs for the correctness and the time bound of the data reduction technique are basically straightforward and are omitted here.

In the following, we use $G' = (V' \cup S, E')$ with $V' \subseteq V \setminus X$ to denote the graph resulting after exhaustive application of the data reduction; note that none of the vertices in S has been removed during the data reduction process. In order to prove that $|V'| \leq 14 \cdot |S|$, we partition V' into three subsets, each of which will have a provable size bound linearly depending on $|S|$—the partition is illustrated in Figure 11.1:

$$F := \{v \in V' \mid |N(v) \cap S| \geq 2\},$$
$$H := \{v \in V' \setminus F \mid |N(v) \cap V'| \geq 3\},$$
$$R := V' \setminus (F \cup H).$$

To upper-bound the number of vertices in F, consider the bipartite subgraph $G_F = (F \cup S, E_F)$ of $G' = (V' \cup S, E')$ with $E_F := (F \times S) \cap E'$. Let us first consider conditions for G_F being acyclic. If there are more than $|S| - 1$ vertices in F, then there is a cycle in G_F: if G_F is acyclic then G_F is a forest and, thus, $|E_F| \leq |S| + |F| - 1$. Moreover, the fact that each vertex in F has at least two incident edges in G_F implies $|E_F| \geq 2|F|$, which further implies that $|F| \leq |S| - 1$ if G_F is acyclic. It follows directly that if $|F| \geq 2|S|$, it is impossible to delete at most $|S|$ vertices chosen from F such that $G'[F \cup S]$ is acyclic.

To upper-bound the number of vertices in H, observe that the subgraph $G'[V']$ of G' induced by V' is a forest. Furthermore, all leaves of the trees in $G'[V']$ are from F since G' is reduced with respect to the above data reduction rules. By definition, each vertex in H has at least three vertices in V' as neighbors. Thus there cannot be more vertices in H than there are in F, that is $|H| < 2|S|$.

Finally, consider the vertices in R. By the definitions of F and H and since G' is reduced, each vertex in R has degree two in $G'[V']$ and exactly one neighbor in S. Hence, the subgraph $G'[R]$ of G' induced by R is a forest consisting of paths and isolated vertices. We now separately bound the number of isolated vertices and those vertices participating in paths.

Each of the isolated vertices in $G'[R]$ connects two vertices from $F \cup H$ in $G'[V']$. Since $G'[V']$ is acyclic, the number of isolated vertices in $G'[R]$ cannot exceed $|F \cup H| - 1 < 4|S|$. The total number of vertices participating in paths in $G'[R]$ can be upper-bounded as follows: consider the subgraph $G'[R \cup S]$ of G' induced by $R \cup S$. Each edge in $G'[R]$ creates a path between two vertices from S, that is, if $|E(G'[R])| \geq |S|$, then there exists a cycle in $G'[R \cup S]$. By an analogous argument to the one that upper-bounded the size of F (and considering that removing a vertex from $G'[R]$ destroys at most two edges), the total number of edges in $G'[R]$ may thus not exceed $|S| + 2|S|$, bounding from above the total number of vertices participating in paths in $G'[R]$ by $6|S|$.

Altogether, we have shown that

$$|V'| = |F| + |H| + |R| < 2|S| + 2|S| + 4|S| + 6|S| = 14|S|,$$

directly giving the claim. □

We conclude with two remarks. First, due to refined analysis based on extended data reduction rules, the running time $O(10.6^k \cdot n^3)$ for Feedback Vertex Set could be shown by basically the same approach. Second, a very simple randomized algorithm—using the described data-reduction rules but a direct approach instead of iterative compression—runs in $O(c \cdot 4^k \cdot k \cdot n)$ time with probability of success at least $(1 - (1 - 4^{-k}))^{c \cdot 4^k}$ for an arbitrary constant c (see Section 15.2.2).

11.4 Greedy localization

In the previous section with iterative compression we described a new method to show fixed-parameter tractability for minimization problems. A new approach attacking maximization problems is now considered; it is called greedy localization. As for iterative compression, the parameter relates to the size of the solution set. Again the basic ideas behind greedy localization are very natural and appealing:

- In a first step greedily compute a *maximal* solution of the problem.
- Then either this solution is already good enough, that is, the solution size is at least as big as the given parameter value, or we can make use of this known solution to construct—if it exists—a larger one.

In a way similar to iterative compression, the "greedy solution" provides some initial information that narrows down our search space. In other words, it "localizes" the search efforts to this initial structure.

We illustrate greedy localization using the *NP*-complete problems Set Splitting and 3-Set Packing as examples. Our emphasis lies on presenting the fundamental concepts and ideas as concisely as possible. Thus the derived upper bounds on the running time—in particular, the combinatorial explosion with respect to the parameters—are not the best known ones from the literature. Although similar in spirit, the greedy localization solution for SET SPLITTING seems significantly easier to grasp than that for 3-SET PACKING, which appears to be the harder problem.

11.4.1 *Set Splitting*

The *NP*-complete SET SPLITTING problem is defined as follows.

> **Input:** A collection $\mathcal{C}$ of n subsets of a finite set S and a nonnegative integer k.
> **Task:** Find a partition of S into two disjoint subsets S_1 and S_2 such that there exist at least k subsets in $\mathcal{C}$ with nonempty intersections with both S_1 and S_2.

Without loss of generality we assume in the following that each subset from $\mathcal{C}$ consists of at least two elements from S. Otherwise, the subset can be discarded from consideration because it is "unsplittable". We say that S_1 and S_2 *split* a subset $C \in \mathcal{C}$ if $S_1 \cap C \neq \emptyset$ and $S_2 \cap C \neq \emptyset$. Before describing the actual work of the greedy localization procedure, we consider a simple data reduction step done in a preprocessing phase. The point is the following.

Lemma 11.15 *If there exists a subset $C \in \mathcal{C}$ with $|C| \geq 2k$ then a solution for* SET SPLITTING*—if it exists—can be found by searching a solution for the reduced instance $(\mathcal{C} \setminus \{C\}, k-1)$ and splitting C into two subsets appropriately.*

Proof Clearly, if $\mathcal{C} \setminus \{C\}$ has no solution splitting $k-1$ sets, then there is no solution splitting k sets for $\mathcal{C}$. So we have only to show that every partition of $\mathcal{C} \setminus \{C\}$ splitting $k-1$ sets suffices to construct a solution for $\mathcal{C}$.

Assume that $(\mathcal{C} \setminus \{C\}, k-1)$ has a solution S_1' and S_2' splitting $k-1$ subsets from $\mathcal{C} \setminus \{C\}$. To split a subset C' we need two elements $a, b \in C'$ with $a \in S_1'$ and $b \in S_2'$. Hence, to split $k-1$ subsets we need a set of at most $2 \cdot (k-1)$ elements from S. Let $S' := S_1' \cup S_2'$. We conclude that $|C \setminus S'| \geq 2$. Let $a', b' \in C \setminus S'$. Then $S_1 := S_1' \cup \{a'\}$ and $S_2 := S_2' \cup \{b'\}$ is a solution for the input instance $(\mathcal{C}, k)$. □

Clearly, one preprocessing step as implied by Lemma 11.15 can be done in linear time. Moreover, at most k rounds of such preprocessing need to be applied. From now on we will assume that we have an input instance where each subset in $\mathcal{C}$ has size less than $2k$.

We are ready to outline the two phases that greedy localization performs to solve SET SPLITTING.

Greedy phase. Greedily compute a maximal solution S_1', S_2' that splits a collection $\mathcal{C}''$ of subsets:

1. Set $\mathcal{C}' := \mathcal{C}$; $\mathcal{C}'' := \emptyset$; $S_1' := \emptyset$; $S_2' := \emptyset$.
2. If $\mathcal{C}' = \emptyset$, then go to Step 10.

3. Choose an arbitrary subset C from $\mathcal{C}'$.
4. Let $a \in C \setminus (S_1' \cup S_2')$ and $b \in C$ with $b \neq a$, and without loss of generality $b \notin S_2'$.
5. Set $S_1' := S_1' \cup \{a\}$; $S_2' := S_2' \cup \{b\}$.
6. Let $\mathcal{D} \subseteq \mathcal{C}'$ be the set of subsets from $\mathcal{C}'$ split by S_1' and S_2'.
7. Set $\mathcal{C}' := \mathcal{C}' \setminus (\mathcal{D} \cup \{C \mid C \subseteq S_1' \vee C \subseteq S_2'\})$.
8. Set $\mathcal{C}'' := \mathcal{C}'' \cup \mathcal{D}$.
9. If $\mathcal{C}' \neq \emptyset$ then go to Step 2.
10. If $|\mathcal{C}''| \geq k$ then we have found a desired solution and we stop; otherwise, go to the localization phase.

Localization phase. Here, given the maximal solution S_1' and S_2' from the greedy phase, we use S_1' and S_2' to "localize"—if it exists—a solution S_1 and S_2 splitting at least k subsets from $\mathcal{C}$.

Let us begin with some fundamental observations. Due to the maximality of S_1' and S_2', we know that

$$\forall C \in \mathcal{C} \setminus \mathcal{C}'' : C \subseteq S_1' \cup S_2'. \tag{11.2}$$

Otherwise, we could further extend the collection $\mathcal{C}''$ and the sets S_1' and S_2' in a straightforward way (see greedy phase) contradicting maximality.

Hence, we proceed as follows: try all possible partitions of $S_1' \cup S_2'$ into new disjoint subsets S_1'' and S_2'' with $S_1'' \cup S_2'' = S_1' \cup S_2'$.

1. If one of these partitions splits at least k subsets from $\mathcal{C}$ then we have found a desired solution and we stop.
2. Otherwise, we consider each of these new partition sets S_1'' and S_2'' and proceed as follows. Let k' be the number of subsets from $\mathcal{C}$ split by S_1'' and S_2''. We want to find $k - k'$ further subsets from $\mathcal{C}$ that can be split. According to the inclusion (11.2), all these subsets are subsets of $S_1' \cup S_2'$. Hence the only way to "extend" the partitioning given by S_1'' and S_2'' is to choose elements from $S \setminus (S_1'' \cup S_2'')$ that are contained in the subsets of the "maximal collection" $\mathcal{C}''$ found by the greedy phase.

Since $|\mathcal{C}''| < k$ and due to the preprocessing there are less than $k \cdot 2k = 2k^2$ elements from S in

$$\left(\bigcup_{C \in \mathcal{C}''} C\right) \setminus (S_1'' \cup S_2'').$$

In this way, we have localized our search to this set with size only depending on parameter k. By simply trying all partitions of this set by brute force, we arrive at $2^{2k^2} = 4^{k^2}$ possibilities. This must be multiplied by the upper bound of $2^{2k} = 4^k$ for all partitions of $S_1' \cup S_2'$, yielding a combinatorial explosion bounded from above by $4^k \cdot 4^{k^2} = 4^{k^2+k}$. Thus we have shown the following result.

Theorem 11.16 SET SPLITTING *is fixed-parameter tractable with respect to parameter k.*

By using problem kernelization by means of crown reduction rules and further refining the above greedy localization method, one can achieve a combinatorial explosion upper-bounded by only 8^k instead of 4^{k^2+k}.

11.4.2 *Set Packing*

The SET PACKING problem is defined as follows.

> **Input:** A collection $\mathcal{C}$ of n subsets of a finite set S and a nonnegative integer k.
> **Task:** Find a subcollection $\mathcal{C}' \subseteq \mathcal{C}$ which consists of at least k mutually disjoint subsets.

By way of contrast to the fixed-parameter tractable SET SPLITTING, SET PACKING is $W[1]$-complete with respect to parameter k. Hence, as for HITTING SET (see Sections 7.5 and 8.4), we focus our attention on restricted versions of SET PACKING where all given subsets have bounded size. In the remainder of this section we focus on the still *NP*-complete special case 3-SET PACKING where subset sizes are bounded by three in order to illustrate the greedy localization technique. Note that 2-SET PACKING is solvable in polynomial time by matching techniques.

Let us sketch the basic ideas behind the fixed-parameter algorithm for 3-SET PACKING using greedy localization. As indicated before, the algorithm begins by greedily computing a *maximal* subcollection $\mathcal{C}'_0$: it does so by starting with $\mathcal{C}'_0$ as the empty set and considering each set in $\mathcal{C}$ in some arbitrary order for addition to $\mathcal{C}'_0$ if it does not intersect with any set already in $\mathcal{C}'_0$. Using a suitable data representation, this phase takes linear time. Now, if $|\mathcal{C}'_0| \geq k$ we are clearly done. Hence, from now on assume that $|\mathcal{C}'_0| < k$. In addition, if $|\mathcal{C}'_0| < k/3$, then the given 3-SET PACKING instance has no solution: with little effort this can be shown by making use of the fact that, due to the maximality of $\mathcal{C}'_0$, for every set S_1—if it exists—in a solution $\mathcal{C}'$ with $|\mathcal{C}'| \geq k$ there has to exist a set S_2 in $\mathcal{C}'_0$ with $S_1 \cap S_2 \neq \emptyset$. From this one can conclude that there must be at least k different elements from S occurring in the sets from $\mathcal{C}'_0$, implying the claim because each subset of the collection contains only three elements.

Without loss of generality, we now can assume the following. Let

$$\mathcal{C}'_0 = \{\{a_1, a_2, a_3\}, \{a_4, a_5, a_6\}, \ldots, \{a_{3j-2}, a_{3j-1}, a_{3j}\}\}$$

with $j < k$ be the greedy solution, with all subsets pairwisely disjoint. If there exists a solution $\mathcal{C}'$ with $|\mathcal{C}'| = k$, then every 3-element set in $\mathcal{C}'$ contains at least one element from $A := \{a_1, a_2, \ldots, a_{3j}\}$. In order to determine for each set in $\mathcal{C}'$ one of its corresponding elements from A, we thus only need to consider $\binom{3j}{k} < \binom{3k}{k}$ possibilities. In this way, for each possibility we obtain a "partial collection"

$$\mathcal{C}^* = \{\{a_{i_1}, *, *\}, \{a_{i_2}, *, *\}, \ldots, \{a_{i_k}, *, *\}\},$$

where $a_{i_1}, a_{i_2}, \ldots, a_{i_k} \in A$ are pairwisely different and "$*$" represents an undetermined element from S. Then the question is whether this partial collection can be completed into a valid solution

$$\mathcal{C}' = \{\{a_{i_1}, b_1, c_1\}, \{a_{i_2}, b_2, c_2\}, \ldots, \{a_{i_k}, b_k, c_k\}\}.$$

A naive way to answer this would be to try all candidates for $b_1, c_1, \ldots, b_k, c_k$, giving $\binom{|S|}{2k}$ possibilities which is clearly not a function solely depending on parameter k. Hence the goal is to restrict the size of the candidate set from which we must choose $b_1, c_1, \ldots, b_k, c_k$. A simple observation is to check for every a_{i_ℓ}, $1 \leq \ell \leq k$, whether the partial set $\{a_{i_\ell}, *, *\}$ has exactly one extension $\{a_{i_\ell}, b_\ell, c_\ell\}$ in $\mathcal{C}$; that is, there is exactly one 3-element set in $\mathcal{C}$ that contains a_{i_ℓ}. Obviously, if there is a solution with respect to $a_{i_1}, a_{i_2}, \ldots a_{i_k}$, all corresponding uniquely determined 3-element sets have to be chosen—if two of them intersect, there is no such solution. Hence, without loss of generality, from now on we may assume that there is no partial set with a uniquely determined extension.

Again, we apply a greedy algorithm as follows.

1. Let $\mathcal{C}^{**} := \mathcal{C}^*$.
2. For each 3-element set $\{d_1, d_2, d_3\}$ from $\mathcal{C}$ check whether there is exactly one partial 3-element set π from $\mathcal{C}^{**}$ which can be extended to $\{d_1, d_2, d_3\}$, and $\{d_1, d_2, d_3\}$ has an empty intersection with any other set from $\mathcal{C}^{**}$. If so, let
$$\mathcal{C}^{**} := (\mathcal{C}^{**} \setminus \{\pi\}) \cup \{\{d_1, d_2, d_3\}\}.$$
3. Let $\mathcal{C}_0^{**}$ denote the set of 3-element sets from $\mathcal{C}$ contained in $\mathcal{C}^{**}$, that is, all sets added in the second step.

Clearly, if $|\mathcal{C}_0^{**}| = k$ we are done. Otherwise, let $h := |\mathcal{C}_0^{**}|$ and, without loss of generality, assume that

$$\mathcal{C}_0^{**} = \{\{a_{i_1}, b_1, c_1\}, \{a_{i_2}, b_2, c_2\}, \ldots, \{a_{i_h}, b_h, c_h\}\}.$$

If $\mathcal{C}^*$ can be completed into an overall solution $\mathcal{C}'$ with $|\mathcal{C}'| = k$, then the following must hold: consider the 3-element set $\{a_{i_{h+1}}, b_{h+1}, c_{h+1}\}$ from $\mathcal{C}'$. Since all sets in $\mathcal{C}'$ have to be disjoint, $\{a_{i_{h+1}}, b_{h+1}, c_{h+1}\}$ has an empty intersection with every set in $\mathcal{C}^*$ except for the one containing $a_{i_{h+1}}$. In step 2 of the above algorithm no partial 3-element set π has been replaced by $\{a_{i_{h+1}}, b_{h+1}, c_{h+1}\}$. As a consequence, when $\{a_{i_{h+1}}, b_{h+1}, c_{h+1}\}$ was considered in the second step, there must have been a non-empty intersection of $\{a_{i_{h+1}}, b_{h+1}, c_{h+1}\}$ with one of the sets from $\mathcal{C}_0^{**}$. More specifically, one of b_{h+1} and c_{h+1} must already occur in one of the sets from $\mathcal{C}_0^{**}$. Moreover, b_{h+1} and c_{h+1} cannot be any of $a_{i_1}, a_{i_2}, \ldots, a_{i_h}$; otherwise, $a_{i_{h+1}}$ would occur together with an a_{i_ℓ}, $1 \leq \ell \leq h$, in a 3-element set, contradicting that $\mathcal{C}^*$ is a partial collection that can be completed into an overall solution $\mathcal{C}'$. This is the decisive point for concluding fixed-parameter tractability because now we have narrowed down the candidate set of elements to

$$\{b_1, c_1, b_2, c_2, \ldots, b_h, c_h\}.$$

The size of this set is smaller than $2k$. By trying all possibilities, that is, branching into less than $2k$ cases in a search tree manner, we will find the right element if it exists and determine the value of one more so far undetermined element "$*$". Thus, with one application of the above algorithm we can determine the at most $2k$ cases to branch into. Here, we start with $\mathcal{C}^*$ where, step by step, more and more undetermined values get resolved and the next round of the algorithm then starts with $\mathcal{C}^{**}$ being set to this new partial instance with one more resolved element. After $2k$ rounds, starting with $\mathcal{C}^{**} = \mathcal{C}^*$, if a solution with k pairwisely disjoint subsets exists, all elements will be resolved in at least one of the search paths of the corresponding search tree.

Summarizing, we start with less than $\binom{3k}{k} = O((3k)^k)$ initial partial solutions and for each of them we have a procedure that constructs a search tree of depth upper-bounded by $2k$ which branches into less than $2k$ children at each inner node. Altogether, the combinatorial explosion in parameter k can be bounded from above by $O((3k)^k \cdot (2k)^{2k})$. Obviously, all other computations can be done in time polynomial in the input size, giving only a polynomial factor for the running time. Hence we have shown:

Theorem 11.17 3-SET PACKING *is fixed-parameter tractable with respect to the parameter k.*

A much more refined analysis can actually show that 3-SET PACKING can be solved by greedy localization in $O((5.7k)^k \cdot n)$ time, see the literature cited in the bibliographical remarks. Also one can find there the generalization to m-SET PACKING for $m > 3$. Meanwhile, using a sophisticated combination of problem kernelization, color coding, and dynamic programming this bound could be improved to $O(2^{O(k)} + n)$. Again, however, one first finds a maximal solution that here helps to bound the problem kernel size. Thus in a sense greedy localization is used for bounding the problem kernel as a tool in the color-coding dynamic programming. In the example given above, the problem was solved directly without searching for a problem kernel.

As a final remark, note that it might be less frequent to have small parameter values for maximization problems than it is for minimization problems with respect to realistic input instances; this might hamper the practical use of greedy localization.

11.5 Graph minor theory

Graph minor theory as developed by Neil Robertson and Paul D. Seymour is considered to be the deepest achievement of modern graph theory. The corresponding results and proofs are far beyond the scope of this book. The importance of graph minor theory for fixed-parameter tractability investigations is of purely theoretical flavor—probably no other technique of this book is so far from practical algorithmics as what we very briefly discuss now. Nevertheless graph minor theory and the long series of corresponding papers has brought about a tremendous amount of important findings of algorithmically relevant concepts.

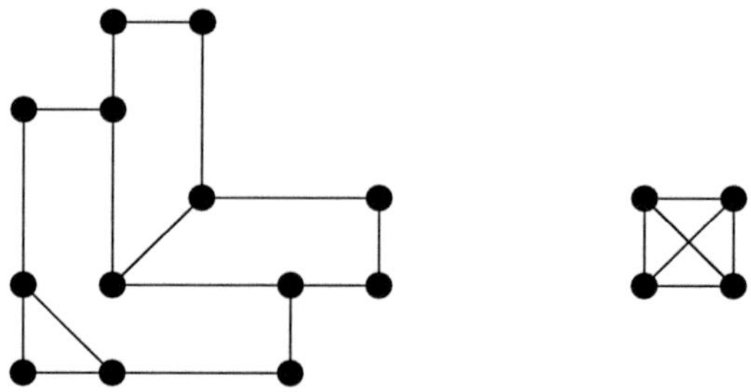

FIG. 11.2. A graph (left) and a minor of it.

Let us only mention here the notion of tree decompositions of graphs that grew out of this theory as a by-product. In what follows, no proofs will appear and we only roughly indicate what graph minor theory may be used for in our field of interest.

Recall from Section 2.3 that planar graphs can be characterized by forbidden subdivisions. Using the concept of minors, we will reformulate this famous result of Kasimir Kuratowski.

Definition 11.18 1. *A graph H is a* minor *of a graph G iff H can be obtained from G by a finite sequence of edge deletions and edge contractions. We then write $H \leq_m G$. Herein, contracting an edge $e = \{v, w\}$ means to replace vertices v and w by one new vertex u which is connected by an edge to all vertices from $(N(v) \cup N(w)) \setminus \{v, w\}$.*

2. *A family $\mathcal{F}$ of graphs is* closed under taking minors *if for graphs $G, H \in \mathcal{F}$ we have*

$$((G \in \mathcal{F}) \wedge (H \leq_m G)) \Rightarrow (H \in \mathcal{F}).$$

Figure 11.2 shows a graph and one of its minors. The family of planar graphs is clearly closed under taking minors. Moreover, it can be characterized by Kuratowski's result saying that a graph is planar iff it does not contain one of K_5 and $K_{3,3}$ as a minor. Here K_5 is the complete graph with five vertices and $K_{3,3}$ is the complete bipartite graph with three vertices on each side. In a sense, the results of graph minor theory we focus on here strongly generalize Kuratowski's characterization of planar graphs. The following two theorems are of core interest to us. The first one is known as *graph minor theorem.*

Theorem 11.19 *Let $\mathcal{F}$ be a family of finite graphs closed under taking minors. Then there exists a finite "obstruction set" of graphs $O_{\mathcal{F}} = \{H_1, H_2, \ldots, H_t\}$ such that*

$$(G \notin \mathcal{F}) \equiv (\exists 1 \leq i \leq t : H_i \leq_m G).$$

In other words, Theorem 11.19, formerly known as Wagner's conjecture, is equivalent to stating that every minor-closed class of graphs has a finite number of "minor-minimal" elements. Note that in Kuratowski's characterization of planar graphs we have $t = 2$ and $H_1 = K_5$ and $H_2 = K_{3,3}$. The minor-closed class of forests has the obstruction set $\{K_3\}$. It is important to observe that Theorem 11.19 is not constructive in the sense that, other than in the case of planar graphs, it is generally not known how to determine the obstruction set for a given graph family. This is a severe drawback from an algorithmic point of view. Under the assumption of having an obstruction set given, we still need a so-called *minor test* in order to draw algorithmic conclusions with respect to fixed-parameter tractability.

Theorem 11.20 *Given an n-vertex graph G and a fixed k-vertex graph H, we can decide in $f(k) \cdot n^3$ time whether $H \leq_m G$. Here f is a function depending only on k but not on f.*

Function f in Theorem 11.20 grows enormously quickly, however, so the result is only useful for classification purposes. Theorem 11.20 implies a cubic-time test for the planarity of graphs—there are linear-time algorithms, though. Let us see how graph minor theory applies to derive the fixed-parameter tractability of VERTEX COVER.

For fixed k, the family $\mathcal{F}$ of graphs with a vertex cover of size at most k is closed under taking minors. According to Theorem 11.19 there must exist a finite obstruction set $O_{\mathcal{F}}$. Using Theorem 11.20, for a given graph G we can check whether there is an $H \in O_{\mathcal{F}}$ with $H \leq_m G$. If no such H exists, we know that G has a vertex cover of size at most k. Since one can also show that all graphs in $O_{\mathcal{F}}$ have size depending only on k, we can conclude that this takes $f(k) \cdot n^3$ time for some function f depending only on k.

By way of contrast, the family of graphs with a dominating set of size at most k is not closed under taking minors, in this sense signalling the fixed-parameter intractability of DOMINATING SET.

In conclusion, the result for VERTEX COVER clearly cannot compete with the trivial depth-bounded search tree algorithm for VERTEX COVER running in $O(2^k \cdot n)$ time. But graph minor theory provides such a general and powerful classification tool that this was not to be expected.

11.6 Summary and concluding remarks

Although the techniques in this chapter have been termed advanced, this does not mean that they are harder to apply than the methods presented in previous chapters. By way of contrast, perhaps, these techniques particularly offer many opportunities for extensions, improvements, and new applications. A fruitful research ground is prepared in this chapter.

- Color-coding—randomized or derandomized—is a very natural way to confine the combinatorial complexity in (graph) search problems. In a way it narrows down the search space by identifying different objects.
- Integer linear programs are more of a classification tool because of the huge associated combinatorial explosions in the parameter "number of variables". Still, as for CLOSEST STRING, this may be a first and perhaps "only" way to decide on fixed-parameter tractability.

- Iterative compression is a very new, promising technique to cope with hard minimization problems. It seems useful for a fresh attack on unsettled fixed-parameter complexity questions. A prominent example of this is the newly achieved fixed-parameter tractability of GRAPH BIPARTIZATION. Moreover, there are concrete indications that iterative compression may lead to practically useful algorithms.
- Greedy localization appears as the counterpart of iterative compression, tailored for maximization problems. Here, however, the ultimate use for practical applications is less clear.
- Graph minor theory again is more or less beyond the scope of this book due, on the one hand, to its great technical demands and difficulties, and on the other to the huge constants that are involved. It seems to be an attractive candidate for classification purposes only.

It is difficult to decide now what must be considered and what will develop into a methodology on its own. Hence the example applications presented here and in previous chapters must be seen as a personally biased snapshot of a dynamic field of research. It is possible that perhaps five or ten years from the writing of this book the picture will be drawn significantly differently. No doubt other authors might already do so now. Still, however, there is some confidence that at least several of the core ingredients of fixed-parameter complexity have been touched on—and they will survive.

11.7 Exercises

1. Describe a simple randomized algorithm for FEEDBACK VERTEX SET—using the described data reduction rules as presented in Section 11.3.2—which runs in $O(4^k \cdot kn)$ time with probability of success at least $(1-(1-4^{-k}))^{c \cdot 4^k}$ for an arbitrary constant c.
2. Provide direct combinatorial algorithms solving CLOSEST STRING for $k=2$ (easy) and $k=3$ (hard).
3. Show how the following problem can be solved using color-coding:
   Input: A graph G and a nonnegative integer k.
   Task: Find out whether there are at least k vertex-disjoint induced stars with one center vertex and five neighbor vertices in G. (Note that the neighbor vertices must form an independent set.)
4. The following is a network configuration problem. Consider a cycle of n vertices. On this cycle are "stacked" multiple rings, each of which has capacity c. On these rings we realize "communication channels" subject to the following conditions.
   - A communication channel between two vertices on the cycle must always use the same ring.
   - Through each edge of each ring there may be at most c communication channels.

- If a vertex of the cycle is the endpoint of a communication channel on a ring, then it must use a so-called ADM (add/drop multiplexer) for this vertex/ring pair.

We end up with the following *NP*-complete problem, RING GROOMING.
Input: Positive integers n (cycle size), c (ring capacity), k, and a list $\{j_1, k_1\}, \{j_2, k_2\}, \ldots, \{j_s, k_s\}$ of unordered pairs from $\{1, 2, \ldots, n\}$ representing communication demands. Note that communication demands between the same endpoints may occur multiple times, but the two endpoints must be different.
Task: Check whether all communication demands can be fulfilled by keeping all capacity conditions and using at most k ADMs.
Show that RING GROOMING is fixed-parameter tractable with respect to parameter k. To this end, design an integer linear program for RING GROOMING with a number of variables that is a function only of k.

5. The *NP*-complete GRAPH BIPARTIZATION problem is defined as follows.
Input: A graph G and a nonnegative integer k.
Task: Find a set of at most k vertices whose deletion makes the graph bipartite.
Using iterative compression, show that GRAPH BIPARTIZATION is fixed-parameter tractable with respect to parameter k. Note: this is a hard exercise.
6. Given that GRAPH BIPARTIZATION is fixed-parameter tractable, show that so-called EDGE BIPARTIZATION is fixed-parameter tractable as well.
Input: A graph G and a nonnegative integer k.
Task: Find a set of at most k edges whose deletion makes the graph bipartite.
7. Use greedy localization to find at least k vertex-disjoint triangles in an arbitrary graph.

11.8 Bibliographical remarks

The color-coding technique is due to Alon *et al.* (1995). A recent application in computational biology appears in Scott *et al.* (2005).

The integer linear program with a bounded number of variables that solves CLOSEST STRING appears in Gramm *et al.* (2003*b*). The central result concerning the feasibility of these programs is due to Lenstra (1983); see also Kannan (1987) and Lagarias (1995).

The iterative compression technique was introduced in Reed *et al.* (2004). For more details and two speed-up methods for iterative compression refer to (Guo, 2005). See Hüffner (2005) for a simplified presentation and encouraging empirical tests with implementing this technique for GRAPH BIPARTIZATION. Further applications appear in Dehne *et al.* (2004), Dehne *et al.* (2005), Marx (2004*a*), and Guo *et al.* (2005). The VERTEX COVER example is due to Jiong Guo and Sebastian Wernicke [personal communication] and the FEEDBACK VERTEX SET example closely follows Guo *et al.* (2005); basically the same algorithm with better analysis was independently achieved in Dehne *et al.* (2005). The randomized algorithm for FEEDBACK VERTEX SET appears in Becker *et al.* (2000).

Greedy localization was introduced in Jia *et al.* (2004) and the presentation here follows this source. The SET SPLITTING application is from Dehne *et al.* (2004). Further applications appear in Dehne *et al.* (2004), Fellows *et al.* (2004*a*), and Prieto and Sloper (2004).

For more information about the graph minor theory due to Robertson and Seymour we refer to Diestel (2005) and Downey and Fellows (1999) as starting points.

12

SUMMARY AND CONCLUDING REMARKS

Part II is the heart of this book. If there is nothing to be learnt here, the other two parts will not be beneficial.

The classification into five methodological chapters as presented here is debatable. One must be aware of the fact that there will be omissions to complain about and focal points that could or should have been chosen differently. Fixed-parameter algorithmics is probably not mature enough, or, to express it more positively, is still developing too dynamically, to decide clearly what the essentials finally will turn out to be. The arsenal of algorithmic methods presented, however, might be sufficient to convince the reader that this is a versatile and prospective field of research.

Among others, particular topics where the last word cannot be pronounced on the current stage of research are iterative compression (Section 11.3) and greedy localization (Section 11.4). Also the role that various graph parameters related to treewidth (Section 10.7) will eventually play seems wide open. By way of contrast, concepts such as reduction to a problem kernel (Chapter 7), depth-bounded search trees (Chapter 8), and dynamic programming (Chapter 9) are solid parts of the "establishment" in fixed-parameter algorithmics. In addition, a technique such as color-coding (Section 11.1) has profitable applications in many settings and is a clear candidate to survive as a fundamental contribution. At this still early stage of development, however, it is open whether everything from this central part of the book will last for a longer time or whether better replacements will eventually be found.

Besides the call for new methods and techniques that arises particularly from this part of the book—recall that an attractive technique such as iterative compression was invented during the writing of this book—there is also a call for further broadening and deepening of "known" methods. For instance, can mechanized analysis (Section 8.8) replace many of the "hand-made" search tree strategies based on extensive and error-prone case distinctions? Is such a form of automation also possible in the context of finding data reduction rules?

We have tried to exhibit many aspects of fixed-parameter algorithmics by using Vertex Cover as the main running example. One may wonder whether one single problem such as Vertex Cover and its variants can be used as a source to base a whole introduction to fixed-parameter algorithms upon it. Discovering structure using only a few fundamental concepts and problem variations is rewarding and would help to achieve a better understanding and more popular presentation of the field.

Having presented several nice algorithmic results from a mainly theoretical point of view, it remains a task for future work to better evaluate and understand their practical benefits and potentials. Algorithm engineering for fixed-parameter algorithms is still at its very beginning.

Part III

Some Theory, Some Case Studies

One reason why parameterized complexity analysis has still not reached a really broad degree of attention in the algorithms and complexity community may lie in the fact that the corresponding complexity theory seems to be somehow unwieldy. As a matter of fact, there are more "parameters" that one has to take into account for developing the theory, making it more technical than classical computational complexity theory. In the first chapter of this third and last part of the book, we try to condense the essential highlights of parameterized complexity theory in around 30 pages. What we provide here are the absolute basics that are needed and that may also be sufficient to appreciate what parameterized complexity theory can perform. It is insufficient to fully explore the richness of the field.

There are two main objectives of our presentation in Chapter 13:

- to provide enough basic working knowledge for a designer of fixed-parameter algorithms to be in a position to see the peculiarities of parameterized intractability results, maybe even deriving such results oneself; and
- to put the designer on a firm footing such that, with the knowledge presented in Chapter 13, it is easier to appreciate, understand, and further explore the world of structural parameterized complexity theory.

Chapter 13 is not a replacement for other, deeper presentations as provided in the cited literature. It is a starting point only.

The second chapter of this part deals with the still neglected relationship between approximation algorithms and fixed-parameter algorithms. Apparently, the interaction between both fields can and should be further strengthened; Chapter 14 is very brief still. A fair competition and interchange between these two main "theory-based" attacks against computational intractability is needed, clearly stating the pros and cons of each approach. We try to initiate a corresponding agenda here.

Finally, the third and last chapter exhibits a selection of more than twenty case studies dealing with various computational problems studied from a parameterized point of view. Also, reverting to the material that has been presented throughout the book, the objective is to show a colorful spectrum of facets of fixed-parameter algorithmics, ranging from computational biology over graph problems to problems from logic and many other fields. To this end, in a concise style mostly newer results are discussed. To inspire future research activities is the ultimate goal here.

13

PARAMETERIZED COMPLEXITY THEORY

After the crash course in Chapter 3 we now delve a little deeper into parameterized complexity theory. So far in this book, we have mainly been concerned with algorithmic methods to show fixed-parameter tractability of various problems. Unfortunately, not all parameterized problems of interest turn out to be fixed-parameter tractable. By way of contrast, there are natural and important problems such as CLIQUE and DOMINATING SET—both parameterized by the size of the solution set—which have resisted all attempts to confine the combinatorial explosion to the parameter. As is to be expected when studying the computational complexity of problems, we are not able to prove fixed-parameter *intractability* unconditionally. A way out of this misery is at least to show that fixed-parameter tractability for some problems could only be achieved if a long series of other hard parameterized problems are fixed-parameter tractable as well. Thus we obtain what are known as "relative lower bounds" which work analogously to *NP*-hardness theory. This is the topic of this chapter.

Where is the borderline—with respect to the current state of knowledge—between fixed-parameter tractability and intractability? Let us gain some intuition by coming back to the drosophila of computational complexity theory—the satisfiability of Boolean formulae we introduced in Section 1.1. Recall the definition of CNF-SATISFIABILITY:

Input: A Boolean formula F in conjunctive normal form.
Question: Is there a satisfying truth assignment for F?

Bounding the maximum clause size by two makes this decision problem polynomial-time solvable, whereas bounding it by three leaves CNF-SATISFIABILITY *NP*-complete. Now let us parameterize this problem by weight. Define the *weight of a truth assignment* as the number of variables that are set true. Then the WEIGHTED CNF-SATISFIABILITY problem is the following:

Input: A Boolean formula F in conjunctive normal form and a nonnegative integer k.
Question: Is there a satisfying truth assignment for F which has weight *exactly* k?

For ease of presentation, in the following we concentrate on WEIGHTED 2-CNF-SATISFIABILITY where the maximum clause size is upper-bounded by two. Surprisingly, WEIGHTED 2-CNF-SATISFIABILITY is not known to be fixed-parameter tractable with respect to parameter k. Indeed, showing that WEIGHTED 2-CNF-SATISFIABILITY is fixed-parameter tractable would imply an unlikely collapse of parameterized complexity classes similarly, as would showing that 3-CNF-SATISFIABILITY is solvable in polynomial time in classical complexity theory. In this sense, the essence of parameterized intractability theory can be exhibited and experienced by studying WEIGHTED 2-CNF-SATISFIABILITY. The border of fixed-parameter tractability is a very sharp one here. As discussed in Section 1.1, the seemingly harmless relaxation of WEIGHTED 2-CNF-SATISFIABILITY where one asks for a satisfying truth assignment of weight *at most* k instead of a weight exactly k or at least k is easily shown to be fixed-parameter tractable.

In what follows, we explore only a few of the facets of parameterized complexity classes. Since this is a book mainly about algorithms, we adopt a fairly pragmatic point of view, omitting several proofs and some insights more related to a structural complexity theory investigation. As a rule of thumb, one may say that parameterized complexity theory—in order to reflect (subtle) parameterization aspects—has to follow a more fine-grained and perhaps more technical methodology than classical complexity theory does.

13.1 Basic definitions and concepts

Historically, VERTEX COVER, INDEPENDENT SET, CLIQUE, and DOMINATING SET were, and to some extent still are, the driving forces for developing the theory of parameterized complexity. The investigations started by trying to better understand and classify what makes VERTEX COVER so "easy" with respect to fixed-parameter algorithms whereas the other problems appear to be so "hard". So far in this book, we have chosen a relatively informal approach when introducing parameterized problems. In order to give a clean and simple presentation of complexity-theoretic considerations, we now become somewhat more strict. First, we only consider *decision problems*. Second, the parameter must be a function of the "basic input". This leads to the following refinement of Definition 3.1 from Section 3.1.

Definition 13.1 *A* parameterized problem *L over an alphabet Σ is a set of pairs (x, k) with $x \in \Sigma^*$ and k a nonnegative integer such that there is no x with $(x, k) \in L$ and $(x, k') \in L$ for some $k' \neq k$.*

In particular, Definition 13.1 describes problems where we search for solutions with parameter value exactly k. As we have seen for WEIGHTED 2-CNF-SATISFIABILITY, the distinction between "at most" and "exactly" can be of decisive importance. Let us reformulate our favorite problems in order to fulfill Definition 13.1. The minimization problem VERTEX COVER then turns into the following.

Input: A graph $G = (V, E)$ and a nonnegative integer k.
Question: Is there a $C \subseteq V$ with exactly k vertices such that each edge in E has at least one of its endpoints in C?

Analogously, INDEPENDENT SET then is defined in the following way.

Input: A graph $G = (V, E)$ and a nonnegative integer k.
Question: Is there an $I \subseteq V$ with exactly k vertices that form an independent set, that is, I induces an edgeless subgraph in G?

The dual problem to INDEPENDENT SET, that is, CLIQUE, is then defined as follows.

Input: A graph $G = (V, E)$ and a nonnegative integer k.
Question: Is there a subset of vertices $C \subseteq V$ with exactly k vertices such that C forms a clique, that is, C induces a complete subgraph of G?

Finally, the "close relative" of VERTEX COVER, that is, DOMINATING SET, where one has to cover vertices instead of edges, is then defined in the following way.

Input: A graph $G = (V, E)$ and a nonnegative integer k.
Question: Is there a $S \subseteq V$ with exactly k vertices such that each vertex in V is contained in S or has at least one neighbor in S?

Note that in the case of these four problems, other than for WEIGHTED 2-CNF-SATISFIABILITY, the distinction between "at least", "at most", and "exactly" is not decisive for our considerations. The reason is that, for instance, in the case of the minimization problems VERTEX COVER and DOMINATING SET, a solution of size $k' < k$ can be turned into a size-k solution by simply adding $k - k'$ arbitrarily chosen further vertices. Observe that, by way of contrast, it is not clear how to transform a weight-k' satisfying assignment with $k' < k$ into a weight-k satisfying assignment.

We will compare the four above problems by means of reducibility among them. Before giving a formal definition of (parameterized) reductions, let us first provide some intuition. To this end, let $G = (V, E)$ be an n-vertex graph. We have the following simple observations.

- G has a size-k vertex cover iff it has a size-$(n - k)$ independent set.
- G has a size-k independent set iff its complement graph $\overline{G}$ has a size-k clique.
- G has a size-k vertex cover iff the following graph G' has a size-k dominating set: $G' := (V \cup V', E \cup E')$ where $V' := \{v_e \mid e \in E\}$, $V' \cap V = \emptyset$, and E' consists of all edges connecting any vertex u that is an endpoint of edge e with v_e.

13.1.1 *Parameterized reducibility*

What can we learn from the above observations? In a nutshell, we learn that VERTEX COVER on the one side and INDEPENDENT SET on the other side are dual problems of each other, INDEPENDENT SET and CLIQUE are "basically the same" problems, and DOMINATING SET is "at least as hard" as VERTEX COVER. We will develop a better picture of these mutual relationships once we have become acquainted with an appropriate concept of reducibility between combinatorial problems. Let us start with many–one reductions as known from classical complexity theory.

Definition 13.2 *We call a function* $g : \Sigma^* \longrightarrow \Sigma^*$ *a polynomial-time* many–one reduction *from a problem* $L_1 \subseteq \Sigma^*$ *to a problem* $L_2 \subseteq \Sigma^*$ *if*

1. $x \in L_1$ *iff* $g(x) \in L_2$ *and*
2. $g(x)$ *is computable in* $|x|^{O(1)}$ *time.*

As an example, consider the reduction of CNF-SATISFIABILITY to 3-CNF-SATISFIABILITY. Here the central idea is as follows. Replace a clause

$$(\ell_1 \vee \ell_2 \vee \ldots \vee \ell_m)$$

by the expression

$$(\ell_1 \vee \ell_2 \vee z_1) \wedge (\overline{z_1} \vee \ell_3 \vee z_2) \wedge \ldots \wedge (\overline{z_{m-3}} \vee \ell_{m-1} \vee \ell_m),$$

where z_1, z_2, ..., z_{m-3} denote newly introduced variables. It is not hard to verify that an original CNF-formula is satisfiable iff the 3-CNF formula constructed by replacing all its size-at-least-four clauses in the above way is satisfiable. Clearly, the reduction is computable in polynomial time, thus also fulfilling the second condition of Definition 13.2. Note that a core observation concerning the use of reductions for the sake of proving relative lower bounds is that if a problem L_1 is reducible to a problem L_2 and L_1 is not solvable in polynomial time, then L_2 is also not solvable in polynomial time.

Would this classical reduction as given in Definition 13.2 suffice to show that if a parameterized problem L_1 is reducible to a parameterized problem L_2 and L_1 is not fixed-parameter tractable, then L_2 is not as well? The answer is no. The problem is that the interrelationship between the respective parameters cannot be modelled by using standard polynomial-time many–one reductions. For instance, using the previous observation, we can trivially reduce INDEPENDENT SET to VERTEX COVER in a many–one fashion—but VERTEX COVER is fixed-parameter tractable and INDEPENDENT SET (probably) is not. Hence, we need a more fine-grained reducibility concept to better characterize the relationship between parameterized problems.

Definition 13.3 *Let* $L, L' \subseteq \Sigma^* \times \mathbb{N}$ *be two parameterized problems. We say that* L reduces to L' by a standard parameterized (many–one-)reduction *if there are functions* $k \mapsto k'$ *and* $k \mapsto k''$ *from* $\mathbb{N}$ *to* $\mathbb{N}$ *and a function* $(x, k) \mapsto x'$ *from* $\Sigma^* \times \mathbb{N}$ *to* Σ^* *such that*

1. $(x, k) \mapsto x'$ *is computable in* $k'' \cdot |(x, k)|^c$ *time for some constant* c *and*
2. $(x, k) \in L$ *iff* $(x', k') \in L'$.

Clearly, the relationship between INDEPENDENT SET and CLIQUE described above yields parameterized reductions in both directions, even computable in polynomial time. The indicated reduction from VERTEX COVER to INDEPENDENT SET, however, does not yield a parameterized reduction: whereas VERTEX

COVER has parameter value k, INDEPENDENT SET receives parameter value $n-k$, a value that does *not exclusively* depend on k but also on the number n of vertices in the input graph.

Finally, let us look back at the reduction from CNF-SATISFIABILITY to 3-CNF-SATISFIABILITY. Could this be extended into a parameterized reduction from WEIGHTED CNF-SATISFIABILITY to WEIGHTED 3-CNF-SATISFIABILITY? This does not seem to be the case. The point is as follows. Assume that the CNF formula has a weight-k satisfying truth assignment, making exactly one literal ℓ_j in clause $(\ell_1 \vee \ell_2 \vee \ldots \vee \ell_m)$ true. To satisfy the corresponding 3-CNF formula associated with $(\ell_1 \vee \ell_2 \vee \ldots \vee \ell_m)$, however, one has has to set all variables z_1, z_2, ..., z_{j-2} to true. Thus, the weight of a satisfying truth assignment for the constructed 3-CNF formula does not depend exclusively on the original parameter k. In other words, an explanation for the difference is that, although the CNF and the 3-CNF formulae are equivalent with respect to satisfiability, they are not "logically equivalent". Indeed, although both WEIGHTED 3-CNF SATISFIABILITY and WEIGHTED CNF-SATISFIABILITY appear to be fixed-parameter *intractable*, they seem to belong to different structural degrees of parameterized intractability. As we shall see later on, these degrees of intractability are directly tied to the "logical depth" required to express these problems as weighted satisfiability problems. In its very simple form here, 3-CNF formulae have logical depth one in the sense that we only have one big (that is, unbounded fan-in) OR-gate whereas CNF formulae have logical depth two because we basically have a big AND of big ORs.

13.1.2 *Parameterized complexity classes*

The enchantment of *NP*-completeness theory is that thousands of combinatorial problems can be interrelated through polynomial-time many–one reducibility. If any *NP*-complete problem is solvable in polynomial time, then all of them are. This is a basic reason for believing that none of them is. From a parameterized point of view, we want to follow an analogous agenda: try to interrelate as many parameterized, seemingly fixed-parameter intractable problems as possible using parameterized reductions. In this context, note that parameterized reductions as well as classical ones are *transitive*, that is, if L_1 reduces to L_2 and L_2 reduces to L_3, then L_1 also reduces to L_3. There are two problems with parameterized complexity theory and the definition of adequate complexity classes:

- Since parameterized reductions are more fine-grained than classical ones, it seems to be unavoidable to split the world of intractability into more classes (although the related worst-case upper bounds on the running times may not differ significantly between these classes).
- The machine characterization is not as nice for parameterized complexity classes as we have for *NP* with nondeterministic Turing machines running in polynomial time.

Losing some elegance because the parameterized world seems to be more complicated, we accept the first problem. We ignore the second problem because we focus on hardness and, to this end, a machine model is not really needed.

NP-hard problems are all those problems that 3-SATISFIABILITY can be reduced to by means of polynomial-time many–one reductions. In complete analogy, we can use weighted satisfiability problems to define complexity classes of "parameterized hardness". The basic degree of parameterized intractability is the class *W*[*1*].

Definition 13.4 1. *The class W[1] contains all problems that can be reduced to* WEIGHTED 2-CNF-SATISFIABILITY *by a parameterized reduction.*

2. *A parameterized problem is called W[1]*-hard *if the parameterized problem* WEIGHTED 2-CNF-SATISFIABILITY *can be reduced to it by a parameterized reduction.*
3. *A problem in W[1] that fulfills both the above properties is W[1]*-complete.
4. *The class W[2] is defined analogously by replacing* WEIGHTED 2-CNF-SATISFIABILITY *with* WEIGHTED CNF-SATISFIABILITY.

Before we continue our discussion of parameterized intractability, we give simple examples for problems in *W*[*1*] and *W*[*2*].

Example 13.5 • INDEPENDENT SET (analogously CLIQUE) is in *W*[*1*]. This is based on the observation that a graph $G = (V, E)$ with $V = \{1, 2, \ldots, n\}$ has an independent set of size k iff the 2-CNF formula

$$\bigwedge_{\{i,j\} \in E} (\overline{x_i} \vee \overline{x_j})$$

has a weight-k satisfying truth assignment. Clearly, the corresponding parameterized reduction works in polynomial time.

• DOMINATING SET is in *W*[*2*]. This is based on the observation that a graph $G = (V, E)$ with $V = \{1, 2, \ldots, n\}$ has a dominating set of size k iff the CNF-formula

$$\bigwedge_{i \in V} \bigvee_{j \in N[i]} x_j$$

has a weight-k satisfying truth assignment. Here, note that the size of the closed neighborhood $N[i]$ of vertex i in general cannot be bounded from above by a constant; hence an unbounded OR might be necessary here.

Clearly, we could analogously show that VERTEX COVER is in *W*[*1*]. This also follows from the inclusion $FPT \subseteq W[1]$ which is a direct consequence of the definition of *W*[*1*]. As we see, *W*[*1*] is defined through a satisfiability problem which employs a formula structure that is a "big AND" of "small ORS" (2-CNF). In fact, we will shortly see that small ORs can be defined by OR-gates having fan-in bounded from above by a *constant.* Moreover, *W*[*2*] is defined through a satisfiability problem which employs a formula structure that is a "big AND" of "big ORS" (CNF). This generalizes into the concept of t-normalized formulae, where the term "product" refers to AND-operators and "sum" refers to OR-operators.

Definition 13.6 *A Boolean formula is called* t-normalized *if it can be written in the form of a "product-of-sums-of-products-..." of literals with $t-1$ alternations between products and sums.*

2-CNF formulae are 1-normalized and CNF formulae are 2-normalized.

With this concept at hand, we may define "higher" classes of parameterized intractability. To this end, we introduce the WEIGHTED t-NORMALIZED SATISFIABILITY problem which is the natural extension of WEIGHTED CNF-SATISFIABILITY (equivalently, WEIGHTED 2-NORMALIZED-SATISFIABILITY) to t-normalized Boolean formulae.

Definition 13.7
1. *$W[t]$ for $t \geq 1$ is the class of all parameterized problems that can be reduced to* WEIGHTED t-NORMALIZED SATISFIABILITY *by a parameterized reduction.*
2. *$W[Sat]$ is the class of all parameterized problems that can be reduced to* WEIGHTED SATISFIABILITY *by a parameterized reduction.*
3. *$W[P]$ is the class of all parameterized problems that can be reduced to* WEIGHTED CIRCUIT SATISFIABILITY *by a parameterized reduction.*

In the definition of $W[Sat]$ we generalize t-normalized Boolean formulae to arbitrary Boolean formulae. In the definition of $W[P]$ we furthermore generalize Boolean formulae to Boolean circuits (of polynomial size)—the main distinguishing factor between circuits and formulae here is that circuits may use "intermediate results" more than once at different places in the circuit.

It is time to clarify what "W" stands for. It is used as a shorthand for the *weft* of Boolean formulae and circuits. The weft denotes the maximum number of alternations between unbounded fan-in AND- and OR-gates used plus one. Thus, a CNF formula has weft two whereas a q-CNF formula for constant q has weft one.

Finally, we introduce one more class of parameterized intractability which, in fact, is known to be a proper superset of FPT.

Definition 13.8 *A parameterized language L belongs to the class XP if it can be determined in $f(k) \cdot |x|^{g(k)}$ time whether $(x, k) \in L$ where f and g are computable functions only depending on k.*

Now, we can draw an overview of parameterized complexity hierarchies.

Theorem 13.9 $FPT \subseteq W[1] \subseteq W[2] \subseteq \ldots \subseteq W[Sat] \subseteq W[P] \subseteq XP$.

Proof The inclusion $FPT \subseteq W[1]$ directly follows from the fact that $W[1]$ allows "FPT-computations" through the use of parameterized reductions. The inclusion $W[Sat] \subseteq W[P]$ is immediate because a formula is a special case of a circuit (the "computation graph" is a directed tree in the first case and it is a directed acyclic graph in the second case). Except for $W[P] \subseteq XP$, all other inclusions are trivial by definition. Finally, $W[P] \subseteq XP$ follows from the fact that in order to solve a weighted circuit satisfiability problem one may simply guess a weight-k truth assignment of the n input variables and then check in polynomial time whether the circuit evaluates to true. Since guessing can be realized by trying all $\binom{n}{k}$ possibilities, we end up with running time $n^{O(k)}$. Every problem in $W[P]$ can be parameterized reduced to a weighted circuit satisfiability problem. Thus, by combining the above algorithm with a parameterized reduction we clearly arrive at an overall running time of $f(k) \cdot n^{g(k)}$ for appropriate functions f and g. □

The borderline between "good" (fixed-parameter tractable) and "bad" (fixed-parameter intractable) here lies between *FPT* and $W[1]$. Whereas problems in *FPT* can be solved in $f(k) \cdot n^{O(1)}$ time, for $W[1]$-hard problems only exact solving algorithms with running time $n^{\Theta(k)}$ for some constant c are known. Indeed, we are not aware of any differences concerning running times for say $W[1]$-hard and $W[2]$-hard problems. Thus, pragmatic as we are, henceforward we concentrate on the distinction between *FPT* and $W[1]$-hard in order to separate between good and bad. Therefore, questions such as showing containment in $W[1]$ or $W[2]$ are basically neglected here. A thorough study of these more structural complexity issues can be found in the literature.

Next, we take a closer look at $W[1]$ and, in particular, at showing $W[1]$-hardness.

13.2 The complexity class $W[1]$

Although we will later become acquainted with the complexity class $M[1]$ which is a candidate for lying "between" *FPT* and $W[1]$, $W[1]$-hardness continues to be the basic degree of parameterized intractability. Thus, proving $W[1]$-hardness for a parameterized problem is our main tool for proving relative lower bounds. It is important to note that, to the best of our knowledge, there are no results showing that $W[1]$-hard problems can be solved by significantly faster exact algorithms than, say, $W[2]$-hard or even $W[P]$-hard problems can.

We begin by learning more about the structure and properties of $W[1]$ and the related $W[1]$-complete and -hard problems. We defined $W[1]$ as the class of problems parameterized reducible to WEIGHTED 2-CNF-SATISFIABILITY, or, in the words of computational complexity theory, the *closure* of WEIGHTED 2-CNF-SATISFIABILITY under parameterized reductions. The following—technically demanding—theorem tells us that we could have defined $W[1]$ by an even "simpler" problem. To this end, call a Boolean formula in CNF *antimonotone* if it exclusively contains negative literals. Then, in the natural way, we obtain WEIGHTED ANTIMONOTONE 2-CNF-SATISFIABILITY as a special case of WEIGHTED 2-CNF-SATISFIABILITY with formulae restricted to be antimonotone. We omit the proof.

Theorem 13.10 WEIGHTED 2-CNF-SATISFIABILITY *reduces to* WEIGHTED ANTIMONOTONE 2-CNF-SATISFIABILITY *by a parameterized reduction.*

Indeed, Theorem 13.10 can be generalized to show that WEIGHTED q-CNF-SATISFIABILITY reduces to WEIGHTED ANTIMONOTONE q-CNF-SATISFIABILITY by a parameterized reduction. The important consequence of this theorem is that WEIGHTED ANTIMONOTONE 2-CNF-SATISFIABILITY is $W[1]$-complete.

Corollary 13.11 *The set of problems parameterized reducible to* WEIGHTED ANTIMONOTONE 2-CNF-SATISFIABILITY *coincides with $W[1]$. This implies that* WEIGHTED ANTIMONOTONE 2-CNF-SATISFIABILITY *is $W[1]$-complete.*

Proof By definition, WEIGHTED 2-CNF-SATISFIABILITY is $W[1]$-complete. Every problem in $W[1]$ can be reduced to WEIGHTED 2-CNF-SATISFIABILITY by a parameterized reduction and WEIGHTED 2-CNF-SATISFIABILITY in turn can be reduced to WEIGHTED ANTIMONOTONE 2-CNF-SATISFIABILITY by a parameterized reduction. Hence the transitivity of parameterized reductions implies the claim. □

Corollary 13.11 is extremely useful for providing $W[1]$-completeness and hardness results. For instance, the $W[1]$-completeness of INDEPENDENT SET can now easily be shown.

Theorem 13.12 INDEPENDENT SET *is $W[1]$-complete.*

Proof In Example 13.5 we have seen a simple parameterized reduction of INDEPENDENT SET to an instance of WEIGHTED ANTIMONOTONE 2-CNF-SATISFIABILITY. This shows that INDEPENDENT SET is in $W[1]$. Moreover, the reduction presented can actually be reversed, that is, every antimonotone 2-CNF formula can be transformed into a graph such that the original formula has a weight-k truth assignment iff the constructed graph has a size-k independent set. Hence, WEIGHTED ANTIMONOTONE 2-CNF-SATISFIABILITY parameterized reduces to INDEPENDENT SET, implying the $W[1]$-hardness of INDEPENDENT SET. □

Due to the close relationship between CLIQUE and INDEPENDENT SET, we also obtain $W[1]$-hardness for CLIQUE.

Corollary 13.13 CLIQUE *is $W[1]$-complete.*

Proof At the beginning of Section 13.1 we presented a trivial parameterized reduction from CLIQUE to INDEPENDENT SET and vice versa; both problems are equivalent in a parameterized sense. The claim then follows from the $W[1]$-completeness of INDEPENDENT SET. □

In this book, which is oriented towards algorithms and not towards structural complexity investigations, we circumvent basically all of the fairly technical constructions in relation to the W-hierarchy and the corresponding complexity classes with their complete problems. For $W[1]$, however, we will at least touch the surface of this world by showing that WEIGHTED q-CNF SATISFIABILITY for every constant $q > 2$ is $W[1]$-complete as well. To this end (see Definition 13.4), the main difficulty is to show that WEIGHTED q-CNF-SATISFIABILITY is contained in $W[1]$ by establishing a one-to-one correspondence (by parameterized reducibility) to WEIGHTED 2-CNF-SATISFIABILITY.

Theorem 13.14 WEIGHTED q-CNF-SATISFIABILITY *is* $W[1]$*-complete for every constant* $q \geq 2$.

Proof By definition, WEIGHTED 2-CNF-SATISFIABILITY is $W[1]$-complete. Let $q > 2$ be any integer constant. First, it is easy to see that WEIGHTED q-CNF-SATISFIABILITY is hard for $W[1]$: consider the case $q = 3$. Replace every clause $(\ell_1 \vee \ell_2)$ of a 2-CNF formula by two new clauses

$$(\ell_1 \vee \ell_2 \vee y) \wedge (\ell_1 \vee \ell_2 \vee \overline{y}),$$

where y is a new variable that does not occur in the 2-CNF formula. Obviously, in this way the constructed 3-CNF-formula has a weight-k satisfying assignment setting $y = 0$ iff the original 2-CNF formula has a weight-k assignment. Moreover, this transformation is doable in polynomial time, hence WEIGHTED 2-CNF-SATISFIABILITY parameterized reduces to WEIGHTED 3-CNF-SATISFIABILITY, implying the $W[1]$-hardness of WEIGHTED 3-CNF-SATISFIABILITY. The construction easily generalizes to $q > 3$, implying the $W[1]$-hardness of WEIGHTED q-CNF-SATISFIABILITY.

It remains to show the containment of WEIGHTED q-CNF-SATISFIABILITY in $W[1]$ for any $q > 3$. By the generalization of Theorem 13.10 to arbitrary q-CNF formulae it suffices to show the following. Let F be an antimonotone formula in q-CNF, that is, all variables occur in negated form. Then there is a 2-CNF formula F' such that F has a weight-k satisfying assignment iff F' has a satisfying assignment of weight

$$k' = f(k) := k2^k + \sum_{i=2}^{q} \binom{k}{i} = k2^k + O(2^k).$$

We first show the construction of F' from F and then show the claimed equivalence.

Let $x_1, \ldots, x_n$ be the variables occurring in F. Consider all possible subsets of sizes between two and q of this variable set. Call these subsets $A_1, \ldots A_p$. A clause thus directly corresponds to one of these subsets. Associate with each A_i, $1 \leq i \leq p$, a new variable v_i in F'. When v_i is true, this shall mean that all variables in A_i are also true. We will replace the clause corresponding to A_i by $\overline{v_i}$ and introduce for each input variable x_j, $1 \leq j \leq n$, exactly 2^k "copies" $x_{j,0}, \ldots, x_{j,2^k-1}$. Then, F' has the following structure. It consists of all clauses from F replaced by their corresponding literal $\overline{v_i}$ (thus giving a one-literal-clause) and the following two-literal clauses (written as implications, to simplify the presentation).

1. $v_i \rightarrow v_{i'}$ if $A_{i'} \subseteq A_i$;
2. $v_i \rightarrow x_{j,0}$ if $x_j \in A_i$;
3. $x_{j,r} \rightarrow x_{j,r+1 \pmod{2^k}}$ for $j = 1, \ldots, m$ and $r = 0, \ldots, 2^k - 1$.

With this construction, the size of F' compared to F increases by a factor $O(2^k)$ and an additive term $\sum_{i=1}^{q} \binom{n}{i} = O(n^q)$ with *constant* q. Hence, this is a parameterized reduction.

The correctness of the construction is seen as follows. First, assume that F has a weight-k satisfying assignment which sets exactly the variables $x_{j_1}, \ldots, x_{j_k}$ true. Now consider F'. Setting variables

$$x_{j_1,0}, \ldots, x_{j_1,2^k-1}, \ldots, x_{j_k,0}, \ldots, x_{j_k,2^k-1}$$

true and all those variables v_i that correspond to an A_i which forms a subset of the variable set $\{x_{j_1}, \ldots, x_{j_k}\}$ then gives a satisfying assignment of F' with weight

$$k2^k + \sum_{i=2}^{q} \binom{k}{i} = k'.$$

Second, for the reverse direction assume that F' has a satisfying assignment of weight $k' = k2^k + \sum_{i=2}^{q} \binom{k}{i}$. Due to the construction of F' we know that if one $x_{j,r}$ is true, then all $x_{j,\ell}$, $1 \leq \ell \leq 2^k - 1$, must be true. Moreover, $2^k > \sum_{i=2}^{q} \binom{k}{i}$. Hence every weight-$k'$ satisfying assignment of F' has to set exactly k "variable copy sets" of size 2^k each to true. These variable copy sets naturally correspond to a satisfying assignment of F by identifying each copy set with the original variable in F it belongs to. □

Once more, observe that the main purpose of providing the proof of Theorem 13.14 here was to give some idea of the typical technical expenditure parameterized complexity theory takes.

We have now encountered several $W[1]$-complete problems and there are numerous others. The decisive point is that all $W[1]$-hard problems probably do not allow for fixed-parameter algorithms. Before we study several more parameterized reductions leading to $W[1]$-hardness results in the next section, let us briefly discuss the relationship between FPT and $W[1]$. Clearly, $FPT \subseteq W[1]$. But how strong in the sense of provable intractability is the statement that a parameterized problem is $W[1]$-hard? Here, the current state of knowledge is as follows. We know that if $P = NP$, then $FPT = W[1]$ because then for instance the $W[1]$-complete and NP-complete problem CLIQUE is in P. Since all problems in $W[1]$ can be reduced to CLIQUE by a parameterized reduction, we obtain fixed-parameter algorithms for all $W[1]$-complete problems by combining the assumed polynomial-time algorithm for CLIQUE with the fixed-parameter algorithm performing the reduction.

Establishing the (not believed) equality $FPT = W[1]$ would have weaker consequences than $P = NP$. Still, however, these consequences give some good reason to believe that $W[1]$-hardness delivers robust statements about parameterized intractability. The fundamental separation hypothesis,

$$FPT \neq W[1],$$

upon which the theory of parameterized intractability is based, is underpinned by the following statement.

Theorem 13.15 *If $W[1] = FPT$ then* 3-CNF-SATISFIABILITY *for a Boolean formula F with n variables can be solved in $2^{o(n)} \cdot |F|^{O(1)}$ time.*

The proof of Theorem 13.15 is beyond the scope of this book. The significance of Theorem 13.15 lies in connecting classical complexity theory with parameterized complexity. An even stronger link can be established when considering the separation hypothesis $FPT \neq W[P]$; here, there are tight connections to the concept of *bounded nondeterminism* in classical complexity theory and, in fact, Theorem 13.15 can be derived from this correspondence between parameterized complexity and bounded nondeterminism. We skip any details here. Let us only remark that solving 3-CNF-SATISFIABILITY in basically $2^{o(n)}$ steps (that is, equivalently, in $2^{o(n)} \cdot |F|^{O(1)}$ steps) seems fairly unlikely from today's point of view. All exact algorithms for 3-CNF-SATISFIABILITY we know to date work in basically c^n steps for some constant $c < 2$ and the (significant) progress achieved in the last twenty years was to bring down the constant from the trivial $c = 2$ to $c = 1.49$ (with a randomized algorithm achieving $c = 1.33$). This seems to indicate that there is reason to believe that $FPT \neq W[1]$. This is the working hypothesis throughout this book. Clearly, from the viewpoint of computational complexity theory there is much more to say and the reader is invited to further explore parameterized complexity theory using the bibliographical remarks as a starting point.

13.3 Concrete parameterized reductions

As explained before, from an algorithmic point of view the main interest for exploring the W-hierarchy lies in showing the "probable non-existence" of fixed-parameter algorithms for particular problems. To this end, our main tool is showing $W[1]$-hardness by giving a parameterized reduction from a $W[1]$-hard problem such as CLIQUE to the problem under consideration. In some cases, however, even stronger statements than $W[1]$-hardness are possible. Let us consider k-COLORING as a simple example:

> **Input:** A graph $G = (V, E)$ and a positive integer k.
> **Question:** Can G be colored with at most k colors, that is, does there exist an assignment of colors to vertices such that no two adjacent vertices have the same color?

It is well-known that k-COLORING and even 3-COLORING are *NP*-complete. This implies that with respect to parameter k the parameterized problem k-COLORING is not fixed-parameter tractable unless $P = NP$, since setting $k = 3$ would result in a polynomial-time algorithm. This statement is clearly stronger than showing that k-COLORING is $W[1]$-hard with respect to parameter k because $P = NP$ would imply $FPT = W[1]$ but the reverse direction is not known.

A second example of fixed-parameter intractability based on the assumption $P \neq NP$ is given by a generalization of VERTEX COVER to the case with vertices having positive real-valued weights—GENERAL WEIGHTED VERTEX COVER:[7]

> **Input:** A graph $G = (V, E)$, a weight function assigning nonnegative real values to vertices, and and a positive real number k.
> **Question:** Does there exist a vertex cover set such that the sum of the weights of its vertices is upper-bounded by k?

GENERAL WEIGHTED VERTEX COVER is not fixed-parameter tractable with respect to parameter k because of the following reduction of (unweighted) VERTEX COVER to it showing NP-completeness even in the case $k = 1$.

Consider an instance of VERTEX COVER with parameter k'. Transform this into an instance of GENERAL WEIGHTED VERTEX COVER by simply assigning weight $1/k$ to every vertex of the original graph. Clearly, the unweighted graph has a vertex cover of size k iff the weighted graph has a vertex cover of total weight 1. Hence, if GENERAL WEIGHTED VERTEX COVER were fixed-parameter tractable we could derive a polynomial-time algorithm for the NP-complete VERTEX COVER problem. We mention in passing that a decisive point in the above reduction is that weights can be arbitrarily close to zero.

Finally, before starting with parameterized hardness proofs, let us end this introduction by showing how parameterized reductions also can lead to fixed-parameter tractability results. Again consider a weighted version of VERTEX COVER—INTEGER WEIGHTED VERTEX COVER:

> **Input:** A graph $G = (V, E)$, a weight function assigning positive integer values to vertices, and and a positive integer k.
> **Question:** Does there exist a vertex cover set such that the sum of the weights of its vertices can be bounded from above by k?

Here we assume that the weights are bounded from above by a polynomial in the graph size.

In what follows, we exhibit that we can easily reduce INTEGER-WEIGHTED VERTEX COVER to unweighted VERTEX COVER via a simple parameterized many–one reduction that does not change the value of the parameter. To prove the following theorem, we may safely assume that the maximum vertex weight is bounded by k. The according preprocessing needs only polynomial time.

Theorem 13.16 INTEGER-WEIGHTED VERTEX COVER *can be solved as fast as unweighted* VERTEX COVER *up to an additive term polynomial in* k.

Proof An instance of INTEGER-WEIGHTED VERTEX COVER is transformed into an instance of VERTEX COVER as follows: replace each vertex v of weight i with a cluster c_v consisting of i vertices. The graph does not contain intra-cluster edges. Furthermore, if $\{u, v\}$ is an edge in the original graph, then we connect

[7]Also positive rational weights are sufficient to show intractability.

every vertex of cluster c_u to every vertex of cluster c_v. Now, it is easy to see that both graphs (the instance for INTEGER-WEIGHTED VERTEX COVER and the new instance for VERTEX COVER) have minimum vertex covers of the same weight/size. Here it is important to observe that the following is true for the constructed instance for VERTEX COVER: either all vertices of a cluster are in a minimum vertex cover or none of them is. Assume that one vertex of cluster c_v is not in the cover but the remaining ones are. Then all vertices in all neighboring clusters have to be included and, hence, it makes no sense to include any vertex of cluster c_v in the vertex cover.

Let $t(k,n)$ be the time needed to solve VERTEX COVER. The running time of the algorithm on the "cluster instance" is clearly bounded by $t(k,wn) \leq t(k,kn)$, where $w \leq k$ is the maximum vertex weight in the given graph. Using the interleaving technique (see Section 8.7), it is possible to show that the running time is not increased by a polynomial factor, but only a polynomial amount of *additional* processing is needed. □

13.3.1 *W[1]-hardness proofs*

Experience tells us that parameterized many–one reductions are, as a rule, harder to achieve than classical many–one reductions. This appears to be plausible because parameterized reductions, taking care of the "parameter structure", have to fulfill more properties. Nevertheless, in this part we present a series of not too difficult examples for parameterized reductions showing *W[1]*-hardness. Far from being exhaustive, we try to exhibit a wide range of different types of reductions and problems.

We start with a generic problem that, since containment in *W[1]* (and, thus, *W[1]*-completeness) can also be shown, is particularly useful for proving containment in *W[1]*. Moreover, it resembles the generic *NP*-complete problem of determining whether a nondeterministic Turing machine has a polynomial-length accepting computation path on a given input word. We focus on *W[1]-hardness* of SHORT TURING MACHINE ACCEPTANCE here:

> **Input:** A nondeterministic Turing machine with its transition table, an input word x for M, and a nonnegative integer k.
> **Question:** Does M on input x have an accepting computation path of length at most k?

Observe that it is of key importance for the following proof that M may have an input alphabet of arbitrary size.

Theorem 13.17 SHORT TURING MACHINE ACCEPTANCE *is W[1]-hard with respect to parameter k denoting the number of computation steps.*

Proof The idea of the proof is to give a parameterized reduction from CLIQUE to SHORT TURING MACHINE ACCEPTANCE. Let $(G = (V, E), k)$ form an input instance of CLIQUE where we want to determine whether G contains k vertices inducing a clique. Without going into the technical but straightforward details, we may clearly assume that the structure of the graph G, for instance its adjacency matrix, can be encoded into the transition table of the Turing machine using alphabet and state sets of unbounded size. Then the Turing machine M is designed to work in two phases. In the first phase, M nondeterministically guesses k vertices from V, each represented by a different alphabet symbol, and writes them on its working tape, using k tape cells. In the second phase, M then makes $\binom{k}{2}$ scans through the work tape, verifying that all pairs of vertices represented on the tape are adjacent in G. Note that the adjacency information is stored in the finite control of M and can thus be "looked up" in constant time. If all scans are successful, then G has a size-k clique. Otherwise, it does not. Clearly, the computation of M as described takes

$$O\left(k + \binom{k}{2}\right) = O(k^2)$$

time, thus showing that we have a parameterized reduction from CLIQUE to SHORT TURING MACHINE ACCEPTANCE computable in polynomial time. □

Next, we come to a more standard parameterized reduction between two graph problems yielding $W[1]$-hardness. We study another generalization of VERTEX COVER known from the literature as PARTIAL VERTEX COVER:

> **Input:** A graph $G = (V, E)$ and two positive integers k and t.
> **Question:** Does there exist a subset $C \subseteq V$ with k vertices such that C covers at least t edges?

Clearly, setting $t = |E|$ gives the standard VERTEX COVER problem. Perhaps surprisingly, PARTIAL VERTEX COVER is $W[1]$-hard.

Theorem 13.18 PARTIAL VERTEX COVER *is $W[1]$-hard with respect to parameter k.*

Proof We give a parameterized reduction from INDEPENDENT SET to PARTIAL VERTEX COVER. Let $(G = (V, E), k)$ be an instance of INDEPENDENT SET. For every vertex $v \in V$, let $\deg(v)$ denote the degree of v in G. We construct a new graph $G' = (V', E')$ in the following way: for each vertex $v \in V$ we insert $|V| - \deg(v)$ new vertices into G and connect each of these new vertices with v. In the following, we show that a size-k independent set in G one-to-one corresponds to a size-k partial vertex cover in G' which covers $t := k \cdot |V|$ edges.

First, a size-k independent set in G also forms a size-k independent set in G'. Moreover, each of these k vertices has exactly $|V|$ incident edges. Then, these k vertices form a partial vertex cover covering $k \cdot |V|$ edges. Second, if we have a size-k partial vertex cover in G' which covers $k \cdot |V|$ edges, then we know that none of the newly inserted vertices in G' can be in this cover. Hence this cover contains k vertices from V. Moreover, a vertex in G' can cover at most $|V|$ edges and two adjacent vertices can together cover only $2|V| - 1$ edges. Therefore no two vertices in this partial vertex cover can be adjacent, which implies that this partial cover forms a size-k independent set in G. □

The next $W[1]$-hardness result shows a simple relation between the graph problem CLIQUE and the set problem SET PACKING. Recall the definition of SET PACKING from Section 11.4.2:

> **Input:** A collection $\mathcal{C}$ of subsets of a finite set S and a nonnegative integer k.
> **Question:** Is there a subcollection $\mathcal{C}' \subseteq \mathcal{C}$ which consists of at least k mutually disjoint subsets?

Theorem 13.19 SET PACKING *is $W[1]$-hard with respect to the desired number k of disjoint sets.*

Proof We give a parameterized reduction from the $W[1]$-hard INDEPENDENT SET problem to SET PACKING. Let $(G = (V, E), k)$ be an instance of INDEPENDENT SET. We construct an instance of SET PACKING as follows. The elements of the base set $S \subseteq 2^V$ are two-element multi-subsets of V.

More precisely, for every vertex $v \in V$ we build two-element sets $\{v, u\}$ for all u that are non-adjacent to v in G. Moreover, for every $v \in V$ we build the two-element set $\{v, v\}$—to simplify the presentation we think of multisets here such that $\{v\} \neq \{v, v\}$. Altogether, for every graph vertex v we have then built the following subset of S:

$$S_v := \{\{v, u\} \mid \{v, u\} \notin E\} \cup \{\{v, v\}\}.$$

The collection $\mathcal{C}$ of sets S_v thus constructed together with the unchanged parameter value k then forms our instance of SET PACKING. Altogether, it is not hard to see that graph G together with k forms a yes-instance of INDEPENDENT SET iff the constructed collection forms a yes-instance of SET PACKING. To this end, observe that an independent set in G corresponds to a subcollection of pairwise disjoint subsets of $\mathcal{C}$ and a vertex v one-to-one corresponds to the subset S_v. □

We end this subsection with a more demanding construction that shows the $W[1]$-hardness of an *NP*-complete string problem arising in computational biology. The purpose is to illustrate the significant technical expenditure that may be needed in order to prove $W[1]$-hardness results. More specifically, we consider the following generalization of the CLOSEST STRING problem (see Sections 8.5 and 11.2), called CLOSEST SUBSTRING.

> **Input:** A set of k strings $s_1, s_2, \ldots, s_k$ over a finite alphabet Σ and nonnegative integers d and L.
> **Question:** Is there a string s of length L, and for $i = 1, \ldots, k$, a substring s'_i of s_i of length L such that, for all $i = 1, \ldots, k$, $d_H(s, s'_i) \leq d$?

Here $d_H(s, s'_i)$ denotes the Hamming distance between s and s'_i.

Observe that we have increased combinatorial complexity compared with CLOSEST STRING because the input strings do not need to be of same length anymore and the desired substrings may start in $|s_i|-|s|+1$ positions individually for *every* of the given k strings. Indeed, whereas we have shown in Section 11.2 that CLOSEST STRING is fixed-parameter tractable with respect to the number of input strings k, we now show that CLOSEST SUBSTRING becomes $W[1]$-hard with respect to this parameterization. To simplify the presentation, we only prove $W[1]$-hardness in the case of unbounded size of the alphabet Σ—by significantly increased mathematical efforts one can also show $W[1]$-hardness even in the case of Σ being a binary alphabet. We refer to the literature for the latter proof, which is based on the proof for unbounded alphabet size as described here.

We derive the $W[1]$-hardness result by a series of intermediate steps, aiming at a reduction from CLIQUE to CLOSEST SUBSTRING.

Let a CLIQUE instance be given by an undirected graph $G = (V, E)$, with a set $V = \{v_1, v_2, \ldots, v_n\}$ of n vertices, a set E of m edges, and a positive integer k denoting the size of the desired clique. We describe how to generate a set S of $\binom{k}{2}$ strings such that G has a clique of size k iff there is a string s of length $L := k+1$ such that every $s_i \in S$ has a substring s'_i of length L with $d_H(s, s'_i) \leq d := k-2$. If a string $s_i \in S$ has a substring s'_i of length L with $d_H(s, s'_i) \leq d$, we call s'_i a *match* for s. We assume $k > 2$, because $k = 1$ and $k = 2$ are trivial cases.

Alphabet. The alphabet of the produced instance is given by the disjoint union of the following sets:

- $\{\, [v_i] \mid v_i \in V \,\}$, i.e. an alphabet symbol for every vertex of the input graph; we call them *encoding symbols*;
- $\{\, [c_{i,j}] \mid i = 1, \ldots, k, j = i+1, \ldots, k \,\}$, that is, a unique symbol for every of the $\binom{k}{2}$ produced strings; we call them *string identification symbols*;
- $\{\#\}$ which we call the *synchronizing symbol*.

This accounts for a total of $n + \binom{k}{2} + 1$ alphabet symbols.

Choice strings. We generate a set of $\binom{k}{2}$ *choice strings* $S_c = \{c_{1,2}, \ldots, c_{1,k}, c_{2,3}, c_{2,4}, \ldots, c_{k-1,k}\}$ and assume that the strings in S_c are ordered as shown. *Every* choice string will encode the whole graph; it consists of m concatenated strings, each of length $k+1$, called *blocks*; from this, we have one block for every edge of the graph. The blocks will be separated by *barriers*, which are length k strings consisting of k identification symbols corresponding to the respective string. A choice string $c_{i,j}$ is given by

$$c_{i,j} := \langle \text{block}(i,j,e_1) \rangle \, ([c_{i,j}])^k \, \langle \text{block}(i,j,e_2) \rangle \, ([c_{i,j}])^k \ldots ([c_{i,j}])^k \, \langle \text{block}(i,j,e_m) \rangle,$$

where $e_1, e_2, \ldots, e_m$ are the edges of G and $\langle \text{block}() \rangle$ will be defined below. The solution string s will have length $k+1$, which is exactly the length of one block.

Block in a choice string. Every block is a string of length $k+1$ and it encodes an edge of the input graph. Every choice string contains a block for every edge of the input graph; different choice strings, however, encode the edges in different positions of their blocks: for a block in choice string $c_{i,j}$, positions i and j are called *active* and these positions encode the edge. Let e be the edge to be encoded and let e connect vertices v_r and v_s, $1 \leq r < s \leq n$. Then the ith position

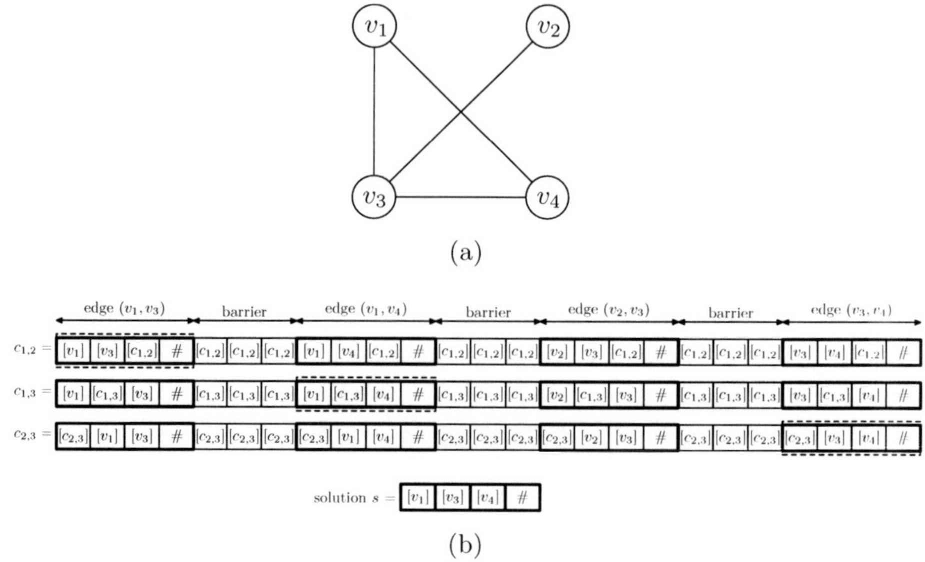

FIG. 13.1. Example of the reduction from a CLIQUE instance G with $k = 3$ (shown in (a)) to a CLOSEST SUBSTRING instance with bounded alphabet (shown in (b)) as explained in Example 13.20. In (b), we display the constructed strings $c_{1,2}$, $c_{1,3}$, and $c_{2,3}$ (the contained blocks are highlighted by bold boxes) and the solution string s that is found, since G has a clique of size $k = 3$; s is a string of length $k + 1 = 4$ such that $c_{1,2}$, $c_{1,3}$, and $c_{2,3}$ have length 4 substrings (indicated by dashed boxes) that have Hamming distance at most $k - 2 = 1$ to s.

of the block is equal to $[v_r]$ in order to encode v_r and the jth position is equal to $[v_s]$ in order to encode v_s. The last position of a block is set to the synchronizing symbol #. All remaining positions in the block are set to the identification symbol $[c_{i,j}]$ of $[c_{i,j}]$. Thus, a block is given by

$$\langle \text{block}(i, j, (v_r, v_s)) \rangle \;:=\; ([c_{i,j}])^{i-1}\, [v_r]\, ([c_{i,j}])^{j-i-1}\, [v_s]\, ([c_{i,j}])^{k-j}\, \#.$$

Values for L and d. We set $L := k + 1$ and $d := k - 2$.

Example 13.20 Let $G = (V, E)$ be an undirected graph with $V = \{v_1, v_2, v_3, v_4\}$ and $E = \{\{v_1, v_3\}, \{v_1, v_4\}, \{v_2, v_3\}, \{v_3, v_4\}\}$ (as shown in Figure 13.1(a)) and let $k = 3$. Using G, we exhibit the above construction of $\binom{k}{2} = 3$ choice strings $c_{1,2}$, $c_{1,3}$, and $c_{2,3}$ (as shown in Figure 13.1(b)). We claim that (which will be proven in the following) there exists a clique of size k in G iff there is a string s of length $L := \binom{k}{2} + 1 = 4$ such that, for $1 \leq i < j \leq 3$, each $c_{i,j}$ contains a length 4 substring $s_{i,j}$ with $d_H(s, s_{i,j}) \leq d := k - 2 = 1$.

The choice strings are over an alphabet consisting of $\{[v_1], [v_2], [v_3], [v_4]\}$ (the encoding symbols, that is, one symbol for every vertex of G), $\{[c_{1,2}], [c_{1,3}], [c_{2,3}]\}$

(the string identification symbols), and $\{\#\}$ (the synchronizing symbol). Every string $c_{i,j}$, $1 \le i < j \le 3$, consists of four blocks, each of which encodes an edge of the graph. Every block is of length $\binom{k}{2} + 1 = 4$ and has # at its last position. The blocks are separated by barriers $([c_{i,j}])^k = ([c_{i,j}])^3$.

In string $c_{1,2}$, positions 1 and 2 within each block are active and encode the corresponding edge (in general, in $c_{i,j}$ positions i and j within a block are active). All of the first k positions of a block in string $c_{i,j}$ which are not active contain the $[c_{i,j}]$ symbol. Thus, for instance, the block in $c_{1,2}$ encoding the edge $\{v_1, v_3\}$ is given by $[v_1]\,[v_3]\,[c_{1,2}]\,\#$. Further details can be found in Figure 13.1.

The closest substring that corresponds to the k-clique in G consisting of vertices v_1, v_3, v_4 is $[v_1]\,[v_3]\,[v_4]\,\#$. The corresponding matches are $[v_1]\,[v_3]\,[c_{1,2}]\,\#$ in $c_{1,2}$ (encoding the edge $\{v_1, v_3\}$), $[v_1]\,[c_{1,3}]\,[v_4]\,\#$ in $c_{1,3}$ (encoding the edge $\{v_1, v_4\}$), and $[c_{2,3}]\,[v_3]\,[v_4]\,\#$ in $c_{2,3}$ (encoding the edge $\{v_3, v_4\}$).

We have now seen the construction. It remains to show its correctness. To prove the correctness of the proposed reduction, we have to show two directions. The easier one is to see that a k-clique implies a closest substring fulfilling the given requirements.

Proposition 13.21 *For a graph with a k-clique, the above construction produces an instance of* Closest Substring *which has a solution, i.e. there is a string s of length L such that every $c_{i,j} \in S_c$ has a substring $s_{i,j}$ with $d_H(s, s_{i,j}) \le d$.*

Proof Let the input graph have a clique of size k. Let $h_1, h_2, \ldots, h_k$ denote the indices of the clique's vertices, $1 \le h_1 < h_2 < \cdots < h_k \le n$. Then we claim that a solution for the produced Closest Substring instance is

$$s := [v_{h_1}]\,[v_{h_2}] \ldots [v_{h_k}]\,\#.$$

Consider the choice string $c_{i,j}$, $1 \le i < j \le k$. As the vertices $v_{h_1}, v_{h_2}, \ldots, v_{h_k}$ form a clique, we have an edge connecting v_{h_i} and v_{h_j}. The choice string $c_{i,j}$ contains a block $s_{i,j} := \langle \text{block}(i, j, (v_{h_i}, v_{h_j})) \rangle$ encoding this edge:

$$s_{i,j} := ([c_{i,j}])^{i-1}\,[v_{h_i}]\,([c_{i,j}])^{j-i-1}\,[v_{h_j}]\,([c_{i,j}])^{k-j}\#.$$

We have $d_H(s, s_{i,j}) = k - 2$, and we can find such a block for every $c_{i,j}$, $1 \le i < j \le k$. □

For the reverse direction, we show in Proposition 13.24 that a solution in the produced Closest Substring instance implies a k-clique in the input graph. For this, we need the following two lemmas which show that a solution to the constructed instance has encoding symbols at its first k positions and the synchronizing symbol # at its last position.

Lemma 13.22 *A closest substring s contains at least two encoding symbols and at least one synchronization symbol.*

Proof Let s be a solution of the CLOSEST SUBSTRING instance produced by the construction above. Let $A_{[c]}(s)$ be the set of string identification symbols from $\{ [c_{i,j}] \mid 1 \le i < j \le k \}$ that occur in s. Let $S'_{[c]}(s) \subseteq S_c$ be the subset of choice strings that do *not* contain a symbol from $A_{[c]}(s)$.

Since s is of length $k+1$, we have $|A_{[c]}(s)| \le k+1$. Therefore, for $k \ge 4$, there are at least $\binom{k}{2} - (k+1)$ choice strings in $S'_{[c]}(s)$. We show that with less than two encoding symbols and no synchronizing symbol, we cannot find matches for s (with maximally allowed Hamming distance $d = k-2$) in the choice strings of $S'_{[c]}(s)$. Observe that in every choice string, because of the barriers, every length $k+1$ substring contains at most two encoding symbols and at most one symbol #. Further observe that, having taken a choice string from $S'_{[c]}(s)$, positions with symbols from $\{ [c_{i,j}] \mid 1 \le i < j \le k \}$ cannot coincide with the corresponding positions in s. Therefore s has a match in such a string only if s has two encoding symbols and one symbol # that all coincide with the corresponding positions in the selected substring. This proves the claim for $k \ge 4$. Regarding $k = 3$, if $|A_{[c]}(s)| < 3$, then the above argument applies here, too. If, however, $|A_{[c]}(s)| = 3$, a length 4 substring in every choice string has at least two positions that do not coincide with the corresponding positions in s.
□

Based on Lemma 13.22, we can now exactly specify the numbers and positions of the encoding and synchronizing symbols in the closest substring.

Lemma 13.23 *A closest substring s contains encoding symbols at its first k positions and a symbol # at its last position.*

Proof Let $n_{\#}(s)$ denote the number of symbols # in s, let $n_{[c]}(s)$ denote the number of string identification symbols in s, and let $n_{[v]}(s)$ denote the number of encoding symbols in s. Let $S'_{[c]}(s) \subseteq S_c$ be the subset of choice strings whose string identification symbol does not occur in s. In the following, we establish a lower bound on the number of strings in $S'_{[c]}(s)$ and an upper bound on the number of strings from $S'_{[c]}(s)$ in which we can find a match for s. By comparing these bounds, we will show that, if $n_{\#}(s) > 1$, there are choice strings in $S'_{[c]}(s)$ in which we cannot find a match; we will conclude that $n_{\#}(s) = 1$. Then we will show that if $n_{[v]}(s) < k$, there are again strings in $S'_{[c]}(s)$ without a match for s; which will lead to the conclusion that $n_{[v]}(s) = k$.

Regarding the size of $S'_{[c]}(s)$, a lower bound on its size is $|S'_{[c]}(s)| \ge \binom{k}{2} - n_{[c]}(s)$. To obtain an upper bound on the number of strings in $S'_{[c]}(s)$ in which we can find a match for s, recall that such matches must contain two encoding symbols and one symbol # that all coincide with the corresponding positions in s. On the one hand, the synchronizing symbol of a block must coincide with a symbol # in s. On the other hand, in all blocks of a choice string, its encoding symbols are in fixed positions relative to the block's synchronizing symbol; for example, in the choice string $c_{1,2}$, the encoding symbols are located only at the first and second positions and # at the last position of a block in $c_{1,2}$. For these two reasons, one symbol $\#$ in s can provide matches in at most $\binom{n_{[v]}(s)}{2}$ choice strings from $S'_{[c]}(s)$. Consequently, $n_{\#}(s)$ many symbols $\#$ in s can provide matches in at most $n_{\#}(s) \cdot \binom{n_{[v]}(s)}{2}$ choice strings from $S'_{[c]}(s)$.

Summarizing, we have at least $\binom{k}{2} - n_{[c]}(s)$ choice strings in $S'_{[c]}(s)$ and we can find matches in at most $n_{\#}(s) \cdot \binom{n_{[v]}(s)}{2}$ of them. Thus, we find matches for s in all choice strings only if

$$n_{\#}(s) \cdot \binom{n_{[v]}(s)}{2} \geq \binom{k}{2} - n_{[c]}(s). \tag{13.1}$$

In order to show that s contains exactly one synchronizing symbol, we assume that $n_{\#}(s) > 1$ (we know that $n_{\#}(s) \geq 1$ by Lemma 13.22) while $k > 2$, and show that then inequality (13.1) is violated.

We know that $k + 1 = n_{[v]}(s) + n_{[c]}(s) + n_{\#}(s)$ and, by Lemma 13.22, that $n_{[v]}(s) \geq 2$. Using this, we conclude on the one hand, that $n_{\#}(s) \cdot \binom{n_{[v]}(s)}{2} \leq n_{\#}(s) \cdot \binom{k+1-n_{\#}(s)}{2}$ and, since $n_{\#}(s) \geq 2$, that $n_{\#}(s) \cdot \binom{k+1-n_{\#}(s)}{2} \leq 2 \cdot \binom{k-1}{2}$. On the other hand, we have $\binom{k}{2} - n_{[c]}(s) \geq \binom{k}{2} - (k - 1 - n_{\#}(s))$ and, since $n_{\#}(s) \geq 2$, also $\binom{k}{2} - (k - 1 - n_{\#}(s)) \geq \binom{k}{2} - (k - 3)$. For $k \geq 3$, however, we have $\binom{k}{2} - (k - 3) > 2 \cdot \binom{k-1}{2}$. Thus,

$$\begin{aligned} n_{\#}(s) \cdot \binom{n_{[v]}(s)}{2} \leq n_{\#}(s) \cdot \binom{k+1-n_{\#}(s)}{2} &< \tbinom{k}{2} - (k - 1 - n_{\#}(s)) \\ &\leq \tbinom{k}{2} - n_{[c]}(s), \end{aligned}$$

that is, there are choice strings in $S'_{[c]}(s)$ which contain no match for s, a contradiction. Since $n_{\#}(s) \geq 1$ (Lemma 13.22), we conclude that $n_{\#}(s) = 1$.

In order to show that s contains exactly k encoding symbols, we assume that $n_{[v]}(s) < k$ while $k > 2$ and $n_{\#}(s) = 1$, and show that inequality (13.1) is violated. Since $k + 1 = n_{[v]}(s) + n_{[c]}(s) + n_{\#}(s) = n_{[v]}(s) + n_{[c]}(s) + 1$, we have $\binom{k}{2} - n_{[c]}(s) = \binom{k}{2} - (k - n_{[v]}(s))$ and, thus,

$$\binom{n_{[v]}(s)}{2} < \binom{k}{2} - (k - n_{[v]}(s)) = \binom{k}{2} - n_{[c]}(s),$$

that is, again, some strings in $S'_{[c]}(s)$ have no match for s, a contradiction. Thus, on the one hand, we have $n_{[v]}(s) \geq k$, and, on the other hand, we have $n_{\#}(s) = 1$ and therefore $n_{[v]}(s) \leq k$.

Note that if an encoding symbol is located *after* the synchronizing symbol in s, then due to the barriers it is not possible that both $\#$ and this encoding symbol will coincide with the respective positions in every choice string from $S'_{[c]}(s)$, e.g. in $c_{1,2}$. Therefore, symbol $\#$ is located at the last position of s. □

Proposition 13.24 *The first k characters of a closest substring correspond to k vertices of a clique in the input graph.*

Proof By Lemma 13.23, a closest substring s has encoding symbols at its first k positions and a synchronizing symbol at its last position. Consequently, the blocks are the only possible matches of s in the choice string. Now, assume that $s = [v_{h_1}]\,[v_{h_2}]\ldots[v_{h_k}]\,\#$ for $h_1, h_2, \ldots, h_k \in \{1, \ldots, n\}$. Consider any two h_i, h_j, $1 \le i < j \le k$, and the choice string $c_{i,j}$. Recall that in this choice string, the blocks encode edges at their ith and jth position, they have # at their last position, and all their other positions are set to a string identification symbol unique for this choice string. Thus we can only find a block that is a match if there is a block with $[v_{h_i}]$ at its ith position and $[v_{h_j}]$ at its jth position. We have such a block only if there is an edge connecting v_{h_i} and v_{h_j}. Summarizing, the closest substring s implies that there is an edge between every pair of $\{v_{h_1}, v_{h_2}, \ldots, v_{h_k}\}$; that is, these vertices form a k-clique in the input graph. □

Propositions 13.21 and 13.24 establish the following hardness result. (Note that hardness for the combination of all three parameters also implies hardness for each subset of the three.)

Theorem 13.25 CLOSEST SUBSTRING *with unbounded alphabet is W[1]-hard for every combination of the parameters L, d, and k.*

As mentioned before, the parameterized reduction above can be strengthened for parameter k in order to show $W[1]$-hardness in the case of binary alphabets. In contrast to the previous construction, here one cannot encode every vertex with its own symbol and one cannot use a unique symbol for every string produced. Also, one has to find new ways to "synchronize" the matches of a solution, a task previously done by the synchronizing symbol "#". To overcome these problems, one can construct an additional "complement string" for the input instance and lengthen the blocks in the produced choice strings considerably. We omit the lengthy technical details, just stating the result.

Theorem 13.26 CLOSEST SUBSTRING *is W[1]-hard with respect to the number of input strings k for binary input alphabet.*

13.3.2 *Further reductions and W[2]-hardness*

We have used WEIGHTED CNF-SATISFIABILITY (see Definition 13.4) as the "defining" complete problem for the class $W[2]$. In Example 13.5 we have seen that DOMINATING SET can be expressed as a weighted CNF-satisfiability problem, showing that DOMINATING SET is contained in $W[2]$. In fact—by a fairly complicated construction—one can also show that, given a Boolean formula F in conjunctive normal form and a nonnegative integer k, a graph G can be constructed such that F has a weight-k satisfying truth assignment iff G has a size-$2k$ dominating set. We only state the result here and refer to the literature for the proof.

Theorem 13.27 DOMINATING SET *is W[2]-complete with respect to the size of the solution set.*

Theorem 13.27 can be used as the starting point for several $W[2]$-hardness results. We provide three simple examples.

Our first example deals with the HITTING SET problem. Recall the definition.

> **Input:** A collection $\mathcal{C}$ of subsets of a base set S and a nonnegative integer k.
> **Question:** Is there a subset $S' \subseteq S$ with $|S'| \leq k$ such that S' contains at least one element from each subset in $\mathcal{C}$?

In Section 8.4 we have observed that the special cases with subset sizes bounded from above by a constant are fixed-parameter tractable. By way of contrast, an unbounded subset size leads to $W[2]$-hardness.

Theorem 13.28 HITTING SET *is $W[2]$-hard with respect to the size of the solution set.*

Proof We give a parameterized reduction from DOMINATING SET to HITTING SET. Let $(G = (V, E), k)$ be an instance of DOMINATING SET. We construct an instance of HITTING SET as follows. Let $S := V$ and let $\mathcal{C} := \{N[v] \mid v \in V\}$, that is, the subsets in $\mathcal{C}$ are exactly the closed neighborhoods of all vertices in V. Based on the fact that a vertex can only be dominated by a vertex from its closed neighborhood, we can immediately conclude that G has a dominating set of size k iff $\mathcal{C}$ has a hitting set of size k. The reduction is clearly computable in polynomial time. □

HITTING SET has a close relative: SET COVER.

> **Input:** A base set $S = \{s_1, s_2, \ldots, s_n\}$, a collection $\mathcal{C} = \{c_1, \ldots c_m\}$ of subsets of S such that $\bigcup_{1 \leq i \leq m} c_i = S$, and a nonnegative integer k.
> **Question:** Is there a subset $\mathcal{C}'$ of $\mathcal{C}$ of size at most k which covers all elements in S, that is, $\bigcup_{c \in \mathcal{C}'} c = S$?

In Section 9.5 we have studied (fixed-parameter) tractable variants of SET COVER. By way of contrast, we show here that the general problem is $W[2]$-hard.

Theorem 13.29 SET COVER *is $W[2]$-hard with respect to the size of the solution set.*

Proof The $W[2]$-hardness of SET COVER follows from a well-known equivalence between SET COVER and HITTING SET. Let $(\mathcal{C}, k)$ be an input instance of HITTING SET over the base set S with $S = \{s_1, s_2, \ldots, s_n\}$ and $\mathcal{C} = \{c_1, c_2, \ldots, c_m\}$. Define

$$\hat{\mathcal{C}} = \{c'_1, c'_2, \ldots, c'_n\},$$

where

$$c'_i := \{t_j \mid 1 \leq j \leq m, s_i \in c_j\}.$$

Here, $S' := \{t_1, t_2, \ldots, t_m\}$ forms the base set in the set cover instance. By considering the analogous inverse reduction one easily obtains the equivalence of both problems in the sense that the HITTING SET instance has a size-k solution iff the SET COVER instance has a size-k solution. This straightforward obviously correct polynomial-time reduction works in both directions. □

Our last example has a more exotic flavor, presenting a fairly new variant of the domination problem in graphs, applied to electric power networks. The problem is called POWER DOMINATING SET and we need two "observation rules" to define it.

Observation Rule 1 (OR1): A vertex in the power dominating set observes itself and all of its neighbors.

Observation Rule 2 (OR2): If an observed vertex v of degree $d \geq 2$ is adjacent to $d - 1$ observed vertices, then the remaining unobserved vertex becomes observed as well.

Thus, the definition of POWER DOMINATING SET reads as follows:

> **Input:** A graph $G = (V, E)$ and a nonnegative integer k.
> **Question:** Is there a subset $C \subseteq V$ with at most k vertices that observes all vertices in V with respect to the observation rules OR1 and OR2?

Note that the classical DOMINATING SET problem can be defined by simply omitting OR2. Subsequently, the following simple lemma is useful—the reader is asked to prove it in the exercises.

Lemma 13.30 *If G is a graph with at least one vertex of degree three or higher, then there is always a minimum power dominating set which contains only vertices with degree as least three.*

The following reduction, as a side result, also gives a simplified NP-hardness proof for POWER DOMINATING SET.

Theorem 13.31 POWER DOMINATING SET *is $W[2]$-hard with respect to the size of the solution set.*

Proof We give a parameterized reduction from DOMINATING SET to POWER DOMINATING SET. Given an instance $(G = (V, E), k)$ of DOMINATING SET, we construct an instance $(G' = (V \cup V_1, E \cup E_1), k)$ of POWER DOMINATING SET by simply attaching newly introduced degree-1 vertices to all vertices from V. Figure 13.2 illustrates this transformation.

For a dominating set D of G, we set $C := D$ to be a power dominating set for G'. By definition, all vertices from V are observed by OR1. Applying OR2 to every vertex in V, the vertices in V_1 become observed as well. Thus, C is a power dominating set of G'.

If G' has a power dominating set C with $|C| = k$, we can assume due to Lemma 13.30 that each vertex in C has degree at least three. This implies that $C \cap V_1 = \emptyset$. The proof is by contradiction. Assume that C is not a dominating set of G. Then there is a vertex $v \in V$ with $N_G[v] \cap C = \emptyset$. Let v' denote the

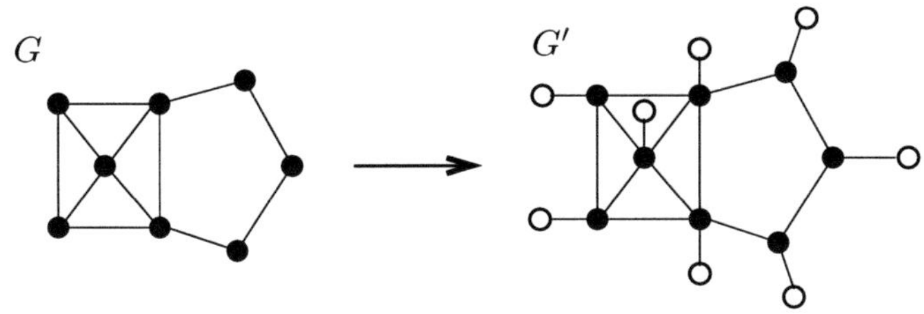

FIG. 13.2. An example of the reduction from DOMINATING SET to POWER DOMINATING SET in the proof of Theorem 13.31. The vertices in V_1 are drawn white.

newly introduced degree-1 neighbor of v in G'. We also have that $N_{G'}[v] \cap C = \emptyset$. Vertex v can get observed in G' only by applying OR2 to one of its neighbors in G'. Denote this neighbor by u. It is easy to see that u can not be v'. Hence, $u \in V$. Furthermore, u has also a degree-1 neighbor $u' \in V_1$ in G'. Since $u \notin C$ and $u' \notin C$, u' can be observed only by applying OR2 to u. However, this is impossible since u has two unobserved neighbors v and u'. Thus, C cannot be a power domination set of G', yielding a contradiction. □

So far, we have seen several examples of proving $W[1]$- and $W[2]$-hardness. Many other (and also more intricate) examples can be found in the literature. We conclude this section by demonstrating that one single problem with its parameterization and slight variations can exhibit the whole scenario of parameterized complexity. To this end, we consider the *NP*-complete LONGEST COMMON SUBSEQUENCE problem, which has numerous applications ranging from computational biology to text processing.

> **Input:** A set of k strings $s_1, s_2, \ldots, s_k$ over an alphabet Σ and a positive integer L.
> **Question:** Is there a string $s \in \Sigma^*$ of length at least L that is a subsequence of every s_i, $1 \leq i \leq k$?

Three parameters appear naturally in LONGEST COMMON SUBSEQUENCE:

- the number k of input strings;
- the length L of the common subsequence; and
- somewhat aside, the size of the alphabet Σ.

In the case of constant-size alphabets it is straightforward to see that LONGEST COMMON SUBSEQUENCE is fixed-parameter tractable with respect to parameter L by a simple brute-force enumerative approach—the combinatorial explosion amounts to $|\Sigma|^L$. By way of contrast, it is known that the problem is $W[1]$-hard with respect to parameter k. In the case of unbounded alphabet size, however, parameterized hardness predominates. LONGEST COMMON SUBSEQUENCE then is

- $W[t]$-hard for all $t \geq 1$ with respect to parameter k,

- $W[2]$-hard with respect to parameter L, and
- $W[1]$-complete with respect to the combined parameters k and L.[8]

We mention in passing that the proofs of all above hardness results are fairly demanding.

13.4 Some recent developments

To fully explore the rich literature of parameterized complexity issues is far beyond the scope of this book. Thus, in what follows we present only a small selection of recent developments that deserve further investigations elsewhere. Our presentation here is completely informal. To obtain a more accurate picture of developments, studying the literature we will point to is essential.

13.4.1 *Lower bounds and the complexity class $M[1]$*

There may be problems lying between *FPT* and $W[1]$. The class $M[1]$ is a proposal for classifying problems in this field, where $FPT \subseteq M[1] \subseteq W[1]$ and it is an open problem of current research whether $M[1] = W[1]$ or not. Whatever the answer to this question turns out to be, $M[1]$ plays an important role in the context of "subexponential lower bounds" for fixed-parameter algorithms. The main technical motivation for introducing $M[1]$ is based on the following observation. Recalling that weighted satisfiability problems play a central role in the definition of the W-hierarchy, consider the following "trick", which leads to reductions between satisfiability and weighted satisfiability problems: specifying a weight-k assignment for a set of n variables requires $k \cdot \log n$ bits for binary encoding. This can be used to reduce the weighted satisfiability problem of a formula with n variables to an unweighted satisfiability problem for a formula with only $k \cdot \log n$ variables.

Based on our current knowledge, there are two natural "routes" to $M[1]$.

- The first is called the *renormalization route*. One "renormalizes" the VERTEX COVER problem by defining the problem $k \log n$-VERTEX COVER.

  Input: An n-vertex graph $G = (V, E)$ and a nonnegative integer k.
  Question: Does G have a vertex cover of size at most $k \cdot \log n$?

  Using parameterized reductions[9], this can be used as a defining $M[1]$-complete problem. Note that $k \log n$-VERTEX COVER can be trivially solved in $n^{O(k)}$ steps simply by translating the size-2^k search tree idea for the standard VERTEX COVER problem into this context.

[8]Observe that in the case of combined parameters one might have a running time such as $2^{O(kL)}$ which would imply fixed-parameter tractability with respect to the combined parameter. By way of contrast, such a running time would neither imply fixed-parameter tractability with respect to the parameter k nor with respect to the parameter L.

[9]Note that instead of the many–one reductions we used before so-called Turing reductions are more suitable here. We refer to the literature for details.

- The second is called the *miniaturization route* and it shows that the basic idea behind the definition of $M[1]$ and classes beyond (again there is a whole M-hierarchy, $M[1]$ being the lowest level as $W[1]$ is for the W-hierarchy) is a translation of a natural hierarchy of satisfiability problems into the parameterized complexity context. Importantly, the supposed fixed-parameter intractability of the corresponding satisfiability problems lays the foundations for these complexity classes. Consider the MINI-3-CNF-SATISFIABILITY problem:

  Input: A Boolean formula F in conjunctive normal form with at most three literals per clause, and two nonnegative integers k and n encoded over a one-element alphabet such that F contains at most $k \cdot \log n$ variables.
  Question: Does F have a satisfying truth assignment?

  By means of parameterized reductions, MINI-3-CNF-SATISFIABILITY can again be used as a defining $M[1]$-complete problem. It can be trivially solved by checking all $2^{k \cdot \log n} = n^{O(k)}$ truth assignments.

We mention in passing that the main combinatorial tool in the development of a $M[1]$-completeness theory is a famous *sparsification lemma* which basically says that the satisfiability problem for d-CNF formulae can be reduced to the satisfiability problem for d-CNF formulae whose size is linear in the number of variables by using a suitable reduction that also preserves subexponential-time solvability.

Notably, whereas $k \log n$-VERTEX COVER and $k \log n$-INDEPENDENT SET are $M[1]$-complete, $k \log n$-CLIQUE turns out to be fixed-parameter tractable because the CLIQUE problem for an n-vertex graph can be solved in $n^{O(\sqrt{n})}$ time. We omit any details.

Recall from Theorem 13.15 that $W[1] = FPT$ would imply that the 3-CNF-SATISFIABILITY problem for a Boolean formula F with n variables can be solved in $2^{o(n)} \cdot |F|^{O(1)}$ time. This can be turned into an equivalence as follows.

Theorem 13.32 *$FPT = M[1]$ iff the* 3-CNF-SATISFIABILITY *problem for a formula with n variables can be solved in basically $2^{o(n)}$ steps.*

Here, as on previous occasions, "basically" means that we ignore polynomial factors in the running time.

Again the proof is beyond the scope of this book. The assumption that 3-CNF-SATISFIABILITY cannot be solved in basically $2^{o(n)}$ steps for an n-variable formula is also known as the *exponential time hypothesis*.

Besides providing a tool for fixed-parameter intractability results, $M[1]$ can also be used in the deployment of subexponential lower bounds. We give two examples in the following theorem, not proven here.

Theorem 13.33 *If $FPT \neq W[1]$, then there is no fixed-parameter algorithm that solves* VERTEX COVER *with a combinatorial explosion $f(k) = 2^{o(k)}$, and* *there is no fixed-parameter algorithm that solves* VERTEX COVER IN PLANAR GRAPHS *with a combinatorial explosion* $f(k) = 2^{o(\sqrt{k})}$.

In summary, one might say that currently the main use of $M[1]$ at this point in time lies in showing subexponential lower bounds such as those in Theorem 13.33.

13.4.2 *Lower bounds and linear FPT reductions*

As in classical complexity theory, the derivation of (relative) lower bounds in parameterized complexity theory is based on a "reduction-and-completeness"-program. So far, we have seen how classes such as $W[1]$ or $M[1]$ are employed for this purpose. In particular, as indicated in the previous subsection, this line of research has led to subexponential (relative) lower bounds for fixed-parameter tractable problems. Now let us return to $W[1]$-hard problems and their presumed fixed-parameter intractability.

By definition, $W[1]$-hardness does not, in principle, exclude the possibility that say CLIQUE in an n-vertex graph can be solved in $n^{O(\log\log k)}$ time. Such an algorithm would be very useful for moderate values of k. The best known algorithms for CLIQUE, however, run in $n^{O(k)}$ time. To close this gap, a refined parameterized reducibility concept has been introduced which we very briefly sketch in the following. To simplify the presentation and to make things more concrete, we consider the CLIQUE problem for a graph with n vertices and overall input size m. We have to find k out of the n vertices—this forms the *search space*—that induce a clique. Thus in the following definition we assume that a parameterized problem is characterized by a triple (m, n, k), where m is the overall input size, n is the size of the search space, and k is the parameter. The central definition then reads as follows.

Definition 13.34 *There is a linear parameterized reduction from a parameterized problem L with an* (m, n, k)*-instance* x *to a parameterized problem* L' *with an* (m', n', k')*-instance* x' *if there is an algorithm that runs in* $f(k) \cdot n^{o(k)} \cdot m^{O(1)}$ *time and produces an* x' *with*

- $m' = m^{O(1)}$,
- $n' = n^{O(1)}$,
- $k' = O(k)$, *and*
- x *is a yes-instance of* L *iff* x' *is a yes-instance of* L'.

Using this refined type of reducibility with linear dependence between the parameters then leads to refined concepts of $W[1]$-and $W[t]$-hardness for $t \geq 2$. The point again is that (relative) lower bounds for weighted satisfiability problems are proven by leading these problems back to the solvability of unweighted satisfiability problems. These lower bounds, together with linear parameterized reductions, then are used to establish lower bounds for $W[1]$-hard problems such as CLIQUE or DOMINATING SET (the latter being even $W[2]$-hard). For instance, the following result has been shown.

Theorem 13.35 *Unless* $W[1] = FPT$, DOMINATING SET *and* WEIGHTED CNF-SATISFIABILITY *cannot be solved in* $f(k) \cdot n^{o(k)} \cdot m^{O(1)}$ *time for any computable function* f.

Note that Theorem 13.35 refers to two $W[2]$-hard problems and deals with only two example problems. In the case of $W[1]$-hardness, under somewhat different conditions referring to the (non-)existence of subexponential-time algorithms, the non-existence of algorithms running in $f(k) \cdot m^{o(k)}$ for any function f has been shown for problems such as CLIQUE and WEIGHTED q-CNF-SATISFIABILITY.

In conclusion, we note that all of the above-mentioned problems, either $W[1]$- or $W[2]$-hard, are solvable by algorithms running in $O(n^k \cdot m^2)$ time by enumerating all size-k subsets of the solution space. New approaches are needed to significantly improve these upper bounds.

13.4.3 *Machine models, limited nondeterminism, and bounded FPT*

All parameterized complexity classes except for FPT have been defined using complete problems—they have not been defined based on natural machine characterizations based, for instance, on certain Turing machine models. This is probably one of the main reasons that makes these complexity classes harder to understand than classical ones. For the class $W[P]$, however, there is a very nice machine characterization. A problem is in $W[P]$ iff it is solvable by a nondeterministic fixed-parameter algorithm whose use of nondeterminism is bounded in terms of a parameter. More precisely, one can show:

Theorem 13.36 *A parameterized problem with parameter* k *and input size* n *is in* $W[P]$ *iff it is solvable in* $f(k) \cdot n^{O(1)}$ *time by a nondeterministic Turing machine that makes at most* $f(k) \cdot \log n$ *nondeterministic steps for some computable function* f.

Theorem 13.36 clearly indicates a strong relationship between parameterized complexity and the concept of *limited nondeterminism.* Moreover, one can also show that a problem is in $W[1]$ iff it is solvable by a nondeterministic fixed-parameter algorithm that makes its nondeterministic decisions only among the last steps of the computation. To make precise what is meant by "among the last steps" the concept of nondeterministic random access machines is useful—we skip any details here. Note, however, that machine characterizations are particularly useful for showing membership in a class, whereas in this book the classes of the W-hierarchy are mainly of interest in terms of showing parameterized hardness results. To this end, machine models are of less importance in our context.

Closer to the focus of an algorithmic point of view on parameterized complexity of problems is the question of how small the combinatorial explosion in a fixed-parameter algorithm can be made. For instance, note that fixed-parameter algorithms obtained from Courcelle's famous theorem (see Section 10.6) are completely impractical due to the huge combinatorial explosions involved. Hence, to better differentiate between the extremely varying combinatorial explosions of fixed-parameter algorithms, the notion of bounded fixed-parameter tractability has been proposed. The idea is to put upper bounds on the growth of the "parameter dependence" f, the two most natural being $f \in 2^{k^{O(1)}}$ and $f \in 2^{O(k)}$. The resulting bounded fixed-parameter classes probably contain all problems that are "fixed-parameter tractable in practice". In this way, one may define natural subclasses of *FPT* and, on top of them, together with a refined reducibility concept, a hierarchy of classes corresponding to various degrees of presumable "bounded fixed-parameter intractability" similar to the W-hierarchy.

In this context note that certain natural model-checking problems that are known to be fixed-parameter tractable in the classical "unbounded sense" have a very high complexity in terms of bounded parameterized complexity, illustrating why these fixed-parameter tractability results might be of little practical value.

13.5 Summary and concluding remarks

Summarizing, this chapter provides an excursion into the deep and rich world of parameterized complexity theory. We have mostly just scratched the surface. For deeper insights into structural complexity, further reading of the cited literature is highly recommended.

A general observation is that it usually takes a while to familiarize oneself with the difficulties and pitfalls that the theory bears. The main reason is that, because the parameters play the key role, all constructions and—in particular—the reductions need to be more fine-grained, thus posing higher technical demands. To some extent, this is similar to the case of approximation theory with its "approximation-preserving" reductions which are also more subtle than standard polynomial-time many–one reductions.

Finally, let us emphasize that parameterized complexity theory is still a highly dynamic research field with numerous recent innovations and ongoing new developments. We end by highlighting five points in this respect.

- With the concept of bounded fixed-parameter tractability a desirable substructuring of the class of fixed-parameter tractable problems has recently been initiated.
- Lower bounds on the running time of exact algorithms for fixed-parameter tractable problems as well as for $W[1]$-hard problems are another interesting research issue that has only recently been addressed.
- The parameterized complexity class $M[1]$ as a candidate for lying between the classes *FPT* and $W[1]$ leads to new challenges such as asking for new parameterized intractability results using $M[1]$-hardness as a basic degree of parameterized intractability or asking whether $M[1] = W[1]$.
- Several further parameterized complexity classes (and corresponding hierarchies) not discussed here are known and lead to numerous challenges concerning structural complexity investigations to clarify their mutual relationships.

- So far, only a few links have been established between parameterized intractability theory and inapproximability theory (see Chapter 14); both theories should have more to say to each other than we currently know.

13.6 Exercises

1. What is the intuitive explanation for why it is not surprising that there exists a parameterized reduction from CLIQUE, which is classically *NP*-complete, to a problem such as VC DIMENSION (also see Section 15.6.4) which is not known to be *NP*-hard but contained in *NP*?
2. Show that the problem to decide whether a 3-CNF formula has a satisfying truth assignment of weight at most k is fixed-parameter tractable with respect to parameter k.
3. Show that parameterized reductions are transitive. That is, for parameterized problems L_1, L_2, L_3, if $L_1 \leq L_2$ and $L_2 \leq L_3$ by parameterized reductions, then also $L_1 \leq L_3$ by a parameterized reduction.
4. Show that the following problems are $W[1]$-hard:
   (a) The variant of PARTIAL VERTEX COVER where we wish to choose at least k vertices such that at most t edges are covered; the parameter being k.
   (b) The introductory example in Section 1.2 with the parameter being the size of the desired set.
   (c) The "exact variant" of MAXIMUM SATISFIABILITY where one only counts clauses where exactly one literal is true; the parameter is the number of clauses with exactly one true literal.
   (d) STEINER TREE IN GRAPHS with respect to the parameter "number of non-terminal vertices", that is, the dual parameterization to the one studied in Section 9.3.
   (e) SET PACKING with respect to the parameter "number of mutually disjoint sets"; see Section 11.4.2.
5. Prove Lemma 13.30.

13.7 Bibliographical remarks

Most material in this chapter is taken from the groundbreaking monograph Downey and Fellows (1999); also refer to this work for older literature not cited here. Some recent surveys also provide insight in latest developments in parameterized complexity theory (Downey, 2003; Fellows, 2003*a*; Fellows, 2003*b*; Flum and Grohe, 2004*a*).

Concerning some of the parameterized reductions, note that WEIGHTED VERTEX COVER problems are studied in Niedermeier and Rossmanith (2003*b*), the $W[1]$-hardness of PARTIAL VERTEX COVER is proven in Guo *et al.* (2005*b*), and the $W[1]$-hardness of CLOSEST SUBSTRING is from Fellows *et al.* (2002). See Marx (2005) for a very recent result showing the $W[1]$-hardness of CLOSEST

SUBSTRING also with respect to the distance parameter d. The LONGEST COMMON SUBSEQUENCE problem is studied in Bodlaender *et al.* (1995), Downey and Fellows (1999), and Pietrzak (2003), the close relationship between SET COVER and HITTING SET was already observed in Ausiello *et al.* (1980), and the $W[2]$-hardness of POWER DOMINATING SET independently appears in Guo *et al.* (2005*a*), and Kneis *et al.* (2004).

The development of the class $M[1]$ goes back to Downey *et al.* (2003), and is further investigated in Chen and Flum (2004). The roots of studying $M[1]$ and lower bounds for fixed-parameter tractable problems lie in Cai and Juedes (2003), with close relations to the earlier work Impagliazzo *et al.* (2001). The study of lower bounds and linear FPT reductions is due to Chen *et al.* (2004*b*). More about machine models for parameterized complexity classes can be found in Chen *et al.* (2005); limited nondeterminism is discussed in Papadimitriou and Yannakakis (1996) and Chen *et al.* (2003). The use of special versions of the TURING MACHINE HALTING problem to show membership in the classes $W[1]$, $W[2]$, and $W[P]$ is advocated in Cesati (2003). Finally, studying the structure inside FPT introducing bounded fixed-parameter tractability has been undertaken in Flum *et al.* (2004).

14

CONNECTIONS TO APPROXIMATION ALGORITHMS

In dealing algorithmically with *NP*-hard problems, polynomial-time approximation algorithms have so far been the main focus of the theoretical computer science community: when we cannot afford to search for optimal solutions, then let us investigate how good an approximate solution we can find in polynomial time. For instance, in the case of VERTEX COVER we can always find—in linear time—a solution that is at most twice as large as an optimal one: instead of having to decide which of the two endpoints of an edge to choose, simply take both. Somewhat surprisingly, in the worst case this is basically the best strategy known; that is, it has been an open problem for more than twenty years whether there exists a factor-$(2-\epsilon)$ polynomial-time approximation algorithm for a constant $\epsilon > 0$. The deep and famous "PCP inapproximability theory" has proceeded to a point that shows that, unless $P = NP$, there is no polynomial-time approximation algorithm for VERTEX COVER with approximation factor better than 1.36.

In contrast to approximation algorithms, fixed-parameter algorithms aim to find optimal solutions at the price of accepting seemingly unavoidable exponential running times. This appears to be reasonable whenever the parameter to which the combinatorial explosion is confined turns out to be small in real applications. Thus, in some sense the concept of fixed-parameter tractability is a competitor of polynomial-time approximability with respect to coping with computational intractability. Indeed, both methodologies have their obvious pros and cons. Notably, polynomial-time approximability is the much deeper explored and better developed field at the current time.

What is there in favor of polynomial-time approximation algorithms?

- No matter what the input is, efficiency in terms of polynomial-time complexity is always guaranteed.[10]
- The approximation factor provides a worst-case guarantee, and in practical applications the actual approximation might be much better, thus turning approximation algorithms into useful heuristic algorithms as well.
- There is a huge arsenal of methods and techniques developed over the years in conjunction with studying approximation algorithms, and there is a strong and deep theoretical foundation for impossibility results particularly concerning lower bounds for approximation factors.

[10] In the author's opinion, however, the analysis of the degree of the polynomial in the running time is too often neglected. To this end, note that depending on the exponent, polynomial running time does not necessarily imply efficiency in a practical sense.

By way of contrast, on the one hand, fixed-parameter algorithms may have unreasonably high running times because the parameters might be too large or the involved combinatorial explosion might be astronomical even for small parameter values. On the other hand, the following speaks in favor of fixed-parameter algorithms:

- There is complete freedom to choose the "right" parameterization, which is not only fixed to measure the size of the solution but can also incorporate structural aspects of the input such as the treewidth parameter does.
- Fixed-parameter algorithms stick to worst-case analysis—in this case concerning the combinatorial explosion—but in practical applications they might turn out to run much faster than the worst-case runtime analysis suggests.
- Although not as developed and mature as inapproximability theory, with the famous PCP theorem, parameterized complexity has already made worthwhile contributions to structural complexity theory, culminating in $W[1]$-hardness theory.

Hence, in conclusion we propose the still young field of parameterized problem analysis and algorithm design as a valuable and interesting alternative to polynomial-time approximation algorithms. It seems that both fields can complement each other very well, thus giving hope for progress based on synergetic effects.

Altogether, however, instead of discussing where which approach might be be superior to the other, it is more productive to see where one approach may serve the other. Interaction between both disciplines is needed. The establishment of links between both fields is still underdeveloped and we hope to see more mutual fertilization in the future.

14.1 Approximation helping parameterization

We started our discussion by mentioning that it is a longstanding open problem to find a better-than-factor-two polynomial-time approximation algorithm for VERTEX COVER. In Section 7.4 we have seen that by a result of Nemhauser and Trotter—a result that originated in the context of approximation algorithms—a $2k$-vertices problem kernel for VERTEX COVER has been developed. From this, our first conclusion is:

- Approximation algorithms may help to prove fixed-parameter tractability results.[11]

There is a perhaps even more important statement we can make concerning lower bounds. Since a $(c \cdot k)$-vertices polynomial-time problem kernelization trivially

[11] This observation is not confined to a scenario dealing with reduction to a problem kernel. For instance, using a linear-time constant-factor approximation algorithm for FEEDBACK VERTEX SET, the iterative compression fixed-parameter algorithm presented in Section 11.3.2 can be turned into a "*linear-time* fixed-parameter algorithm"; that is, this algorithm runs in linear time for a constant parameter value k.

gives a polynomial-time factor-c approximation algorithm, we can conclude for VERTEX COVER that a problem kernel with less than $2k$ vertices seems unlikely. More specifically, since the PCP inapproximability theory tells us that there is no hope of a polynomial-time better-than-factor-1.36 approximation algorithm, we can infer that there is also no hope of a $1.36k$-vertices problem kernel. Hence our second conclusion is:

- Lower bound results for approximation algorithms directly yield lower bounds for problem kernel sizes; thus, PCP theory also helps parameterized complexity theory.

It remains for future work to extend these observations by perhaps also making deeper connections between both fields. In any case, there is no doubt that techniques developed in the context of approximability may provide tools and starting points which parameterized complexity can make use of.

14.2 Parameterization helping approximation

The field of parameterized complexity is younger and so far much less work-time has been invested in its exploration. Still it may help to derive new and fruitful insights concerning approximability. So far, these are mostly based on the $W[1]$-hardness theory and concern "impossibility results" in context with polynomial-time approximation schemes.

We become a little more formal here. The central starting point is the observation that every optimization problem that has an "efficient" polynomial-time approximation scheme is fixed-parameter tractable with respect to the parameter being the optimization value (which is the size of the solution set searched for): for ease of presentation, let us focus on minimization problems—the case of of maximization problems works in completely analogous fashion. Then, informally speaking, a minimization problem has

- a *polynomial-time approximation scheme* (*PTAS*) if, for any constant $\epsilon > 0$, there is a factor-$(1+\epsilon)$ approximation algorithm running in polynomial time;
- an *efficient polynomial-time approximation scheme* (*EPTAS*) if, for any constant $\epsilon > 0$, there is a factor-$(1+\epsilon)$ approximation algorithm running in $f(1/\epsilon)\cdot|x|^{O(1)}$ time for any computable function f only depending on $1/\epsilon$; and
- a *fully polynomial-time approximation scheme* (*FPTAS*) if, for any constant $\epsilon > 0$, there is a factor-$(1+\epsilon)$ approximation algorithm running in $(1/\epsilon)^{O(1)} \cdot |x|^{O(1)}$ time.

The decisive point above is that in the case of a PTAS the degree of the polynomial in the running time bound may depend on $1/\epsilon$, whereas in the case of an FPTAS or an EPTAS it may not. (Clearly, every FPTAS is an EPTAS.)

To state the central result, we need to define what we understand by *the standard parameterization* of an optimization (here only minimization) problem.

Definition 14.1 *Let x be an input instance of an optimization problem where we want to minimize the goal function $m(x)$. Then its* standard parameterization *is the pair (x, k), which means that we ask whether $m(x) \leq k$ for a given parameter value k.*

We are now ready for the main theorem.

Theorem 14.2 *If a minimization problem has an EPTAS, then its standard parameterization is fixed-parameter tractable.*

Proof Let x be an input instance with goal function $m(x)$ and let k be the threshold value of the standard parameterization, that is, the parameter. Choose $\epsilon := 1/(2k)$. Since there is an EPTAS, we know that there is an algorithm running in $f(2k) \cdot x^{O(1)}$ time that produces a solution with approximation factor $1 + 1/(2k)$. Hence if there is a solution for x such that $m(x) \leq k$, then we know that the above approximation algorithm can find a solution with goal value $(1 + 1/(2k)) \cdot k = k + 1/2$. Considering only integer-valued optimization goals, we may conclude that the constructed solution leads to the goal value k. This implies that the standard parameterization is fixed-parameter tractable. □

Clearly, the above considerations hold completely analogously for maximization problems. The point with Theorem 14.2 is not so much that it gives fixed-parameter tractability results using approximability results—usually a direct fixed-parameter approach yields faster algorithms. More important is the conclusion we can draw from reformulating the statement as follows.

Corollary 14.3 *If the standard parameterization of an optimization problem is not fixed-parameter tractable, then there is no EPTAS for this optimization problem.*

Hence parameterized complexity analysis may help to explore the borders of feasible approximability. Polynomial-time approximation schemes have been intensively studied in the approximation literature. Note that a PTAS may have nothing to do with practical feasibility. For instance, a PTAS may have a running time of $O(|x|^{3000/\epsilon})$ or $O(|x|^{(1/\epsilon)!})$. Thus, for practically interesting small ϵ the associated running times can become astronomically large. The problem is that the quality of approximation, expressed by the constant ϵ, is tied to the degree of the polynomial running time bound. This is not the case for an EPTAS. Hence it is useful to distinguish between optimization problems that allow for an EPTAS and those that only allow for a PTAS. Here, $W[1]$-hardness theory enters the stage, giving a clear theory-based indication of fixed-parameter intractability, and thus of the non-existence of an EPTAS. In fact, there are already examples in the literature of concrete $W[1]$-hard parameterized problems where the non-existence of an EPTAS is indicated whereas a PTAS is known.

Finally, we mention in passing that very recent developments along the lines sketched above have led to results such as stating that an optimization problem has no approximation scheme running in $f(1/\epsilon) \cdot |x|^{o(1/\epsilon)}$ time unless an unlikely collapse in parameterized complexity theory occurs.

14.3 Further (non-)relations

May one expect that a problem that has no "good" polynomial-time approximation algorithm will turn out to be well-behaved from a parameterized point of view? Or may one expect that a problem which is parameterized intractable is hard to approximate? Even if we restrict our attention to standard parameterizations (see Definition 14.1), the general answer is both times "no":

- There are problems such as GRAPH BIPARTIZATION (see Section 15.2.3) for which only factor $O(\log n)$ polynomial-time approximation algorithms are known but which have efficient and practical fixed-parameter algorithms.
- There are $W[1]$-hard problems (such as the so-called DISTINGUISHING SUBSTRING SELECTION problem, a generalization of CLOSEST SUBSTRING) that have PTASs or efficient constant-factor approximations.

Finally, we mention in passing that it also makes sense to combine both concepts. We give two examples here.

- A practically efficient factor-4 approximation algorithm to compute the treewidth of a graph and to construct a corresponding tree decomposition is actually a fixed-parameter algorithm in terms of running time (see also Section 10.2).
- One may easily combine depth-bounded search trees with approximation algorithms—stop the search at a higher level and replace the "undone search tree" part by an approximation algorithm. This may yield an approximation algorithm with an improved approximation factor and a fixed-parameter running time better than in the exact case. A concrete application has been described in the literature for the MAXIMUM SATISFIABILITY problem.

The above discussion only scratches the surface of a world of numerous fruitful interactions between parameterization and approximation—perhaps it will find more attention in future research.

14.4 Discussion and concluding remarks

There still is a gap between the communities of approximation and parameterized algorithmics which needs to be (better) bridged. This is also reflected in the brevity of this chapter. It seems clear that more and deeper interactions between both fields await discovery. Whereas approximation is the more mature field, parameterization seems to offer more opportunities for studying one and the same problem from different angles. This is due to the freedom of parameter choice. By way of contrast, the algorithmic tools and techniques as well as the corresponding lower bound theory currently form the larger and more attractive toolbox of approximation in comparison with parameterization. Perhaps future research will bring some progress such that the fields will grow further into each other.

14.5 Bibliographical remarks

The books Ausiello *et al.* (1999), Hochbaum (1997), and Vazirani (2001) provide an excellent basis for studying approximation algorithms. A few words about the connections between approximation and parameterized algorithmics can be found in Downey and Fellows (1999), Downey (2003), and Fellows (2003*b*). The lower bound for the approximation factor of VERTEX COVER is due to Dinur and Safra (2002). The deep and famous PCP theory is surveyed in Ausiello *et al.* (1999), Hochbaum (1997), and Vazirani (2001).

The mentioned use of a linear-time constant-factor approximation algorithm for FEEDBACK VERTEX SET in order to achieve linear-time fixed-parameter tractability is described in Guo *et al.* (2005). The connection between fixed-parameter tractability and EPTAS is established in Cesati and Trevisan (1997). The efficient fixed-parameter algorithm for GRAPH BIPARTIZATION is due to Reed *et al.* (2004). The "fixed-parameter treewidth approximation algorithm" appears in Reed (1993). The $W[1]$-hardness result for DISTINGUISHING SUBSTRING SELECTION and the thus seemingly first "proof of non-existence" of an EPTAS for a natural problem appears in Gramm *et al.* (2003). Finally, the combination of an approximation and an exact search tree algorithm for MAXIMUM SATISFIABILITY appears in Dantsin *et al.* (2001).

15

SELECTED CASE STUDIES

The objective of this chapter is to provide some further arguments for the versatility of the fixed-parameter complexity agenda. To this end, we enumerate several case studies from various fields and very briefly exhibit why and how parameterized complexity analysis is relevant here. In doing so, after some general statements concerning planar and related graphs, from Section 15.2 on we follow always the same structure:

1. we define and motivate the problem;
2. we make some general considerations concerning the fixed-parameter tractability of the problem;
3. we concisely state the known results;
4. we discuss its relevance and potential for future research; and
5. we point to basic references in the literature.

Here, for different problems, we choose different foci and descriptions with different degree of detail; the individual presentations are limited to at most two pages. Clearly, in all cases it is absolutely necessary to refer to the cited literature in order to obtain more complete pictures of the discussed problems.

Note that most fixed-parameter tractability results discussed in this chapter so far have combinatorial explosions that seem to make their practical usefulness doubtful. The selection of problems was done with the idea of presenting results that carry particular challenges for future research, namely to shrink the combinatorial explosions.

15.1 Planar and more general graphs

The study of the parameterized complexity of *NP*-complete problems in planar graphs was the incentive for findings concerning subexponential-time fixed-parameter algorithms. In follow-up work these results were generalized to larger classes of graphs.

15.1.1 *Planar graphs*

In this monograph we have encountered planar graphs in many places.

- In Section 7.6 we described a sophisticated kernelization algorithm for Dominating Set in Planar Graphs.
- In Section 8.6 we discussed a depth-bounded search tree for Dominating Set in Planar Graphs.
- In Section 10.3 we explored connections between planar graphs and the construction of tree decompositions.

The reasons why planar graphs play such a prominent role in this work are as follows.

- Planar graphs derive as natural abstractions from real-life problems.
- Many *NP*-complete graph problems remain *NP*-complete when restricted to planar graphs. However, they behave much better with respect to approximate (polynomial-time) or exact (exponential-time) solvability.
- Planar graphs have several nice properties:
  - ∗ they are *sparse*, that is, the number of edges is bounded from above by a number less than three times the number of vertices;
  - ∗ there always exists a vertex of maximum degree five;
  - ∗ they are characterized by forbidding the complete graph K_5 and the complete bipartite graph $K_{3,3}$ as minors; and
  - ∗ they have nice decomposition properties to be further discussed in what follows.

There are basically two routes to cope with *NP*-complete problems in planar graphs using fixed-parameter algorithms.

1. Through the use of classical separator theorems. Let $G = (V, E)$ be an undirected graph. A *separator* $S \subseteq V$ of G divides V into two parts $A_1 \subseteq V$ and $A_2 \subseteq V$ such that
   - $A_1 \cup S \cup A_2 = V$, and
   - no edge joins vertices in A_1 and A_2.

   Informally speaking, the point is that for planar n-vertex graphs there is always a graph separator S with $|S| = O(\sqrt{n})$ that divides V into two sets A_1, A_2 each with at most $2n/3$ vertices. Hence, a divide-and-conquer approach applies, yielding algorithms with exponential running time factor $c^{\sqrt{n}}$ for some constant c. In the case of linear-size problem kernels, this can be extended into fixed-parameter algorithms with subexponential combinatorial explosion $c^{\sqrt{k}}$ for parameterized graph problems. An example is DOMINATING SET where parameter k denotes the size of the solution set.
2. A closely related but conceptually different route is that through the concept of k-outerplanarity as sketched in Section 10.3. This has been developed into a fairly general methodology applying to graph problems with the so-called “Layerwise Separation Property”: in this case one can run a general algorithm which quickly computes a tree decomposition of guaranteed small width. Again, this leads to fixed-parameter algorithms with combinatorial explosion $c^{\sqrt{k}}$ for some constant c.

Literature. The literature is rich with studies of subexponential fixed-parameter algorithms for problems such as VERTEX COVER or DOMINATING SET in planar graphs. First results were derived in Alber *et al.* (2002), Alber *et al.* (2003), and Alber *et al.* (2004), with later improvements concerning running times such as the ones in Fomin and Thilikos (2003). The results all suffer from constants c that are still too big in the exponential terms $c^{\sqrt{k}}$.

15.1.2 *More general graphs*

The encouraging results for planar graphs have stirred interest in possible extensions to more general classes of graphs. Under these, graph classes such as the following have been successfully investigated:

- graphs of bounded genus (note that planar graphs have genus 0);
- disk graphs;
- map graphs;
- $K_{3,3}$-minor-free or K_5-minor-free graphs;
- minor-closed families for bounded local treewidth;

and several others. We particularly remark that very recently a "bidimensional theory" of bounded-genus graphs (somewhat analogous to the famous graph minor theory) has been developed. In this context, bidimensionality is considered as a general approach for obtaining treewidth-parameter bounds and therefore subexponential fixed-parameter algorithms. Roughly speaking, a parameterized graph problem is bidimensional if the parameter is large in "grid-like graphs", that is, linear in the number of vertices, and if it is (for instance) closed under taking minors. Examples of bidimensional problems are VERTEX COVER, FEEDBACK VERTEX SET, DOMINATING SET, and many others.

Note, however, that all of the above considerations appear to be of mainly theoretical interest because of the large constants involved.

Literature. The bidimensionality theory is developed in Demaine and Hajiaghayi (2005), Demaine *et al.* (2004*a*), and Demaine *et al.* (2004*c*). Other results for generalizations of planar graphs are to be found in, among others, Alber and Fiala (2004), Demaine *et al.* (2003), Demaine *et al.* (2005), Fomin and Thilikos (2004*b*), and Chen *et al.* (2003).

15.2 Graph modification problems

Graph modification problems provide a very natural setting for studying parameterized complexity in terms of a parameterization by the number of modification operations. There is a long list of problems arising in various fields of applications that fall into the category of graph modification. For instance, one may ask

- to delete the minimum number of vertices of a graph to transform it into a forest—this is the *NP*-complete FEEDBACK VERTEX SET problem[12] (see Section 11.3.2;
- to delete the minimum number of vertices or edges to transform it into a bipartite graph—these are the *NP*-complete problems GRAPH BIPARTIZATION and EDGE BIPARTIZATION;
- to add a minimum number of edges to transform a graph into a chordal one, that is, a graph in which every induced cycle of length at least four has a chord—this is the *NP*-complete MINIMUM FILL-IN problem;

[12] In the case of deleting edges instead of vertices, the problem is solvable in polynomial time—this is equivalent to find a spanning forest with a maximum number of edges.

- to add and delete a minimum number of edges to transform a graph into a collection of disjoint cliques—this is the *NP*-complete CLUSTER EDITING problem (see Sections 7.3 and 8.2); and
- to add and delete a minimum number of edges to transform a graph into a so-called 3-leaf power which is a special case of a graph power problem restricted to trees—this is the *NP*-complete CLOSEST 3-LEAF POWER problem.

We have already seen (Sections 7.3 and 8.2) how to attack CLUSTER EDITING with fixed-parameter techniques using the parameter "number of edge additions and deletions". In the following case studies we will sketch how the other four problems are dealt with by fixed-parameter methods. In this way, we revisit the FEEDBACK VERTEX SET problem, complementing the results using the iterative compression technique from Section 11.3.2. Before that, we give a general classification of graph modification problems into four natural categories:

- *Edge deletion* problems where the goal is to find a minimum set of edges to be removed in order to transform the graph as wanted.
- *Edge addition* problems where the goal is to find a minimum set of edges to be added in order to transform the graph as wanted.
- *Edge editing* problems where the goal is to find a minimum set of edges to be removed or to be added in order to transform the graph as wanted.
- *Vertex deletion* problems where the goal is to find a minimum set of vertices to be removed (together with their incident edges) in order to transform the graph as wanted.

We have omitted vertex addition problems since this, in its general form, can be seen as a general graph construction instead of a graph modification method.

Clearly, vertex deletion can be combined with any other of the three edge modification mechanisms. Often graph modification problems turn out to be *NP*-complete. In many application scenarios behind, however, it appears to be natural that the number of modification operations—often considered as some form of "error compensation"—should be relatively small. This makes graph modification problems prime candidates for parameterized complexity studies.

Before presenting four case studies for the above mentioned concrete problems, we begin with a very general framework dealing with graph modification problems from a parameterized point of view.

15.2.1 *Graph modification and hereditary properties*

Definition and motivation. A *graph property* is a set of graphs. A graph property is *hereditary* if every induced subgraph of a graph with the property has the property as well. For instance, planarity and bipartiteness clearly are hereditary whereas regularity (all vertices shall have same degree) is not. A property Π has a *finite forbidden set characterization* if there is a finite set F of graphs such that any graph has property Π iff it does not contain any of the graphs in F as an induced subgraph. Note that a graph property is hereditary iff it has a (not necessarily finite) forbidden set characterization. We shall see that a very general version of graph modification problems is fixed-parameter tractable if the goal graph class of the modification process has a finite forbidden set characterization. To this end, we formulate the following problem $\Pi_{i,j,k}$ GRAPH MODIFICATION for a particular graph property Π:

> **Input:** A graph $G = (V, E)$ and nonnegative integers i, j, k.
> **Question:** Can we delete at most i vertices, delete at most j edges, and add at most k edges such that G is transformed into a graph with property Π?

Many of the subsequently mentioned problems are special cases of this very general, *NP*-complete graph modification problem.
General considerations. We will see that $\Pi_{i,j,k}$ GRAPH MODIFICATION in case of a finite forbidden set characterization of graph property Π is fixed-parameter tractable with respect to the combined parameters i, j, and k. The basic idea of the approach is to—if it exists—determine an induced subgraph in the given graph G that coincides with a forbidden subgraph. If the finitely many forbidden subgraphs all have constant size, then this step is doable in polynomial time by a simple exhaustive approach. Then the crucial idea is to modify such a found induced forbidden subgraph in G, exhaustively trying all of finitely many edge modifications (deletions or additions) or vertex deletions on this subgraph structure in order to destroy the forbidden subgraph. In a natural way, this leads to a depth-bounded search tree strategy: in each search tree node, find a forbidden subgraph and branch into all (of finitely many) cases that lead to a destruction of the considered forbidden induced structure.
Result. $\Pi_{i,j,k}$ GRAPH MODIFICATION can be decided in $O(N^{i+2j+2k} \cdot |G|^{N+1|})$ time for any graph property with finite forbidden set characterization. Here N denotes the maximum number of vertices over all graphs in the forbidden set of graphs. Thus, $\Pi_{i,j,k}$ GRAPH MODIFICATION is fixed-parameter tractable with respect to the combined parameters i, j, and k.
Discussion. The above result is clearly of mainly theoretical interest. There are several special cases of the problem for which more efficient fixed-parameter algorithms can be achieved. Altogether, this indicates that graph modification problems are a fertile ground for fixed-parameter research.
Literature. The basic reference for the result on $\Pi_{i,j,k}$ GRAPH MODIFICATION is Cai (1996).

15.2.2 *Feedback Vertex Set revisited*

Definition and motivation. The FEEDBACK VERTEX SET problem is defined as follows.

> **Input:** A graph $G = (V, E)$ and a nonnegative integer k.
> **Question:** Is there a subset $V' \subseteq V$ with k or fewer vertices such that each cycle in G contains at least one vertex from V'?

Thus removing the vertices in V' from G results in a forest.

We have seen in Section 11.3.2 how to solve FEEDBACK VERTEX SET by a fixed-parameter algorithm running in $O(c^k \cdot m \cdot n)$ time for some constant c. Actually, this can be turned into a linear-time fixed-parameter algorithm by combining it with a linear-time constant-factor approximation algorithm, further increasing constant c. By relaxing our paradigm of considering deterministic algorithms and moving to randomized ones—which means to accept some likelihood of errors—a fairly simple and efficient randomized fixed-parameter algorithm for FEEDBACK VERTEX SET can be derived.

General considerations. The key observations are as follows. First, it is easy to eliminate vertices of degree at most two from the graph. Thus, assuming without loss of generality a graph with minimum vertex degree three, the algorithm is based on the easy-to-prove observation that if one picks an edge at random then there is a probability of at least $1/2$ that at least one of its two endpoints belongs to any feedback vertex set searched for. This implies that with probability $1/4$ a vertex chosen uniformly at random is part of the desired feedback vertex set. This leads to a simple randomized algorithm that terminates with a feedback vertex set of size k with probability at least $1/4^k$.

Result. Repeating the above algorithm independently $c \cdot 4^k$ times results in an algorithm that after $O(c \cdot 4^k \cdot n)$ steps finds a size-at-most-k feedback vertex set—if it exists—with probability at least $1 - (1 - 1/4^k)^{c \cdot 4^k}$.

Discussion. The above randomized algorithm gives one of the few known (see also the color-coding method in Section 11.1) concrete examples where randomization is employed in the design of fixed-parameter algorithms. This randomized algorithm for FEEDBACK VERTEX SET appears to be superior to its deterministic counterparts from a practical viewpoint. Bringing the run time of the deterministic algorithm closer to that of the randomized algorithm appears to be an interesting challenge for future research.

Literature. The randomized fixed-parameter algorithm is due to Becker *et al.* (2000).

15.2.3 *Graph Bipartization*

Definition and motivation. Deleting as few vertices as possible from a graph to make it bipartite is an *NP*-complete problem with applications in diverse fields such as VLSI design and computational biology. Formally, GRAPH BIPARTIZATION, also known as MAXIMUM BIPARTITE SUBGRAPH or ODD CYCLE TRANSVERSAL, is defined as follows.

> **Input:** A graph $G = (V, E)$ and a nonnegative integer k.
> **Question:** Is there a subset $V' \subseteq V$ with k or fewer vertices such that each odd-length cycle in G contains at least one vertex from V'?

Thus removing the vertices in V' from G results in a bipartite graph.

General considerations. The iterative compression technique (see Section 11.3) was first introduced in conjunction with showing that GRAPH BIPARTIZATION is fixed-parameter tractable. At the current time, however, the associated compression lemma seems to be a little more technical for GRAPH BIPARTIZATION than it appears to be for the closely related FEEDBACK VERTEX SET (see Section 11.3.2). Whereas in the latter we need to destroy all cycles by a minimum number of vertex deletions, in the first case we "only" have to destroy the cycles of odd length. Importantly, experiments demonstrate that iterative compression is in fact a worthwhile alternative to integer linear programming approaches for solving GRAPH BIPARTIZATION in practice. To this end, some useful "heuristics" improving the running time have been explored.

Result. GRAPH BIPARTIZATION can be solved in $O(3^k \cdot m \cdot n)$ time, where m denotes the number of edges and n denotes the number of vertices.

Discussion. The closely related EDGE BIPARTIZATION problem can be solved in $O(2^k \cdot m^2)$ time. Here the task is to delete edges instead of vertices. Again, the algorithm is based on iterative compression. Note that iterative compression can be employed to "compress" a non-optimal solution until an optimal one is found. Initial experiments indicate that with GRAPH BIPARTIZATION such a strategy finds an optimal solution very quickly, even when starting with $C = V$, but then takes a long time to actually prove the optimality.

Literature. The basic reference for graph bipartization and iterative compression is Reed *et al.* (2004). The algorithm engineering paper Hüffner (2005) gives a somewhat more accessible presentation of the underlying algorithm together with numerous practical improvements. EDGE BIPARTIZATION is studied in Guo *et al.* (2005).

15.2.4 *Minimum Fill-In*

Definition and motivation. Recall that a graph is *chordal* (or triangulated) if every cycle of length four or more contains a chord, that is, an edge between nonadjacent vertices on the cycle. The MINIMUM FILL-IN problem, also known as CHORDAL COMPLETION, is an *NP*-complete graph completion problem.

> **Input:** A graph $G = (V, E)$ and a nonnegative integer k.
> **Question:** Can G be made chordal by adding at most k edges?

MINIMUM FILL-IN is important in the context of sparse matrix computations.

General considerations. Note that it can be decided in linear time whether a graph is chordal. A simple enumerative approach can solve MINIMUM FILL-IN in $O(n^{2k} \cdot m)$ time where n denotes the number of vertices and m denotes the number of edges. Related *NP*-complete problems are to find the minimum number of vertex deletions to make a graph chordal and the minimum number of edge deletions to make a graph chordal.

Results. MINIMUM FILL-IN can be decided in $O(c^k \cdot m)$ time for $c \approx 4$ or in $O(6^k \cdot k^6 + k^2 \cdot m \cdot n)$ time.

Discussion. The first result means that MINIMUM FILL-IN is linear-time fixed-parameter tractable with respect to parameter k. It is based on a relatively simple search tree strategy. The second result is based on a more involved algorithm.

Both algorithms can actually enumerate all minimal triangulations obtained by adding at most k edges within the same time bounds. Fixed-parameter tractability has recently been stated for the problem to delete at most k vertices to make a graph chordal based on a complicated use of the iterative compression technique, tree decompositions, and Courcelle's theorem for monadic second-order logic.

Literature. The central reference point is Kaplan *et al.* (1999). See also Cai (1996) for related results and Marx (2004*a*) for the result concerning the deletion of vertices.

15.2.5 *Closest 3-Leaf Power*

Definition and motivation. For an unrooted tree T with leaves one-to-one labeled by the elements of a set V, the *k-leaf power* of T is a graph, denoted T^k, with $T^k := (V, E)$, where

$$E := \{\{u, v\} \mid u, v \in V \text{ and } d_T(u, v) \leq k\}.$$

The tree T is called a *k-leaf root* of T^k. The recognition problem then is, given a graph G, is there a tree T such that $T^k = G$. In practical applications motivated by studying phylogenies in computational biology, errors occur. This motivates the definition of the CLOSEST k-LEAF POWER problem.

> **Input:** A graph $G = (V, E)$ and a nonnegative integer l.
> **Question:** Is there a tree T such that T^k and G differ by at most l edges, that is, $|E(T^k) \triangle E(G)| \leq l$?

Here, for two sets A and B, $A \triangle B$ denotes the *symmetric difference* $(A \setminus B) \cup (B \setminus A)$.

General considerations. CLOSEST k-TREE POWER is *NP*-complete for $k \geq 2$. In fact, CLOSEST 2-LEAF POWER is equivalent to CLUSTER EDITING. The recognition problem is polynomial-time solvable for $k \leq 4$ and its complexity is open for $k \geq 5$. To the best of our knowledge, except for the "simpler" case $k = 2$ no results concerning polynomial-time approximation or nontrivial exact algorithms are known. In contrast, here we discuss the first positive algorithmic results, stating fixed-parameter tractability with respect to the number l of edge modifications for CLOSEST 3-LEAF POWER. The key tool here is a forbidden subgraph characterization of graphs that are 3-leaf powers: a graph is a 3-leaf power iff it is chordal and it contains none of the 5-vertex graphs bull, dart, and gem as an induced subgraph (see Figure 15.1). A simpler characterization of graphs that are 2-leaf powers is already known by forbidding an induced path of three vertices. This characterization finds direct applications in corresponding fixed-parameter algorithms (see Section 8.2), whereas the above characterization of 3-leaf powers requires a more sophisticated approach.

According to the above characterization of 3-leaf powers, the fixed-parameter algorithm has two tasks to fulfill:

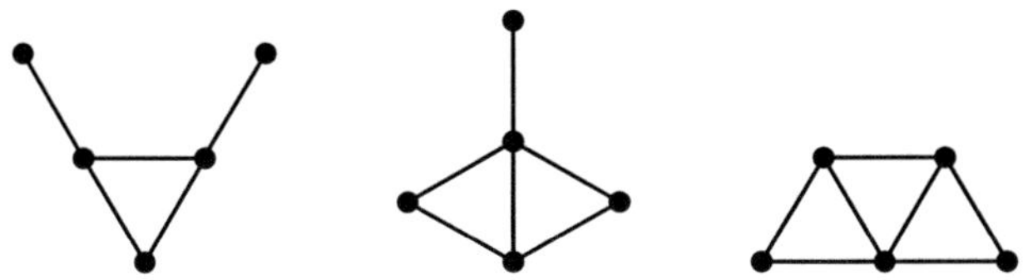

FIG. 15.1. 5-vertex graphs that occur as forbidden induced subgraphs.

1. Edit the input graph G to get rid of the forbidden subgraphs bull, dart, and gem.
2. Edit G to make it chordal.

Results. CLOSEST 3-LEAF POWER on a graph with l edge editing operations allowed is fixed-parameter tractable with respect to l. More precisely, the overall running time is $O((2l+8)^l \cdot m \cdot n)$. Fixed-parameter tractability can also be shown for CLOSEST 4-LEAF POWER.

Discussion. Compared with the corresponding result for CLUSTER EDITING, here the additional difficulty arises that we have no characterization by a *finite* number of forbidden subgraphs. The problem remains fixed-parameter tractable in the case of studying the corresponding vertex deletion, edge deletion, and edge addition problems. It is open whether there is a non-trivial reduction to a problem kernel for CLOSEST 3-LEAF POWER or even CLOSEST 4-LEAF POWER.

Literature. See Dom *et al.* (2004) for the basic reference. The results on the recognition problem are due to Nishimura *et al.* (2002). The CLOSEST 4-LEAF POWER problem is investigated in Dom *et al.* (2005).

15.3 Miscellaneous graph problems

Graphs are the mathematical objects that appear most often in this book. Indeed, the book could almost have been called "parameterized graph algorithms". Here, we provide some further examples from the rich field of *NP*-hard graph problems together with their consideration from the viewpoint of parameterized complexity analysis.

15.3.1 *Capacitated Vertex Cover*

Definition and motivation. VERTEX COVER, the drosophila of fixed-parameter algorithmics, has a number of significant generalizations and variants. We discuss one of them here. CAPACITATED VERTEX COVER is motivated by applications in drug design. To define the problem, for a graph $G = (V, E)$, assume that each vertex $v \in V$ is assigned a *capacity* $c(v) \in \mathbb{N}^+$. For each vertex, this capacity limits the number of edges that it can cover when being part of the vertex cover. Then, given a "capacitated graph" $G = (V, E)$ and a vertex cover C for G, we call C *capacitated vertex cover* if there exists a mapping $g : E \to C$ which maps each edge in E to one of its two endpoints such that the total number of edges mapped by g to any vertex $v \in C$ does not exceed $c(v)$.

Input: A vertex-weighted (with positive real numbers) and capacitated graph $G = (V, E)$, a nonnegative integer k, and a real number $W \geq 0$.
Question: Is there a capacitated vertex cover C for G that contains k or fewer vertices such that $\sum_{c \in C} w(c) \leq W$?

Capacitated Vertex Cover is *NP*-complete.
General considerations. Obviously, ordinary Vertex Cover is the special case of Capacitated Vertex Cover, where each vertex has capacity equal to its degree. It is possible to show fixed-parameter tractability for Capacitated Vertex Cover and both its soft and hard variants. The easiest way to do so is by means of a reduction to a problem kernel. First, assume uniform vertex weights.

Let $u, v \in V$, $u \neq v$, and $\{u, v\} \notin E$. The simple observation that lies at the heart of the data reduction rule needed to attain the claimed kernelization is that if the open neighborhoods coincide, that is, $N(u) = N(v)$, and $c(u) < c(v)$, then u is part of a minimum capacitated vertex cover only if v is as well. We can generalize this finding to a data reduction rule: let $\{v_1, v_2, \ldots, v_{k+1}\} \subseteq V$ with the induced subgraph $G[\{v_1, v_2, \ldots, v_{k+1}\}]$ being edgeless, and $N(v_1) = N(v_2) = \cdots = N(v_{k+1})$. Then delete from G a vertex $v_i \in \{v_1, v_2, \ldots, v_{k+1}\}$ which has minimum capacity. The correctness of this rule is given by the fact that any size-k capacitated vertex cover C containing v_i can be modified by replacing v_i with a vertex from $\{v_1, v_2, \ldots, v_{k+1}\}$ which is not in C.

Based on this data reduction rule, the reduced graph can be computed from G by the following two steps:

1. Use the straightforward linear-time factor-2 approximation algorithm to find a vertex cover S for G of size at most $2k'$ (where k' is the size of a minimum vertex cover for G and hence $k' \leq k$). If $|S| > 2k$, then we can stop because then no size-k (capacitated) vertex cover can be found. Note that $V \setminus S$ induces an edgeless subgraph of G.
2. By examining $V \setminus S$, check whether there is a subset of $k + 1$ vertices that fulfill the premises of the above rule. Repeat application of the data reduction rule until the rule is no longer applicable. Note that this process continuously shrinks $V \setminus S$.

In the above computation the number of all possible neighbor sets can be at most 2^{2k} (the number of different subsets of S). For each neighbor set, there can be at most k neighboring vertices in $V \setminus S$; otherwise, the reduction rule would apply. Hence in the worst case we can have at most $2^{2k} \cdot k$ vertices in the remaining graph $\tilde{G}$. The generalization to non-uniform vertex weights yields a problem kernel with $2^{2k} \cdot 2k^2$ vertices.
Results. For an n-vertex graph $G = (V, E)$ and an integer $k \geq 0$ as part of an input instance for Capacitated Vertex Cover one can construct an $O(4^k \cdot k^2)$-vertex graph $\tilde{G}$ such that G has a size-k solution for Capacitated Vertex Cover iff $\tilde{G}$ has a size-k solution for Capacitated Vertex Cover. In the special case of uniform vertex weights, $\tilde{G}$ has only $O(4^k \cdot k)$ vertices. The construction of $\tilde{G}$ can be performed in $O(n^2)$ time. Altogether, by combining the

above problem kernelization with an enumerative approach, CAPACITATED VERTEX COVER can be solved in $O(1.2^{k^2} + n^2)$ time.
Discussion. The exponential time bounds for CAPACITATED VERTEX COVER are far from the upper bounds for VERTEX COVER and clearly are not practical. Perhaps future research may lead to significant savings in the combinatorial explosion involved. Note, however, that fixed-parameter tractability is far less obvious for CAPACITATED VERTEX COVER than it was for VERTEX COVER. Recall that the PARTIAL VERTEX COVER problem as studied in Section 13.3.1 turned out to be $W[1]$-hard with respect to the cover size. This is worth emphasizing because all VERTEX COVER, CAPACITATED VERTEX COVER, and PARTIAL VERTEX COVER have polynomial-time factor-two approximation algorithms but behave completely differently from a parameterized point of view.
Literature. The basic reference is Guo *et al.* (2005*b*). CAPACITATED VERTEX COVER has been introduced and motivated by Guha *et al.* (2003), studying it in terms of polynomial-time approximability.

15.3.2 *Constraint Bipartite Vertex Cover*

Definition and motivation. VERTEX COVER restricted to bipartite graphs is equivalent to the polynomial-time solvable maximum matching problem in bipartite graphs. The situation changes drastically when two parameters come into play, leading to the definition of CONSTRAINT BIPARTITE VERTEX COVER.

> **Input:** A bipartite graph $G = (V_1, V_2, E)$ and two nonnegative integers k_1 and k_2.
> **Question:** Are there two subsets $C_1 \subseteq V_1$ and $C_2 \subseteq V_2$ of sizes $|C_1| \leq k_1$ and $|C_2| \leq k_2$ such that each edge in E has at least one endpoint in $C_1 \cup C_2$?

The existence of *two* parameters and two vertex sets makes CONSTRAINT BIPARTITE VERTEX COVER (CBVC) quite different from the original VERTEX COVER problem. Thus, whereas classical VERTEX COVER restricted to bipartite graphs is solvable in polynomial time, by a reduction from CLIQUE it follows that CBVC is *NP*-complete. CBVC is motivated by applications in reconfigurable VLSI design modelling a fault coverage scenario.
General considerations. CBVC can be solved by a depth-bounded search tree algorithm: to cover an edge, we have to put at least one of its two endpoints into the (optimal) vertex cover set. Thus, starting with an arbitrary edge, we can make a binary decision between its two endpoints. In each subcase, we delete the corresponding vertex chosen and its incident edges and repeat this until we have built a search tree of size $2^{k_1+k_2}$. As a consequence, it is easy to obtain an algorithm running in time $O(2^{k_1+k_2} \cdot (n + m))$, where n denotes the number of vertices and m denotes the number of edges in the graph. The exponential base can be significantly improved, however.

The improved algorithm consists of three pieces:

1. a reduction to a problem kernel;

2. a depth-bounded search tree; and
3. a special treatment of graphs consisting of vertices with maximum degree two and some slightly more general graphs.

Only the second part of the algorithm has exponential time complexity. We achieve a reduction of the search tree size by distinguishing between the degree of graph vertices. Since for CBVC we have to minimize with respect to two parameters, this gets significantly harder than in the classical VERTEX COVER case. For instance, in the classical instance, taking the neighbor of a degree-one vertex will always lead to an optimal vertex cover. Thus a branching in the search tree is avoided. This is no longer possible in the CBVC case because the neighbor belongs to the second vertex set in the bipartite graph and we have to minimize with respect to two vertex cover set sizes. In particular, since the *signature* $s := (|C_1|, |C_2|)$ of a vertex cover C_1 and C_2 is a tuple of numbers instead of simply one number, there are in terms of size incomparable solutions. The known fixed-parameter algorithm provides for each minimal s a corresponding minimal solution.

A reduction to a problem kernel in the style of Buss (see beginning of Chapter 7) also works for CBVC. Moreover, isolated components of a maximum vertex degree two are easily dealt with in polynomial time. As for VERTEX COVER, the improved search tree algorithm relies on an extensive case distinction with respect to vertex degrees. In fact, the branching becomes more intricate here. Part of this increased difficulty arises from the fact that we have to minimize with respect to two parameters.

Result. CONSTRAINT BIPARTITE VERTEX COVER is solvable in $O(1.40^{k_1+k_2} + (k_1 + k_2) \cdot n)$ time.

Discussion. The above result derives from a complicated case distinction. It is doubtful whether the high complexity in implementing all these cases pays off from a practical viewpoint. It might well be the case that a somewhat worse (in terms of worst-case time complexity) but much simpler search tree algorithm is superior in applications.

There is a simplified version of CBVC which allows for a more efficient fixed-parameter algorithm. The CONSTRAINED MINIMUM VERTEX COVER problem is defined as follows. The input consists of a bipartite graph $G = (V_1, V_2, E)$ and two nonnegative integers k_1 and k_2. The task is to find a vertex cover of G with at most k_1 vertices in V_1 and at most k_2 vertices in V_2. Note that the decisive difference from CBVC as considered before is that here one asks for a *minimum* vertex cover (that is, the sum of the numbers of vertices from the two sides of the bipartite graph) under the given "constraints" k_1 and k_2 whereas CBVC minimizes with respect to the signature. In particular, the previously defined term signature does not make sense for CONSTRAINED MINIMUM VERTEX COVER.

The nice thing about CONSTRAINED MINIMUM VERTEX COVER is that due to its somewhat simpler combinatorial structure it allows for simpler and more efficient algorithms than CBVC does. In particular, classical results from matching theory become applicable and allow for a much simpler search tree structure.

The key tool is the so-called Gallai–Edmonds structure theorem from matching theory which implies a reduction to problem kernel. More precisely, based on this theorem it can be shown that there is a linear problem kernel consisting of only $2(k_1 + k_2)$ vertices, and, moreover, the corresponding kernel graph has a perfect matching. Then the so-called Dulmage–Mendelsohn decomposition for graphs with a perfect matching is applied. This leads to a much simpler search tree procedure than the one known for CBVC. In summary, CONSTRAINED MINIMUM VERTEX COVER thus can be solved in time $O(1.26^{k_1+k_2} + kn)$, where n is the number of graph vertices.
Literature. The basic reference is Fernau and Niedermeier (2001). A better fixed-parameter tractability of CONSTRAINED MINIMUM VERTEX COVER is shown in Chen and Kanj (2003).

15.3.3 *Graph Coloring*

Definition and motivation. GRAPH COLORING is a classic problem in computer science and discrete mathematics with numerous applications. It is one of the hardest *NP*-complete problems under a variety of measures.

> **Input:** A graph $G = (V, E)$ and a nonnegative integer k.
> **Question:** Can G be colored by assigning each vertex one of at most k colors such that two adjacent graph vertices do not have the same color?

It is well-known that GRAPH COLORING is already *NP*-complete for $k = 3$. Thus, parameterization with respect to k seems fruitless.
General considerations. GRAPH COLORING has been subject of recent fixed-parameter studies from different viewpoints.

- Special graph classes with known coloring number (for instance, bipartite graphs have coloring number two) have been studied under the viewpoint that it is asked whether the coloring number can be still determined for graphs that are d edge or vertex modifications away from the given class of graphs. The parameter is d.
- The PRECOLORING EXTENSION problem where a graph with some of the vertices having preassigned colors is given and it must be decided whether the whole graph can be colored extending the given coloring. Parameters are either the number of precolored vertices or the number of colors used in the precoloring.
- The original GRAPH COLORING problem in an n-vertex graph is studied by choosing the parameter to be k and asking whether the graph can be colored using at most k colors. The opposite problem asks the same question with $n - k$ colors instead; this $(n - k)$-GRAPH COLORING thus studies the "dual parameterization" when compared with the original question.

Results. $(n-k)$-GRAPH COLORING can be decided in $O(c^k + k^2 + kn^2)$ time for $c \approx 14$. PRECOLORING EXTENSION is $W[1]$-hard with respect to both parameterizations even when restricted to chordal graphs. For a graph which differs from a bipartite graph by deleting only two vertices or deleting only three edges GRAPH COLORING is *NP*-complete. Choosing so-called split graphs—here the set of vertices can be partitioned into two sets such that one induces an independent set and the other induces a clique—instead of bipartite graphs, the deletion or addition of ℓ edges leads to fixed-parameter tractability of GRAPH COLORING with respect to parameter ℓ, whereas the deletion of ℓ vertices makes the problem $W[1]$-hard.
Discussion. The two last of the above results fall into the category of parameters measuring "distance from triviality". PRECOLORING EXTENSION is an extension of GRAPH COLORING, hence hardness is not surprising. Still, a new parameterization comes into play here. The result for $n-k$-GRAPH COLORING is particularly interesting because it employs a kernelization algorithm based on the crown data reduction rule (see Section 7.4).
Literature. The above results are drawn from Cai (2003), Chor *et al.* (2004), and Marx (2004*b*).

15.3.4 *Crossing Number*

Definition and motivation. Deciding whether a graph is planar can be done in linear time. Relax planarity by admitting a small number of edge-crossings in a drawing of the graph in the plane. Thus, the *crossing number* is the indexcrossing number minimum number of edge crossings needed in a drawing of the graph. The CROSSING NUMBER problem is *NP*-complete.

> **Input:** A graph $G = (V, E)$ and a nonnegative integer k.
> **Question:** Is the crossing number of G at most k?

General considerations. By an exhaustive search approach, one can decide in $n^{O(k)}$ time (where $n = |V|$) whether the crossing number is at most k. This is done by guessing at most k pairs of edges that cross. It is far from obvious to decide whether or not the problem is fixed-parameter tractable with respect to k. Note that the class of graphs of crossing number at most k is not closed under taking minors, hence, Robertson and Seymour's graph minor theory does not apply. Based on sophisticated bounded treewidth machinery, results by Robertson and Seymour (excluded grid theorem), and Courcelle's theorem the fixed-parameter tractability of CROSSING NUMBER can be shown by means of monadic-second order logic machinery.
Result. CROSSING NUMBER can be decided in $f(k) \cdot n^2$ time where f is at least doubly exponential.
Discussion. The above result is of purely theoretical interest—the associated constants are huge. It remains open to give a more practical fixed-parameter algorithm. We remark that the related *NP*-complete problem of deciding about the genus of a graph actually has a linear-time fixed-parameter algorithm. Here,

with respect to the parameter "graph genus", fixed-parameter tractability also follows by application of Robertson and Seymour's graph minor theory, which, by way of contrast, does not apply to CROSSING NUMBER.
Literature. The above result is due to Grohe (2004). For the linear-time fixed-parameter algorithm for the graph genus problem refer to Mohar (1999).

15.3.5 *Power Dominating Set*

Definition and motivation. The DOMINATING SET problem plays a prominent role in this book. In contrast to DOMINATING SET, POWER DOMINATING SET carries some form of *non-locality*: here, the correctness of a dominating set cannot be decided by locally checking every constant-size neighborhood. A vertex may dominate vertices at *arbitrary* distance when certain conditions are fulfilled. POWER DOMINATING SET is motivated by applications in monitoring electrical networks. We already encountered it in Section 13.3.2 as a problem that seems at least as hard as DOMINATING SET. Recall the formal definition. We need two observation rules.

Observation Rule 1 (OR1): A vertex in the power dominating set observes itself and all its neighbors.

Observation Rule 2 (OR2): If an observed vertex v of degree $d \geq 2$ is adjacent to $d - 1$ observed vertices, then the remaining unobserved vertex becomes observed as well.

Thus the definition of POWER DOMINATING SET reads as follows:

> **Input:** A graph $G = (V, E)$ and a nonnegative integer k.
> **Question:** Is there a subset $C \subseteq V$ with k or fewer vertices that observes all vertices in V with respect to the two observation rules OR1 and OR2?

General considerations Whereas DOMINATING SET is trivially linear-time solvable in trees, although true, this is not so obvious in the case of POWER DOMINATING SET. In Section 13.3.2 we showed that POWER DOMINATING SET is $W[2]$-hard with respect to the parameter "size of the solution set". Thus, in generalizing the result for trees it is natural to ask for the "tractability" of POWER DOMINATING SET in graphs of bounded treewidth, that is, is POWER DOMINATING SET fixed-parameter tractable with respect to the parameter treewidth?
Result. For an n-vertex graph with given width-w tree decomposition, POWER DOMINATING SET can be solved in $O(c^{w^2} \cdot n)$ time for a constant c.
Discussion. Clearly, this result is of purely theoretical interest. POWER DOMINATING SET should be further studied for special graph classes. So far, little is known here. One may also ask whether there is a "significant computational" difference between DOMINATING SET and POWER DOMINATING SET. How do fixed-parameter tractability results for DOMINATING SET in planar graphs transfer to POWER DOMINATING SET? Are there nontrivial data reduction rules for POWER DOMINATING SET similar to those we have for DOMINATING SET (see Section 7.6)?

Literature. Power Dominating Set was introduced in Haynes *et al.* (2002). The result discussed here is from Guo *et al.* (2005*a*). Independent work in the same direction (using monadic second-order logic, however) appears in Kneis *et al.* (2004).

15.4 Computational biology problems

Dealing with biologically motivated problems has become a vast area of algorithmic research. Even specialized topics such as the reconstruction of phylogenetic trees, the analysis of gene expression data (also closely related to clustering problems), the comparison and structure prediction of protein, RNA, and DNA molecules, or the search for motifs and signals in sequences (closely related to string searching problems) form subfields that are hard to overview. Thus the considerations in this section build a very small selection of biologically motivated problems in the fixed-parameter context. In addition, observe that some of the problems discussed, such as those related to string algorithms, may also have applications in other fields, for instance information retrieval or coding theory.

Before we come to some more concrete examples of parameterized complexity analysis in computational biology, however, we want to take up the cudgels for fixed-parameter algorithms with the following communicated by an anonymous referee:

> ...fixed-parameter algorithms do seem laudable approaches to *NP*-hard problems in biology, better than approximation methods in most cases.

What makes computational biology a particularly fruitful area for fixed-parameter studies is the fact that often there are several—and frequently all of them "reasonable" at the same time—parameters and "usually" at least one of them can be considered to carry small values. For instance, recall Closest String—with applications in "primer design" and "motif search"—from Sections 8.5 and 11.2. Here, two very natural parameters are the number of input strings k as well as the maximally allowed Hamming distance d to the closest string that is to be found. Both k as well as d in practice are small numbers (for instance, $k \approx 10$ and $d \approx 5$); hence parameterizations in both directions make sense and both actually lead to fixed-parameter algorithms (see Sections 8.5 and 11.2).

Last but not least it goes without saying that computational biology offers a vast amount of *NP*-hard problems, thus triggering research for approximative or heuristic algorithms and now increasingly also fixed-parameter algorithms. In the four case studies we present here, we discuss a tiny selection of fixed-parameter results for biologically motivated problems that have been achieved in recent years. Covering all the material in this direction might easily make up a book on its own.

We begin our considerations with two case studies related to the computation of phylogenetic trees.

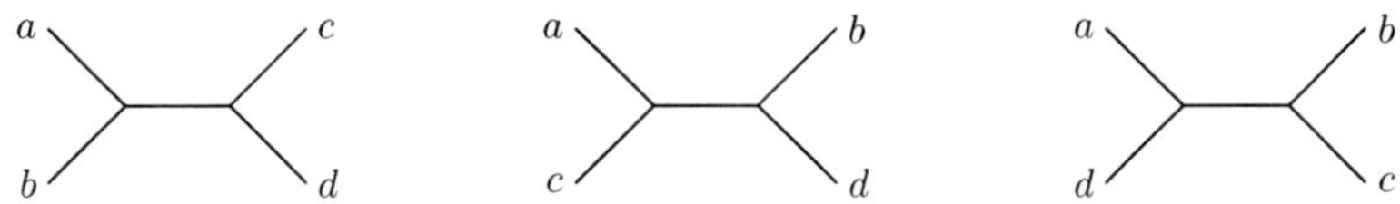

FIG. 15.2. Possible quartet topologies for quartet $\{a, b, c, d\}$, which are (from left to right) $[ab|cd]$, $[ac|bd]$, and $[ad|bc]$.

15.4.1 *Minimum Quartet Inconsistency*

Definition and motivation. To determine the evolutionary relationship of a set of taxa, say, based on DNA or protein sequence data, is an important question in computational biology. A common model for this relationship is an *evolutionary tree*, an unrooted binary tree T in which the leaves are one-to-one labeled by the taxa. In recent years, quartet methods for reconstructing evolutionary trees have received considerable attention. Here, a *quartet* is a size four subset $\{a, b, c, d\}$ of the set of taxa and the *quartet topology* for $\{a, b, c, d\}$ induced by T is simply the four leaves subtree of T for $\{a, b, c, d\}$. The three possible quartet topologies for $\{a, b, c, d\}$ are $[ab|cd]$, $[ac|bd]$, and $[ad|bc]$. They are shown in Figure 15.2.[13] The fundamental objective of quartet methods is, given a set of quartet topologies, to reconstruct the corresponding evolutionary tree. Here, the given set of quartet topologies can be incomplete, may contain errors or more than one topology for one quartet. Hence to reconstruct (a good estimation of) the original evolutionary tree becomes an optimization problem.

We focus on the MINIMUM QUARTET INCONSISTENCY (MQI) problem; see also Section 5.2.

> **Input:** A set S of n taxa and a set Q_S of $\binom{n}{4}$ quartet topologies such that there is exactly one topology for *every* quartet corresponding to S and a nonnegative integer k.
> **Question:** Is there an evolutionary tree T where the leaves are one-to-one labelled by the elements from S such that the set of quartet topologies induced by T differs from Q_S in at most k quartet topologies?

There are several reasons why quartet methods are widely used in practice. They are founded on the fact that an evolutionary tree is uniquely characterized by the quartet topologies for its size-four sets of taxa. From this set of topologies, we can efficiently compute the tree in polynomial time $O(n^4)$. Quartet methods clearly divide the tree construction process into two stages—one can use an arbitrary, even computationally expensive tree construction method to infer the quartet topologies, while the recombination of topologies can be handled independently of the method chosen for inference. Another reason to use quartet methods is data disparity. In practice, one often does not have the same amount of data for all considered taxa.

[13] A fourth possible topology is the star topology, which is not considered here because it is not binary.

Limitations of quartet methods in practice are caused by the process of quartet inference which can be erroneous. Therefore, one cannot be sure that there exists a tree inducing the inferred set of quartet topologies. Assuming that the number of errors is small compared to the number of correct topologies, one may overcome this problem by searching for a tree that matches the inferred topologies as "closely" as possible.

General considerations. MQI is *NP*-complete. It is important to note that there is a so-called "quartet cleaning algorithm" that finds finds the optimal solution for instances with $k < (n-3)/2$. Therefore, MQI is *NP*-hard only for $k \geq (n-3)/2$. It is known that MQI is polynomial time approximable with a factor n^2. Heuristics for the problem include semidefinite programming and the widely used *quartet puzzling*. For the case that the number k of "wrong" quartet topologies is small in comparison with the total number of given quartet topologies, MQI is fixed-parameter tractable with respect to parameter k. The more general variant of MQI where the set Q_S is not required to contain a topology for every quartet is *NP*-complete even if $k = 0$.

The key to developing a fixed-parameter solution for MQI with respect to parameter k is that it is sufficient to examine the size-three sets of quartet topologies and to recursively branch on local conflicts. Roughly speaking, this means that the fixed-parameter algorithm solving MQI can process as follows: if the given set of quartet topologies is non-conflicting and if there is exactly one topology for each possible quartet then one can construct the corresponding evolutionary tree in time $O(n^4)$. Otherwise, as long as the given set of quartet topologies is conflicting, one can deduce that there must exist a "contradicting" subset of three quartet topologies whose set of taxa altogether contains five elements. These three topologies contradict each other in the sense that there is no binary tree with five leaves labeled by the given taxa such that it induces these three topologies. In addition, one can efficiently maintain a conflict list containing all current "local" conflicts of the above kind. Thus the idea of a depth-bounded search tree algorithm with respect to parameter k becomes immediately clear. The only thing one still has to observe is that there are four ways to get rid of such a local conflict by changing one of the three given quartet topologies. This leads to a branching into four cases, each of which decreases the parameter k of maximally allowed quartet topology changes by one.

Result. MINIMUM QUARTET INCONSISTENCY can be solved in $O(4^k \cdot n + n^4)$ time. Observe that the input size is $O(n^4)$.

Discussion. The above algorithm can be sped up in practice by adding several heuristic improvements that do not violate the optimality of the solution obtained. A simple addition is to guarantee that no quartet topology is changed more than once. Another is the fact that if there is a topology t that is involved in more than $3k$ local conflicts (each consisting of three quartet topologies) then t *has* to be changed, a so-called "forced change".

Since MQI can be solved in polynomial time for $k < (n-3)/2$, one may ask—in the spirit of parameterizing above guaranteed values (see Section 5.2)—

whether it is fixed-parameter tractable with respect to parameter k' to find a tree that violates at most $(n-3)/2 + k'$ quartet topologies. This is an open problem.

A problem closely related to MQI is MINIMUM TRIPLET INCONSISTENCY (MTI). Here, a *triplet* is a size-three subset of the taxa set and a *triplet topology* is a *rooted* leaf-labeled three-leaves tree. In comparison to MQI, the input of MTI is a set of triplet instead of quartet topologies, and it asks for a *rooted* evolutionary tree that induces all except k of the given topologies. In contrast to MQI, MTI is solvable in cubic time for $k = 0$ even if there is not a topology for every triplet. Whether the latter general case is fixed-parameter tractable with respect to k is an open question.

Literature. The basic references are Gramm (2003) and Gramm and Niedermeier (2003). A survey of quartet methods can be found in Chor (1998). The polynomial-time algorithm for MQI instances with $k < (n-3)/2$ is due to Berry *et al.* (1999). The $O(n^4)$ time algorithm for non-conflicting topologies is described in Berry and Gascuel (2000). The MTI problem has been communicated by Benny Chor (Tel Aviv). The cubic-time algorithm for the case $k = 0$ is due to Aho *et al.* (1981).

15.4.2 *Compatibility of Unrooted Phylogenetic Trees*

Definition and motivation. The combinatorics of phylogenetic trees yields many interesting and generally *NP*-hard problems. A central problem in this context is that of compatibility. Informally speaking, the problem is, given a collection of phylogenetic trees for overlapping sets of species, is there a "supertree" such that all given trees can be "inferred" from the larger tree. Here, we focus on COMPATIBILITY OF UNROOTED PHYLOGENETIC TREES (CUP):

> **Input:** A collection $T_1, T_2, \ldots, T_k$ of unrooted, leaf-labeled trees.
> **Question:** Is there a tree T such that each tree T_i can be obtained from T by deleting leaves and contracting edges?

If such a tree T exists, then $T_1, T_2, \ldots, T_k$ are said to be compatible. Herein, an unrooted tree simply is an undirected graph without cycles. Deleting a leaf means also deleting the incident edge and contracting an edge means to identify its both endpoints and melting these into one new node, keeping all other adjacency relations to other nodes. CUP is *NP*-hard.

It is important to note that if we replace unrooted trees with rooted ones—that is, directed trees with the root node having indegree zero—then the problem is solvable in polynomial time. By way of contrast, CUP remains *NP*-hard even if all input trees contain only four leaves. The derived results only assume that for all input trees the inner nodes have at least three neighbors.

General considerations. The question for fixed-parameter tractability of CUP with respect to parameter k is very natural and practically relevant. The problem can be solved in $O(n^k)$ time, using the polynomial-time algorithm for the rooted case.

The basic idea behind showing the fixed-parameter tractability of CUP lies in two key steps.

1. The definition of a so-called display graph that provides a simple way to amalgamate the given phylogenetic trees into one graph. It can be shown that if the trees are compatible then the display graph has bounded treewidth.
2. Usage of monadic second-order logic to solve CUP on graphs of bounded treewidth.

Result. COMPATIBILITY OF UNROOTED PHYLOGENETIC TREES is decidable in $O(f(k) \cdot n)$ time, where f is an enormously growing exponential function and n is the total number of leaves. Hence, the problem is linear-time fixed-parameter tractable.

Discussion. The result, due to the huge combinatorial explosion hidden in $f(k)$ which goes along with the usage of monadic second-order logic, is of purely theoretical interest. Nevertheless it means a significant breakthrough result and the race is open to make CUP fixed-parameter tractable in practical terms also.

Literature. The basic reference is Bryant and Lagergren (2005). The polynomial-time solvability of the unrooted case follows from Aho *et al.* (1981), the *NP*-hardness is due to Steel (1992).

15.4.3 *Longest Arc-Preserving Common Subsequences*

Definition and motivation. Structure comparison for RNA and for protein sequences has become a central computational problem in biology. In this context, the LONGEST ARC PRESERVING COMMON SUBSEQUENCE problem (LAPCS) has recently received considerable attention. It is a sound and meaningful mathematical formalization of comparing the secondary structures of molecular sequences:[14] For a sequence s, an *arc annotation* A of s is a set of unordered pairs of positions in s. Focusing on the case of two given arc-annotated input sequences, LAPCS in its general version is defined as follows.

> **Input:** Two arc-annotated sequences s_1 and s_2 and nonnegative integers k_1 and k_2.
> **Question:** Can one delete at most k_1 letters (also called *bases*) from s_1—when deleting a letter at position i, then *all* arcs with endpoint i are also deleted—and at most k_2 letters from s_2 such that in both cases the same arc-annotated sequence t emerges?

Thus, t forms an arc-annotated subsequence of s_1 as well as s_2.

Whereas the LONGEST COMMON SUBSEQUENCE problem for two sequences *without* arc annotations is solvable in quadratic time—it becomes *NP*-complete

[14] As usual in computational biology, we identify the terms "sequence" and "string" here. Note, however, that the terms "subsequence" and "substring" have to be clearly distinguished from each other, the first concept being the much more general one: every substring forms a subsequence but not vice versa.

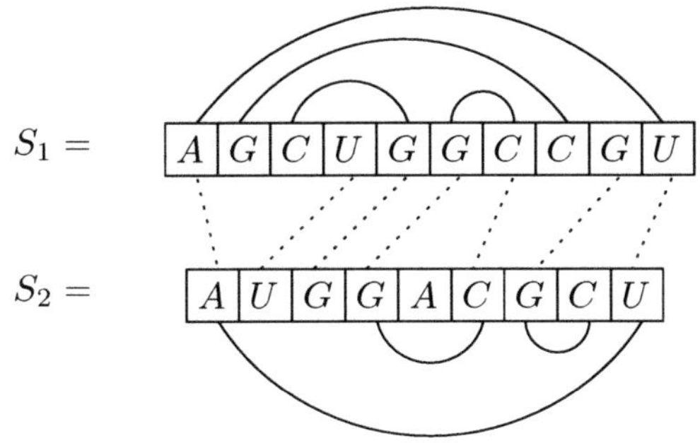

FIG. 15.3. Example of a longest arc-preserving common subsequence for two arc-annotated input sequences s_1 and s_2. The common subsequence is obtained by deleting three bases in s_1 and two bases in s_2.

when allowing for an *arbitrary* number of input sequences—LAPCS for two sequences is *NP*-complete. There is a statement in the literature that LAPCS for *nested arc annotations* is "generally thought of as the most important variant of the LAPCS problem". Here, one requires that no two arcs share an endpoint and no two arcs cross each other, referred to by LAPCS(NESTED,NESTED). An example of a longest common subsequence for two sequences with nested arc annotations is given in Figure 15.3. LAPCS(NESTED,NESTED) remains *NP*-complete.
General considerations. LAPCS(NESTED,NESTED) is studied in terms of two natural parameterizations.

- The first is by the combined parameter $k_1 + k_2$, the number of deletions allowed in sequences s_1 and s_2.
- The second is by the parameter ℓ which denotes the length of the common arc-preserving subsequence.

The first fixed-parameter algorithm with respect to the parameter $k_1 + k_2$ employs a depth-bounded search tree. The algorithm works through both given sequences from left to right, deletes already treated bases of the sequences (when deleting a base adjacent arcs are also deleted). By a case distinction depending on the bases at the currently first positions in s_1 and s_2, we decide how to continue recursively. This case distinction also takes into account arcs which are adjacent to these bases.

The fixed-parameter algorithm with respect to the parameter ℓ employs an enumerative approach. A core tool here is a polynomial-time dynamic programming algorithm that, given two arc-annotated sequences s_1 and s_2, decides in $O(|s_1| \cdot |s_2|)$ time whether s_2 is an arc-preserving subsequence of s_1. The idea behind the algorithm is then simply to try all length-ℓ sequences with nested arc structure as candidates for being a length-ℓ arc-preserving common subsequence. In the case of nested arc annotations and base alphabet set Σ, the number of these candidates is roughly bounded from above by $O((3 \cdot |\Sigma|)^\ell)$. Clearly, each candidate can then simply be tested using the above-mentioned polynomial-time pattern matching algorithm.

Results LAPCS(NESTED,NESTED) is decidable in $O(3.31^{k_1+k_2} \cdot n)$ time where n is the maximum input sequence length. LAPCS(NESTED,NESTED) is decidable in $O(|\Sigma|^{\ell} \cdot \ell \cdot n)$ time when parameterized by the length of the arc-preserving common subsequence. In both cases, n denotes the maximum input sequence length and both cases mean linear-time fixed-parameter tractability.

Discussion. There are numerous related problems in this context. For instance, in protein structure comparison the closely related concept of "contact maps" has been investigated. Moreover, when allowing more complicated structures than nested arcs, with respect to the parameter "length of the common subsequence" $W[1]$-hardness can usually be shown. In addition, still other parameterizations of the LAPCS problem which parameterize properties of the arc structure have been explored. Refer to the extensive literature.

Literature. The basic reference is Alber *et al.* (2004). The polynomial-time pattern matching algorithm appears in Gramm *et al.* (2002); also see Gramm (2004). The first parameterized hardness and tractability results appear in Evans (1999*a*) and Evans (1999*b*). Further investigations of classical complexity and approximation algorithms concerning LAPCS(NESTED,NESTED) appear in Jiang *et al.* (2004) and Lin *et al.* (2002).

15.4.4 *Incomplete Perfect Path Phylogeny Haplotyping*

Definition and motivation. Haplotyping problems currently are a major theme in computational biology. Roughly speaking, haplotype information needs to be inferred from genotype data in order to study the association between genomic variation and medical condition. Haplotype inference methods can be based on so-called *perfect phylogeny principles.* When stripped of the biological context, the haplotype inference problem is a purely combinatorial problem. Consider an $n \times m$-matrix with entries from $\{0, 1, 2\}$. Call this the *genotype matrix.*[15] A *haplotype matrix* then is a $2n \times m$-matrix with entries from $\{0, 1\}$. For a given genotype matrix A, one searches for a haplotype matrix B such that rows $2i - 1$ and $2i$ in B correspond to row i in A in the following way: for every position j in these rows, if there is a 2-entry in position j in the "A-row" then the corresponding entries in the "B-rows" differ; if there is a 0 or 1 in position j in the A-row then these entries in the two B-rows coincide with this value. Then we say that a haplotype matrix B admits a *perfect path phylogeny* if there exists a path P_B with a dedicated root node (which may be one of the leaves) such that

1. every row of B labels exactly one node of P_B;

[15] In fact, each row vector in this matrix is a *genotype*, each entry corresponding to a so-called SNP site (single nucleotide polymorphism). SNPs account for the majority of the variation between DNA sequences of different individuals. Each genotype has two underlying haplotypes, where 0- and 1-entries imply that the underlying haplotypes are identical at that position and a 2-entry implies that they are different at that position.

2. for every column of B, the rows carrying a 1 (or a 0) in these columns form a connected subpath in P_B; and
3. the root node of P_B is labeled with the all-0 row.

Note that the existence of a root node naturally implies two *sides* of a perfect path phylogeny.

The PERFECT PATH PHYLOGENY HAPLOTYPING (PPPH) problem is, given a genotype $\{0, 1, 2\}$-matrix A, decide whether A admits a perfect path phylogeny. Note that PPPH is a special case of the more general case where one has trees instead of paths—the consideration of the simpler path version is motivated by observations that around 70% of real data sets that admit a perfect (tree) phylogeny also allow a perfect path phylogeny. In practice, however, due to experimental noise, one has to work with *incomplete* genotype matrices. That is, genotype matrices lack some entries. These missing entries are represented by a fourth value "?". We arrive at the following central problem INCOMPLETE PERFECT PATH PHYLOGENY HAPLOTYPING (INCOMPLETE PPPH):

> **Input:** An $n \times m$-matrix with entries from $\{0, 1, 2, ?\}$.
> **Question:** Can the ?-entries be replaced by entries from $\{0, 1, 2\}$ such that the resulting $\{0, 1, 2\}$-matrix admits a perfect path phylogeny?

INCOMPLETE PPPH is *NP*-complete.

General considerations. It is natural to assume that the number of ?-entries may be not too large, motivating the study of the parameterization of INCOMPLETE PPPH by the maximum number k of ?-entries per column of the given $\{0, 1, 2, ?\}$-matrix. In the case $k = 0$, the (complete) PPPH can be efficiently solved in linear time. To derive a fixed-parameter algorithm for INCOMPLETE PPPH with respect to parameter k, a two-phase approach is performed, consisting of

1. preprocessing; and
2. dynamic programming.

In the preprocessing phase, the input instance is simplified by collapsing multiple columns into one "consensus column": suppose that several columns can become identical by replacing some ?-entries, and we replace the columns with this consensus column. Clearly, if we can find a perfect path phylogeny for this new matrix, we can also find one for the original matrix. Notably, the reverse implication is also true if the number of columns that formed the consensus is large enough. Altogether, this gives a problem kernel with size exponential in k.

In the dynamic programming phase, one makes use of a certain partial ordering of the columns of the preprocessed input matrix. One iterates over the columns in the given order and makes use of the fact that the perfect path phylogenies for columns up to a certain order value can be constructed from the already computed information of a constant-size range of order values for columns preceding the current column. The details are technical, though. It should be noted, however, that the corresponding combinatorial explosion with respect to parameter k is far from practical sizes.

Result. INCOMPLETE PERFECT PATH PHYLOGENY HAPLOTYPING is fixed-parameter tractable with respect to the parameter "maximum number of ?-entries in a column".
Discussion. One obvious point of interest is to improve the upper bounds on the combinatorial explosion of the above algorithm. Another point of interest is to extend the fixed-parameter tractability result from *path* phylogenies to more general, "bounded width" structures or even (tree) phylogenies.
Literature. The basic reference is Gramm *et al.* (2005).

15.5 Logic and related problems

Logic problems are tightly connected with the development of computational complexity theory. This is best reflected by the role that SATISFIABILITY played in the development of *NP*-completeness theory. Logic problems also play a core role in parameterized complexity theory. Thus weighted satisfiability problems are the key to understanding the theory of parameterized intractability as captured by the *W*-hierarchy (see Chapter 13). Logic and related problems are of fundamental importance in practice as well, having applications in areas such as model checking, database theory, or system verification. In particular, a whole conference series is devoted to satisfiability checking and related problems. In the following we want to briefly survey some of the opportunities and challenges that logic-like problems offer in the fixed-parameter context.

15.5.1 *Satisfiability*

Definition and motivation. In Section 1.1 we have already shown several aspects of the fundamental SATISFIABILITY problem in computational complexity and algorithm theory. Recall the definition.

> **Input:** A Boolean formula F in conjunctive normal form.
> **Question:** Does there exist a truth assignment for the Boolean variables in F such that F evaluates to true?

As pointed out in Section 1.1, there are numerous parameterizations that make sense. For instance, there are the parameters

- number of variables in F;
- number of clauses in F; or
- numbers that restrict the structure of F.

Here, we overview in a little more detail some fixed-parameter tractability results with respect to several structural formula parameters.
General considerations. Our main goal is to spot the great variety of possible structural parameters. To this end, consider formulae in conjunctive normal form simply as sets of clauses. A decisive point in this context is to associate graphs or hypergraphs with clause sets. For instance,

- the *primal graph* of a formula F in conjunctive normal form is the graph whose vertices are joined by an edge if both variables occur together in a clause;

- the *incidence graph* of a formula F in conjunctive normal form is the bipartite graph where one vertex side consists of the variables in F and the other vertex side consists of the clauses of F—a variable and a clause are joined by an edge if the variable occurs in the clause;
- the *directed incidence graph* derives from the incidence graph by orienting edges according to whether a variable occurs positively or negatively in a clause; and
- a *hypergraph* can naturally be associated with a formula F in conjunctive normal form by letting the vertices be the variables and by identifying every clause with a hyperedge formed by the variables occurring in the clause.

By means of the above graph structures associated with Boolean formulae in conjunctive normal form it is possible to employ well-known structural graph parameters such as treewidth and branchwidth in order to quickly solve the corresponding satisfiability problems in the case of small parameter values.

Note that besides drawing such direct connections to graphs and hypergraphs other structural parameterizations are also possible. For instance, the deficiency of a formula F in conjunctive normal form on n variables and m clauses is $\delta(F) := m - n$; its *maximum deficiency* is

$$\delta^*(F) := \max_{F' \subseteq F} \delta(F').$$

Here each F' above simply denotes a subformula of F.

Results. SATISFIABILITY is fixed-parameter tractable with respect to the parameters

- treewidth of the primal graph;
- treewidth of the incidence graph;
- branchwidth of the hypergraph; and
- maximum deficiency.

Discussion. As indicated above, there are numerous fixed-parameter tractability results for natural structural parameterizations of SATISFIABILITY. Some of these involve combinatorial explosions that are so high that there is an urgent need for improvements in this direction. Also, the practical consequences of these fixed-parameter tractability results have been left mostly unexplored. Moreover, it seems likely that there is still more to say about natural parameterizations of SATISFIABILITY. Parameterized complexity studies of satisfiability problems seem to have the potential to develop into a broad research topic.

Literature. The basic reference for this subsection is Szeider (2004*b*), which surveys several more results. The fixed-parameter tractability results mentioned appear in Gottlob *et al.* (2002), Alekhnovich and Razborov (2002), and Szeider (2004*a*). See also Fellows *et al.* (2004*c*) for a recent achievement in this field.

15.5.2 *Maximum Satisfiability*

Definition and motivation. MAXIMUM SATISFIABILITY is a natural generalization of SATISFIABILITY—whereas for SATISFIABILITY one asks to satisfy all clauses, in the case of MAXIMUM SATISFIABILITY one asks to satisfy a prespecified (or maximum) number of clauses. Recall the definition.

> **Input:** A Boolean formula in conjunctive normal form and a nonnegative integer k.
> **Question:** Is there a truth assignment satisfying at least k clauses?

In Section 7.2 we studied the parameterization of MAXIMUM SATISFIABILITY by the number k of satisfied clauses, and a problem kernel of size $O(k^2)$ is easily seen. As for SATISFIABILITY, however, there are numerous other ways for parameterizing MAXIMUM SATISFIABILITY. Little work has been done in this direction and there is much potential for future research. Parameterized complexity studies have mainly focussed on the above parameter k. Here, we will briefly discuss how to complement the problem kernel with further data reduction rules combined with intricate branching strategies yielding a good upper bound on the size of the depth-bounded search tree involved.
General considerations. A particularly notable thing about the fixed-parameter algorithm with respect to the parameter "number of satisfied clauses" is that, besides a number of branching rules, so-called transformation rules play a decisive role. In a sense, transformation rules are nothing but data reduction, as we encountered many times in Chapter 7, but with a different focus here. Here these rules are not used to prove a (better) problem kernelization but are used in order to achieve good branchings in conjunction with good branching vectors. In this context, a transformation rule replaces a formula simply by a single formula (without losing the guarantee of finding an optimal solution) whereas a branching rule replaces it by at least two formulae, at least one of which guarantees a path to an optimal solution. Among others, we have the following transformation rules.

- *Pure literal rule.* If a variable occurs either only positively or only negatively in the formula, then its truth value can be set accordingly, neglecting the second possibility for the complementary truth value.
- *Complementary unit clause rule.* If for variable x the formula contains both the one-literal clauses $\{x\}$ and $\{\overline{x}\}$, then these two can be removed and the parameter counting the number of clauses still to be satisfied can be decremented by one.
- *Resolution rule.* If there are two clauses, one containing literal x and one containing literal $\overline{x}$, then one can merge these two clauses into one with x and $\overline{x}$ removed from this new clause. The parameter counting the number of clauses still to be satisfied can be decremented by one.

These are only three examples and there are several more transformation rules of similar style. These are combined with branching rules which, for instance, distinguish between the cases:

- there is a variable that occurs at least five times in the formula;
- each variable occurs three or four times and some variable occurs exactly three times; and
- all variables occur exactly four times in the formula.

Here it is important that transformation rules are to be applied whenever possible and branching rules are only applied when this is not the case. In this way, one may guarantee favorable branching subcases and corresponding branching vectors because this allows the exclusion of unfavorable branching situations which are "covered" by transformation rules.

Result. MAXIMUM SATISFIABILITY can be decided in $O(1.37^k + |F|)$ time.

Discussion. MAXIMUM SATISFIABILITY together with parameter k is an example where parameterization above guaranteed values (see Section 5.2)—for any formula (more than) half of its clauses can be satisfied—is highly recommendable. Thus the above parameterization by k is doubtful in the sense that k is usually big. Still, the known fixed-parameter tractability result for the guaranteed value "half of all clauses" is based on the above parameterization. The results achieved for parameterizing above guaranteed values are not yet completely satisfactory. Moreover, several alternative parameterizations of MAXIMUM SATISFIABILITY should be the subject of future research (see also Section 1.1).

Literature. The parameterization of MAXIMUM SATISFIABILITY discussed here has been studied in a series of papers (Mahajan and Raman, 1999; Niedermeier and Rossmanith, 2000*b*; Bansal and Raman, 1999; Chen and Kanj, 2004), the mentioned upper bound appearing in Chen and Kanj (2004). Analogous studies have also been undertaken for the *NP*-complete special case MAXIMUM 2-SATISFIABILITY in Gramm *et al.* (2003*a*).

15.5.3 *Constraint satisfaction problems*

Definition and motivation. A *Boolean constraint* is a function

$$f : \{0,1\}^r \longrightarrow \{0,1\}.$$

A *Boolean constraint family* is a finite set of Boolean constraints. The goal is to satisfy all constraints by choosing an appropriate truth assignment. Thus, SATISFIABILITY is a special case of constraint satisfaction where all Boolean constraints simply are OR-functions applied to a number of literals. Constraint satisfaction is a well-studied field. Recently, study of the parameterized complexity of constraint satisfaction has begun. Here, weighted satisfiability problems, as in Chapter 13, are studied, leading to the following base problem WEIGHTED $\mathcal{F}$-SATISFIABILITY.

Input: A formula based on a finite family $\mathcal{F}$ of Boolean constraints.
Question: Does there exist a truth assignment of weight exactly k that satisfies all constraints in $\mathcal{F}$?

Recall that the weight of an assignment is the number of variables it is setting to true. Depending on what $\mathcal{F}$ looks like, the problem may have varying complexity.
General considerations. It can be shown that WEIGHTED $\mathcal{F}$-SATISFIABILITY is in $W[1]$ for every Boolean constraint family $\mathcal{F}$. The point of interest here is to decide for which $\mathcal{F}$ WEIGHTED $\mathcal{F}$-SATISFIABILITY becomes fixed-parameter tractable and when it becomes $W[1]$-complete. In fact, a "dichotomy theorem" holds, that is, for every constraint family $\mathcal{F}$, WEIGHTED $\mathcal{F}$-SATISFIABILITY is either fixed-parameter tractable or $W[1]$-complete. To obtain this classification, one needs the concept of *weakly separable constraints.* To this end, consider an assignment as a Boolean vector $s \in \{0,1\}^r$ and represent it as a subset of $\{1,2,\ldots,r\}$ by putting $i \in \{1,2,\ldots,r\}$ into the subset representing s iff the i-th position in s is set 1. Thus, we may interpret assignments as sets. A Boolean constraint f is *weakly separable* if

1. whenever there are two satisfying assignments for f, their intersection is satisfying as well; and
2. whenever for three satisfying assignments s_1, s_2, s_3 of f we have $s_1 \subseteq s_2 \subseteq s_3$, then $(s_2 \setminus s_1) \cup s_3$ is satisfying as well.

Result. If every constraint in a set of constraints $\mathcal{F}$ is weakly separable, then WEIGHTED $\mathcal{F}$-SATISFIABILITY is fixed-parameter tractable, and it is $W[1]$-complete otherwise.
Discussion. A dichotomy theorem such as the above one was first considered in the classical context for Boolean constraint satisfaction problems showing that every problem in this family of problems is either polynomial-time solvable or *NP*-complete. Also in this context, further structural restrictions on the formula related to incidence graphs, as we discussed for SATISFIABILITY (see Section 15.5.1) have been studied. WEIGHTED $\mathcal{F}$-SATISFIABILITY becomes fixed-parameter tractable with respect to the weight parameter if the incidence graph has treewidth bounded by a constant or is planar.
Literature. The basic reference is Marx (2004*c*). The famous classical dichotomy theorem for Boolean constraint satisfaction goes back to Schaefer (1978).

15.5.4 *Database queries*

Definition and motivation. Database theory makes a very natural setup for studying parameterized complexity. Here we focus on database queries—one of the main issues of database theory. When considering the evaluation of a query on a database instance, one has to distinguish between two sorts of complexity:

- *data complexity* measures the complexity of evaluating a query solely in terms of the size of the database and ignores the query size; and
- *combined complexity* measures the complexity in terms of the size of the query plus the size of the database.

The query is typically much smaller than the database, making this a natural subject for parameterized complexity analysis. Among other things, one analyzes the complexity of relational database queries for two parameters:

- the query size; and
- the number of variables that appear in the query.

Herein, a *database* $d = \{D; R_1, \ldots, R_m\}$ consists of a domain D and a set of relations $R_1, \ldots, R_m$ over D. A query Q is a function that maps a database d to a relation $Q(d)$ over the same domain D. Several classes of query can be distinguished, such as conjunctive queries or first-order queries—refer to the literature. The basic DATABASE QUERY problem then is defined as follows.

Input: A database d and a tuple t.
Question: Is $t \in Q(d)$?

General considerations. As mentioned before, two natural parameterizations of DATABASE QUERY naturally come to mind, that is, the query size and the number of variables appearing in the query. Another point to distinguish is whether the set of relations and their arity is fixed or can vary. Clearly, considering the parameter "number of variables" together with a variable set of relations with variable arity forms the most general problem. It turns out, however, that in most cases the assumption of fixed or variable set of relations and arity makes no difference in the following. Unfortunately, in their general form both of the above-mentioned parameterizations turn out to be $W[1]$-hard. Fixed-parameter tractability can only be achieved by further restricting the structure of queries.
Result. For various query languages such as "conjunctive", "positive", or "first-order" DATABASE QUERY is $W[1]$-hard for parameter "query size" as well as for parameter "number of variables".
Discussion. As mentioned, to achieve practical fixed-parameter tractable cases of DATABASE QUERY one must consider special forms of queries. "Acylicity" plays a central role here. Thus, the so-called class of "acyclic conjunctive queries with inequalities" leads to fixed-parameter tractable cases of DATABASE QUERY with respect to both discussed parameters. By way of contrast, the so-called class of "conjunctive queries with comparisons" remains $W[1]$-hard with respect to both parameterizations. There are many other methods and frameworks for studying parameterized query complexity; refer to the cited literature to gain an overview. There are few fixed-parameter tractability results for DATABASE QUERY for restricted classes of input database instances where—analogously to SATISFIABILITY in Section 15.5.1—one again refers to a graph structure associated with a relational database instance.
Literature. The basic reference for this section is Papadimitriou and Yannakakis (1999). Grohe (2002) provides a brief survey on the parameterized complexity of database query evaluation. Moreover, Frick and Grohe (2001) and Flum and Grohe (2005) provide deeper insights into the closely related field of model checking and descriptive complexity.

15.6 Miscellaneous problems

We conclude our small selection of case studies with a set of miscellaneous problems arising in various contexts. The purpose is eventually to underpin the leitmotif of this work stating the ubiquity of fruitful problem parameterizations. Hence parameterized complexity analysis makes sense almost everywhere.

15.6.1 *Two-dimensional Euclidean TSP*

The TRAVELING SALESPERSON problem (TSP) is one of the most prominent *NP*-complete problems with numerous applications. Unfortunately, the problem is hard to approximate but several successful heuristic strategies have been developed to solve it efficiently. Here, we exhibit a novel sort of parameterization—"distance from triviality" (see Section 5.4)—for an interesting special case of TSP that remains *NP*-complete; that is, we study the TWO-DIMENSIONAL EUCLIDEAN TSP where the set of points under study lies in the Euclidean plane.

> **Input:** A set of n points in the Euclidean plane.
> **Question:** Choosing Euclidean distances as the basis for the cost measure, is there a roundtrip through all n points that has length (resp. cost) at most k?

General considerations. This is once more an example of a problem where the considered parameter has nothing to do with the size or quality of the solution but it refers to "structural aspects" of the input. Thus the parameter is not given explicitly but is an input property as follows. It is well known that TSP is trivial when all n points are in convex position, that is, they are vertices of a convex polygon. Hence the question arises what happens if the number of "inner points" is small: given n points in the Euclidean plane, compute their convex hull and determine the set of points inside the convex hull. Let the number of these inner points be k and consider this as the problem parameter.

The basic strategy of attack is dynamic programming. Two fundamental and basically obvious facts necessary to derive the results are as follows:

1. Every shortest tour has no crossing when drawn in the plane.
2. As a consequence, in every shortest tour all points lying on the convex hull appear in a cyclic order, that is, every two consecutive points in the order are also consecutive on the convex hull.

Results. It can be shown that the problem is solvable in $O(k! \cdot k \cdot n)$ time using $O(k)$ space or in $O(2^k \cdot k^2 \cdot n)$ time using $O(2^k \cdot k \cdot n)$ space.
Discussion. The employed dynamic programming strategies are standard and do not need any complicated definitions or reasoning. They are also applicable to certain variants of TSP such as PRIZE-COLLECTING TSP and PARTIAL TSP. It is tempting to study the parameterization by the "number of inner points" for more problems from computational geometry.
Literature. The basic reference is Deĭneko *et al.* (2004). A parameterization by the number of inner points was also studied for the MINIMUM WEIGHT TRIANGULATION problem (Hoffmann and Okamoto, 2004; Spillner, 2005).

15.6.2 *Multidimensional matching*

Definition and motivation. Matching problems are a key component of many algorithms. In particular, MULTIDIMENSIONAL MATCHING can be used as a subroutine in many other fixed-parameter algorithms. The problem is, given a set of r-tuples, to find at least k "independent" tuples. For two r-tuples T, T' we write $T \sim T'$ if there is an $i \in \{1, 2, \ldots, r\}$ such that T and T' coincide in the ith coordinate; otherwise, we write $T \not\sim T'$.

> **Input:** A set $S \subseteq A_1 \times A_2 \times \ldots \times A_r$ for some collection $A_1, A_2, \ldots, A_r$ of pairwise disjoint sets and a nonnegative integer k.
> **Question:** Is there a matching P for S of size at least k, that is, is there a subset P of S where for any $T, T' \subseteq P$, $T \not\sim T'$, and $|P| \geq k$?

General considerations. The strategy for solving MULTIDIMENSIONAL MATCHING consists of two steps:

1. find a problem kernel; and
2. use color-coding in combination with dynamic programming on a subset of colors.

The idea behind the kernelization is that if a large number of tuples match a particular "pattern" then some of them can be removed. Observe that fixed-parameter tractability with respect to the combined parameter (k, r) already follows from the existence of a size-$O(k^r)$ kernel by simply performing exhaustive search. To gain a more efficient fixed-parameter algorithm, color-coding and dynamic programming are employed. To this end, one begins by greedily computing a *maximal* matching in the kernel instance in $O(k^r)$ time.
Result. MULTIDIMENSIONAL MATCHING can be decided in $O(c^{rk} + n)$ time for some constant c, where n denotes the size of the overall input.
Discussion. Note that the technique used to deploy a fixed-parameter algorithm for MULTIDIMENSIONAL MATCHING makes use of the fact that the goal is to obtain at least k disjoint objects. It generalizes to solving related packing problems in an analogous way with similar time bounds. It remains open to justify the relevance of the fixed-parameter algorithm in terms of implementation and experiments. As MULTIDIMENSIONAL MATCHING derives from a maximization problem, it is unclear whether small parameter values appear in practical applications. Note, however, that the first fixed-parameter algorithms for this problem only had a running time of the form $O(k^{O(k)} + n)$ for constant r, thus the above result means a significant breakthrough.
Literature The basic reference is Fellows *et al.* (2004*b*). Color-coding is due to Alon *et al.* (1995); see also Section 11.1.

15.6.3 *Matrix Domination*

Definition and motivation. MATRIX DOMINATION is a problem closely related to so-called edge domination problems in graphs and also to CONSTRAINT BIPARTITE VERTEX COVER which we discussed in Section 15.3.2. Applications appear in the context of telephone switching or VLSI design. The problem is defined as follows.

> **Input:** An $n \times m$-matrix with entries from $\{0, 1\}$ and a nonnegative integer k.
> **Question:** Is there a set C of at most k entries with value 1 such that all other 1-entries are in the same row or column with at least one entry from C?

The set C may be considered as *matrix dominating set.* MATRIX DOMINATION is *NP*-complete.
General considerations. The basic strategy to derive a fixed-parameter algorithm for MATRIX DOMINATION consists of:

1. deriving a problem kernel; and
2. employing a depth-bounded search tree, particularly making use of the algorithm for CONSTRAINT BIPARTITE VERTEX COVER.

For instance, the following three data reduction rules apply:

- Remove "all-0" rows and columns.
- If there are more than $k + 1$ identical rows (columns), then all of these except for $k + 1$ many can be removed.
- If there are more than k identical rows (columns) that contain more than k 1-entries, then the problem has no solution.

Based on these rules, one can prove a problem kernel of size $O(2^k \cdot k)$.
Result. MATRIX DOMINATION is decidable in $O(1.96^k \cdot k^{5/2} + n^3)$ time where, without loss of generality, $n \geq m$.
Discussion. MATRIX DOMINATION provides an example of a problem where—so far—only a problem kernel of exponential size is known. Another problem where only an exponential-size problem kernel is known is given by MULTICUT IN TREES, as discussed in Section 1.3.
Literature. The basic reference is Weston (2004). The CONSTRAINT BIPARTITE VERTEX COVER problem is studied in Fernau and Niedermeier (2001) (see also Section 15.3.2). The exponential-size problem kernel for MULTICUT IN TREES is derived in Guo and Niedermeier (2005*b*).

15.6.4 *Vapnik–Chervonenkis Dimension*

Definition and motivation. The VAPNIK–CHERVONENKIS DIMENSION (VC DIMENSION for short) problem is of particular importance in learning theory. It is also famous for being a candidate problem being neither polynomial-time solvable nor being *NP*-complete. When n denotes the overall input size, it is known to be solvable in $O(n^{\log n})$ time. The formal definition of VC DIMENSION reads as follows.

> **Input:** A family $\mathcal{C}$ of subsets of some universe U and a nonnegative integer k.
> **Question:** Is the *VC dimension* of $\mathcal{C}$ at least k, that is, is there an $S \subseteq U$ with $|S| \geq k$ with the property that for all subsets T of S there is a set C in $\mathcal{C}$ such that $C \cap S = T$?

The VC dimension is intuitively a measure of the "variability" of $\mathcal{C}$. It has been shown that it is a reasonably precise estimate of the complexity of "learning $\mathcal{C}$" if $\mathcal{C}$ is thought of as a class of concepts to be learned.
General considerations. The question for the complexity of the VC dimension has led to the introduction of natural complexity classes potentially lying between P and NP. These are founded on "limited nondeterminism", a concept seemingly closely related to fixed-parameter tractability. The important thing about VC DIMENSION—which also implies solvability in $n^{O(\log n)}$ time—is that it can be shown that for universe U of size m the VC dimension of $\mathcal{C}$ can be at most $\log m$. This indicates that it is unlikely to be NP-complete. By way of contrast, VC DIMENSION turns out to be intractable with respect to parameter k by reducing the $W[1]$-complete CLIQUE problem to it. Observe that the reduction involves parameter functions that are exponential in the parameter.
Result. VC DIMENSION is $W[1]$-complete with respect to parameter k.
Discussion. The fact that VC DIMENSION is $W[1]$-hard clearly shows that classical and parameterized complexity analysis are in a sense "orthogonal" to each other. So, whereas for instance the NP-complete VERTEX COVER with respect to the parameter "solution size" is fixed-parameter tractable, the presumably not NP-hard VC DIMENSION is $W[1]$-hard with respect to parameter "solution size". This indicates that classical complexity studies do not necessarily always "transfer" into the parameterized context.
Literature. The proof for the $W[1]$-completeness of VC DIMENSION can be found in Downey and Fellows (1999). A systematic study of limited nondeterminism in connection with the complexity of the VC DIMENSION is undertaken in Papadimitriou and Yannakakis (1996). The connections between learnability and the VC dimension are described in Blummer *et al.* (1989).

15.7 Summary and concluding remarks

With the relatively small selection of case studies given in this chapter the main purpose is to briefly sketch various areas and concrete problems where parameterized complexity analysis fits naturally. Through personal bias and ignorance, however, a number of important fields and topics have been left unaddressed. Among these omissions are problems related to

- graph drawing;
- computational geometry;
- model checking problems in connection with descriptive complexity;
- machine learning;
- cryptography;

- formal languages and linguistics;
- social sciences;
- operations research and economy;

and many other fields offering challenging combinatorial problems. It seems hard to overview and reflect all opportunities for studying fixed-parameter algorithms. For an invitation to study fixed-parameter algorithms the above selection may perhaps be enough—for complete coverage hundreds of more pages must be written.

16

ZUKUNFTSMUSIK

The good prospects for fixed-parameter algorithms as predicted in the 1990s have come true. Fixed-parameter tractability will surely continue to prosper in various ways.

The emphasis of this work was on algorithms. Both the algorithmic and the structural complexity theory side of parameterized complexity analysis are flourishing and developing rapidly.

We avoid listing concrete algorithmic challenges here since numerous examples are spread all over the text. The reservoir of challenges for parameterized algorithm design and analysis is almost inexhaustible. This is due to the continuously growing list of *NP*-hard problems and to the fact that fixed-parameter algorithmics encourages the study of one and the same problem under different parameterizations.

A point that has been neglected so far is that fixed-parameter tractability may also, in some cases, be an alternative to polynomial-time solvability. Imagine that one only has an algorithm with a high-degree polynomial running time for some problem. A fixed-parameter algorithm for that problem with a perhaps linear-time component in the overall input size and an exponential time component exclusively depending on a small parameter (so-called "linear fixed-parameter tractability") might still be beneficial in this case. In concrete terms, an $O(2^k \cdot n)$ time algorithm for small parameter k may be superior to an $O(n^3)$ time algorithm.

Among others, one may identify the following five basic themes as important components in ongoing research on fixed-parameter algorithms:

- implementation and experimentation;
- improved mathematical analysis;
- new algorithmic techniques;
- new (structural) problem parameterizations; and
- combination with approximation algorithms.

We conclude with three (of many) central challenges for parameterized algorithm design.

- Is the FEEDBACK VERTEX SET problem—destroy all cycles with at most k vertex deletions—in *directed* graphs fixed-parameter tractable with respect to parameter k?
- Is it fixed-parameter tractable to delete at most k clauses from a Boolean formula in 2-CNF such that the remaining formula becomes satisfiable?

(Note that this can be seen as a generalization of the fixed-parameter tractable GRAPH BIPARTIZATION problem.)

- Is the problem to find an independent set of size $\lceil n/4 \rceil + k$ in an n-vertex planar graph fixed-parameter tractable with respect to the parameter k?

There are many options to explore and extend the range of applicability of fixed-parameter algorithms. Be invited!

REFERENCES

Abu-Khzam, Faisal N., Collins, Rebecca L., Fellows, Michael R., Langston, Michael A., Suters, W. Henry, and Symons, Christof T. (2004). Kernelization algorithms for the Vertex Cover problem: theory and experiments. In *Proc. ALENEX04*, pp. 62–69. ACM/SIAM.

Abu-Khzam, Faisal N., Langston, Michael A., Shanbhag, Pushkar, and Symons, Christopher T. (To appear in *Algorithmica*, 2005). Scalable parallel algorithms for FPT problems.

Aho, Alfred V., Hopcroft, John E., and Ullman, Jeffrey D. (1974). *The Design and Analysis of Computer Algorithms.* Addison-Wesley.

Aho, Alfred V., Sagiv, Yehoshua, Szymanski, Thomas G., and Ullman, Jeffrey D. (1981). Inferring a tree from lowest common ancestors with an application to the optimization of relational expressions. *SIAM Journal on Computing*, **10**(3), 405–421.

Alber, Jochen (2003). *Exact Algorithms for NP-hard Problems on Networks: Design, Analysis, and Implementation.* Ph. D. thesis, WSI für Informatik, Universität Tübingen, Germany.

Alber, Jochen, Betzler, Nadja, and Niedermeier, Rolf (2003). Experiments on data reduction for optimal domination in networks. In *Proc. International Network Optimization Conference (INOC 2003)*, Evry/Paris, France, pp. 1–6. Long version to appear in *Annals of Operations Research.*

Alber, Jochen, Bodlaender, Hans L., Fernau, Henning, Kloks, Ton, and Niedermeier, Rolf (2002). Fixed parameter algorithms for Dominating Set and related problems on planar graphs. *Algorithmica*, **33**(4), 461–493.

Alber, Jochen, Dorn, Frederic, and Niedermeier, Rolf (2005*a*). Empirical evaluation of a tree decomposition based algorithm for Vertex Cover on planar graphs. *Discrete Applied Mathematics*, **145**(2), 219–231.

Alber, Jochen, Fan, Hongbing, Fellows, Michael R., Fernau, Henning, Niedermeier, Rolf, Rosamond, Fran, and Stege, Ulrike (2005*b*). A refined search tree technique for Dominating Set on planar graphs. *Journal of Computer and System Sciences*, **71**(4), 385–405.

Alber, Jochen, Fellows, Michael R., and Niedermeier, Rolf (2004). Polynomial time data reduction for Dominating Set. *Journal of the ACM*, **51**(3), 363–384.

Alber, Jochen, Fernau, Henning, and Niedermeier, Rolf (2003). Graph separators: A parameterized view. *Journal of Computer and System Sciences*, **67(4)**, 808–832.

Alber, Jochen, Fernau, Henning, and Niedermeier, Rolf (2004). Parameterized complexity: exponential speed-up for planar graph problems. *Journal of Algorithms*, **52**(1), 26–56.

Alber, Jochen and Fiala, Jiří (2004). Geometric separation and exact solutions for the parameterized independent set problem on disk graphs. *Journal of Algorithms*, **52**(2), 134–151.

Alber, Jochen, Gramm, Jens, Guo, Jiong, and Niedermeier, Rolf (2004). Computing the similarity of two sequences with nested arc annotations. *Theoretical Computer Science*, **312**(2–3), 337–358.

Alber, Jochen, Gramm, Jens, and Niedermeier, Rolf (2001). Faster exact solutions for hard problems: a parameterized point of view. *Discrete Mathematics*, **229**, 3–27.

Alber, Jochen and Niedermeier, Rolf (2002). Improved tree decomposition based algorithms for domination-like problems. In *Proc. 5th LATIN*, Volume 2286 of *LNCS*, pp. 613–628. Springer.

Alekhnovich, Michael and Razborov, Alexander A. (2002). Satisfiability, branch-width and Tseitin tautologies. In *Proc. 43rd FOCS*, pp. 593–603. IEEE Computer Society.

Alon, Noga, Yuster, Raphael, and Zwick, Uri (1995). Color-coding. *Journal of the ACM*, **42**(4), 844–856.

Appel, K. and Haken, W. (1977*a*). Every planar map is four colorable. i. discharging. *Illinois J. Math.*, **21**, 429–490.

Appel, K. and Haken, W. (1977*b*). Every planar map is four colorable. ii. reducibility. *Illinois J. Math.*, **21**, 491–567.

Arnborg, Stefan, Corneil, Derek G., and Proskurowski, Andrzej (1987). Complexity of finding embeddings in a k-tree. *SIAM Journal on Algebraic Discrete Methods*, **8**, 277–284.

Arvind, V. and Raman, Venkatesh (2002). Approximation algorithms for some parameterized counting problems. In *Proc. 13th ISAAC*, Volume 2518 of *LNCS*, pp. 453–464. Springer.

Atallah, Michael J. (ed.) (1999). *Algorithms and Theory of Computation Handbook.* CRC Press.

Ausiello, Giorgio, Crescenzi, Pierluigi, Gambosi, Giorgio, Kann, Viggo, Marchetti-Spaccamela, Alberto, and Protasi, Marco (1999). *Complexity and Approximation.* Springer.

Ausiello, Giorgio, D'Atri, Alessandro, and Protasi, Marco (1980). Structure preserving reductions among convex optimization problems. *Journal of Computer and System Sciences*, **21**(1), 136–153.

Baker, Brenda S. (1994). Approximation algorithms for NP-complete problems on planar graphs. *Journal of the ACM*, **41**(1), 153–180.

Balasubramanian, R., Fellows, Michael R., and Raman, Venkatesh (1998). An improved fixed-parameter algorithm for vertex cover. *Information Processing Letters*, **65**(3), 163–168.

Bansal, Nikhil, Blum, Avrim, and Chawla, Shuchi (2004). Correlation clustering. *Machine Learning*, **56**(1), 89–113.

Bansal, Nikhil and Raman, Venkatesh (1999). Upper bounds for MaxSat: further improved. In *Proc. 10th ISAAC*, Volume 1741 of *LNCS*, pp. 247–258. Springer.

Bar-Yehuda, Reuven and Even, Shimon (1981). A linear-time approximation algorithm for the weighted vertex cover problem. *Journal of Algorithms*, **2**, 198–203.

Bar-Yehuda, Reuven and Even, Shimon (1985). A local-ratio theorem for approximating the weighted vertex cover problem. *Annals of Discrete Mathematics*, **25**, 27–45.

Becker, Ann, Bar-Yehuda, Reuven, and Geiger, Dan (2000). Randomized algorithms for the Loop Cutset problem. *Journal of Artificial Intelligence Research*, **12**, 219–234.

Bellman, Richard (1962). Dynamic programming treatment of the Traveling Salesman Problem. *Journal of the ACM*, **9**(1), 61–63.

Berry, Vincent and Gascuel, Olivier (2000). Inferring evolutionary trees with strong combinatorial evidence. *Theoretical Computer Science*, **240**(2), 271–298.

Berry, Vincent, Jiang, Tao, Kearney, Paul E., Li, Ming, and Wareham, Todd (1999). Quartet cleaning: Improved algorithms and simulations. In *Proc. 7th ESA*, Volume 1643 of *LNCS*, pp. 313–324. Springer.

Betzler, Nadja, Niedermeier, Rolf, and Uhlmann, Johannes (2004). Tree decompositions of graphs: saving memory in dynamic programming. In *CTW04*, Volume 17 of *Electronic Notes in Discrete Mathematics*. Elsevier. Long version to appear in *Discrete Optimization*.

Biggs, Norman L. (1989). *Discrete Mathematics* (Revised edn). Oxford Science Publications. Clarendon Press.

Blummer, Anselm, Ehrenfeucht, Andrzej, Haussler, David, and Warmuth, Manfred K. (1989). Learnability and the Vapnik–Chervonenkis dimension. *Journal of the ACM*, **36**, 929–965.

Bodlaender, Hans L. (1993). A tourist guide through treewidth. *Acta Cybernetica*, **11**(1–2), 1–22.

Bodlaender, Hans L. (1996). A linear time algorithm for finding tree-decompositions of small treewidth. *SIAM Journal on Computing*, **25**, 1305–1317.

Bodlaender, Hans L. (1997). Treewidth: Algorithmic techniques and results. In *Proc. 22nd MFCS*, Volume 1295 of *LNCS*, pp. 19–36. Springer.

Bodlaender, Hans L. (1998). A partial k-arboretum of graphs with bounded treewidth. *Theoretical Computer Science*, **209**, 1–45.

Bodlaender, Hans L. (2005). Discovering treewidth. In *Proc. 31st SOFSEM*, Volume 3381 of *LNCS*, pp. 1–16. Springer.

Bodlaender, Hans L., Downey, Rodney G., Fellows, Michael R., and Wareham, H. Todd (1995). The parameterized complexity of the longest common subsequence problem. *Theoretical Computer Science*, **147**, 31–54.

Boppana, Ravi B. and Sipser, Michael (1990). The complexity of finite functions. In *Handbook of Theoretical Computer Science, Volume A* (ed. J. van Leeuwen), pp. 757–804. Elsevier.

Brandstädt, Andreas, Le, Van Bang, and Spinrad, Jeremy P. (1999). *Graph*

Classes: a Survey. SIAM Monographs on Discrete Mathematics and Applications.

Bryant, David and Lagergren, Jens (2005, May). Compatibility of unrooted phylogenetic trees is FPT. To appear in *Theoretical Computer Science*.

Buss, Jonathan F. and Goldsmith, Judy (1993). Nondeterminism within *P*. *SIAM Journal on Computing*, **22**(3), 560–572.

Cai, Leizhen (1996). Fixed-parameter tractability of graph modification problems for hereditary properties. *Information Processing Letters*, **58**, 171–176.

Cai, Leizhen (2003). Parameterized complexity of Vertex Colouring. *Discrete Applied Mathematics*, **127**(1), 415–429.

Cai, Liming, Chen, Jianer, Downey, Rodney G., and Fellows, Michael R. (1997). Advice classes of parameterized tractability. *Annals of Pure and Applied Logic*, **84**(1), 119–138.

Cai, Liming and Juedes, David (2003). On the existence of subexponential parameterized algorithms. *Journal of Computer and System Sciences*, **67(4)**, 789–807.

Caprara, Alberto (1999). Formulations and hardness of multiple sorting by reversals. In *Proc. 3d RECOMB*, pp. 84–94. ACM.

Cesati, Marco (2003). The Turing way to parameterized complexity. *Journal of Computer and System Sciences*, **67**(4), 654–685.

Cesati, Marco (2004, January). Compendium of parameterized problems. Available through the author's webpage.

Cesati, Marco and Trevisan, Luca (1997). On the efficiency of polynomial time approximation schemes. *Information Processing Letters*, **64**(4), 165–171.

Chandran, L. Sunil and Grandoni, Fabrizio (2005). Refined memorisation for vertex cover. *Information Processing Letters*, **93**(3), 125–131.

Cheetham, James, Dehne, Frank, Rau-Chaplin, Andrew, Stege, Ulrike, and Taillon, Peter J. (2003). Solving large FPT problems on coarse-grained parallel machines. *Journal of Computer and System Sciences*, **67**(4), 691–706.

Chekuri, Chandra, Mydlarz, Marcelo, and Shepherd, F. Bruce (2003). Multicommodity demand flow in a tree. In *Proc. 30th ICALP*, Volume 2719 of *LNCS*, pp. 410–425. Springer.

Chen, Jianer, Fernau, Henning, Kanj, Iyad A., and Xia, Ge (2005). Parametric duality and kernelization: Lower bounds and upper bounds on kernel size. In *Proc. 22d STACS*, Volume 3404 of *LNCS*, pp. 269–280. Springer.

Chen, Jianer, Friesen, Donald K., Jia, Weijia, and Kanj, Iyad A. (2004*a*). Using nondeterminism to design efficient deterministic algorithms. *Algorithmica*, **40**(2), 83–97.

Chen, Jianer, Huang, Xiuzhen, Kanj, Iyad A., and Xia, Ge (2004*b*). Linear FPT reductions and computational lower bounds. In *Proc. 28th STOC*, pp. 212–221. ACM Press.

Chen, Jianer and Kanj, Iyad A. (2003). Constrained minimum vertex cover in bipartite graphs: complexity and parameterized algorithms. *Journal of Computer and System Sciences*, **67**, 833–847.

Chen, Jianer and Kanj, Iyad A. (2004). Improved exact algorithms for Max-Sat. *Discrete Applied Mathematics*, **142**(1–3), 17–27.

Chen, Jianer, Kanj, Iyad A., and Jia, Weijia (2001). Vertex Cover: Further observations and further improvements. *Journal of Algorithms*, **41**, 280–301.

Chen, Jianer, Kanj, Iyad A., Perkovic, Ljubomir, Sedgwick, Eric, and Xia, Ge (2003). Genus characterizes the complexity of graph problems: Some tight results. In *Proc. 30th ICALP*, Volume 2719 of *LNCS*, pp. 845–856. Springer.

Chen, Jianer, Kanj, Iyad A., and Xia, Ge (2002). Labeled search trees and amortized analysis: Improved upper bounds for NP-hard problems. In *Proc. 14th ISAAC*, Volume 2906 of *LNCS*, pp. 148–157. Springer. Long version to appear in *Algorithmica*.

Chen, Jianer, Kanj, Iyad A., and Xia, Ge (2005, April). Simplicity is beauty: improved upper bounds for Vertex Cover. Manuscript.

Chen, Yijia and Flum, Jörg (2004). On miniaturized problems in parameterized complexity theory. In *Proc. IWPEC*, Volume 3162 of *LNCS*, pp. 108–120. Springer. Long version to appear in *Theoretical Computer Science*.

Chen, Yijia, Flum, Jörg, and Grohe, Martin (2003). Bounded nondeterminism and alternation in parameterized complexity theory. In *Proc. 18th CCC*, pp. 13–29. IEEE Computer Society.

Chen, Yijia, Flum, Jörg, and Grohe, Martin (2005). Machine-based methods in parameterized complexity theory. *Theoretical Computer Science*, **339**(2–3), 167–199.

Chlebík, Miroslav and Chlebíková, Janka (2004). Improvement of Nemhauser–Trotter theorem and its applications in parameterized complexity. In *Proc. 9th SWAT*, Volume 3111 of *LNCS*, pp. 174–186. Springer.

Chor, Benny (1998). From quartets to phylogenetic trees. In *Proc. 25th SOFSEM*, Volume 1521 of *LNCS*, pp. 36–53. Springer.

Chor, Benny, Fellows, Mike, and Juedes, David W. (2004). Linear kernels in linear time, or how to save k colors in $o(n^2)$ steps. In *Proc. 30th WG*, Volume 3353 of *LNCS*, pp. 257–269. Springer.

Chor, Benny and Sudan, Madhu (1998). A geometric approach to betweenness. *SIAM Journal on Discrete Mathematics*, **11**(4), 511–523.

Coppersmith, Don and Winograd, Shmuel (1990). Matrix multiplication via arithmetic progressions. *Journal of Symbolic Computations*, **9**, 251–280.

Cormen, Thomas H., Leiserson, Charles E., Rivest, Ronald L., and Stein, Clifford (2001). *Introduction to Algorithms*. MIT Press.

Crescenzi, Pierluigi and Kann, Viggo (1998). How to find the best approximation results—a follow-up to Garey and Johnson. *ACM SIGACT News*, **29**(4), 90–97.

Damaschke, Peter (2004). Parameterized enumeration, transversals, and imperfect phylogeny reconstruction. In *Proc. IWPEC*, Volume 3162 of *LNCS*, pp. 1–12. Springer. Long version to appear in *Theoretical Computer Science*.

Dantsin, Evgeny, Gavrilovich, Michael, Hirsch, Edward A., and Konev, Boris (2001). MAX SAT approximation beyond the limits of polynomial-time approximation. *Annals of Pure and Applied Logic*, **113**(1–3), 81–94.

Dantsin, Evgeny, Goerdt, Andreas, Hirsch, Edward A., Kannan, Ravi, Kleinberg, Jon, Papadimitriou, Christos, Raghavan, Prabhakar, and Schöning, Uwe (2002). A deterministic $(2 - 2/(k+1))^n$ algorithm for k-SAT based on local search. *Theoretical Computer Science*, **289**(1), 69–83.

Dantsin, Evgeny, Hirsch, Edward A., Ivanov, Sergei, and Vsemirnov, Maxim (2001). Algorithms for SAT and upper bounds on their complexity. Technical Report TR01-012, Electronic Colloquium on Computational Complexity.

Dehne, Frank, Fellows, Michael R., Langston, Michael A., Rosamond, Frances A., and Stevens, Kim (2005). An $\mathcal{O}^*(2^{O(k)})$ FPT algorithm for the undirected feedback vertex set problem. In *Proc. 11th COCOON*, Volume 3595 of *LNCS*, pp. 859–869. Springer.

Dehne, Frank, Fellows, Michael R., Rosamond, Frances A., and Shaw, Peter (2004). Greedy localization, iterative compression, and modeled crown reductions: New FPT techniques, an improved algorithm for set splitting, and a novel $2k$ kernelization for Vertex Cover. In *Proc. IWPEC*, Volume 3162 of *LNCS*, pp. 271–280. Springer.

Deĭneko, Vladimir G., Hoffmann, Michael, Okamoto, Yoshio, and Woeginger, Gerhard J. (2004). The traveling salesman problem with few inner points. In *Proc. 10th COCOON*, Volume 3106 of *LNCS*, pp. 268–277. Springer.

Demaine, Erik D., Fomin, Fedor V., Hajiaghayi, Mohammad Taghi, and Thilikos, Dimitrios M. (2003). Fixed-parameter algorithms for the (k, r)-center in planar graphs and map graphs. In *Proc. 30th ICALP*, Volume 2719 of *LNCS*, pp. 829–844. Springer.

Demaine, Erik D., Fomin, Fedor V., Hajiaghayi, Mohammad Taghi, and Thilikos, Dimitrios M. (2004*a*). Bidimensional parameters and local treewidth. *SIAM Journal on Discrete Mathematics*, **18**(3), 501–511.

Demaine, Erik D., Fomin, Fedor V., Hajiaghayi, Mohammad Taghi, and Thilikos, Dimitrios M. (2004*b*). Subexponential parameterized algorithms on graphs of bounded-genus and H-minor-free graphs. In *Proc. 15th SODA*, pp. 830–839. ACM/SIAM.

Demaine, Erik D. and Hajiaghayi, Mohammad Taghi (2005). Bidimensionality: New connections between FPT algorithms and PTASs. In *Proc. 16th SODA*, pp. 590–601. ACM/SIAM.

Demaine, Erik D., Hajiaghayi, Mohammad Taghi, and Thilikos, Dimitrios M. (2004*c*). The bidimensional theory of bounded-genus graphs. In *Proc. 29th MFCS*, Volume 3153 of *LNCS*, pp. 191–203. Springer.

Demaine, Erik D., Hajiaghayi, Mohammad Taghi, and Thilikos, Dimitrios M. (2005). Exponential speedup of fixed-parameter algorithms for classes of graphs excluding single-crossing graphs as minors. *Algorithmica*, **41**(4), 245–267.

Deng, X., Li, G., Li, Z., Ma, B., and Wang, L. (2003). Genetic design of drugs without side-effects. *SIAM Journal on Computing*, **32**(4), 1073–1090.

Diestel, Reinhard (2005). *Graph Theory* (3 edn). Springer-Verlag, New York.

Dinur, Irit and Safra, Shmuel (2002). On the importance of being biased. In *Proc. 34th STOC*, pp. 33–42. ACM Press.

Dom, Michael, Guo, Jiong, Hüffner, Falk, and Niedermeier, Rolf (2004). Error compensation in leaf root problems. In *Proc. 15th ISAAC*, Volume 3341 of *LNCS*, pp. 389–401. Springer. Long version to appear in *Algorithmica*.

Dom, Michael, Guo, Jiong, Hüffner, Falk, and Niedermeier, Rolf (2005). Extending the tractability border for closest leaf powers. In *Proc. 31st WG*, Volume 3787 of *LNCS*. Springer.

Downey, Rodney G. (2003). Parameterized complexity for the skeptic. In *Proc. 18th IEEE Annual Conference on Computational Complexity*, pp. 147–169.

Downey, Rodney G., Estivill-Castro, Vladimir, Fellows, Michael R., Prieto, Elena, and Rosamond, Frances A. (2003). Cutting up is hard to do: the parameterized complexity of k-cut and related problems. *Electronic Notes in Theoretical Computer Science*, **78**.

Downey, Rod G. and Fellows, Michael R. (1995). Parameterized computational feasibility. In *Feasible Mathematics II*, pp. 219–244. Birkhäuser.

Downey, Rodney G. and Fellows, Michael R. (1999). *Parameterized Complexity*. Springer.

Downey, Rod G., Fellows, Michael R., and Stege, Ulrike (1999). Parameterized complexity: A framework for systematically confronting computational intractability. In *Contemporary Trends in Discrete Mathematics: From DIMACS and DIMATIA to the Future*, Volume 49 of *AMS-DIMACS*, pp. 49–99. AMS.

Dreyfus, S. E. and Wagner, R. A. (1972). The Steiner problem in graphs. *Networks*, **1**, 195–207.

Ellis, John, Fan, Hongbing, and Fellows, Michael R. (2004). The Dominating Set problem is fixed parameter tractable for graphs of bounded genus. *Journal of Algorithms*, **52**(2), 152–168.

Eppstein, David (2000). Diameter and treewidth in minor-closed graph families. *Algorithmica*, **27**(3), 275–291.

Eppstein, David (2004). Quasiconvex analysis of backtracking algorithms. In *Proc. 15th SODA*, pp. 788–797. ACM/SIAM.

Evans, Particia A. (1999*a*). *Algorithms and Complexity for Annotated Sequence Analysis*. Ph. D. thesis, University of Victoria, Canada.

Evans, Particia A. (1999*b*). Finding common subsequences with arcs and pseudoknots. In *Proc. 10th CPM*, Volume 1645 of *LNCS*, pp. 270–280. Springer.

Fedin, Sergey and Kulikov, Alexander (2004). Automated proofs of upper bounds on the running time of splitting algorithms. In *Proc. IWPEC*, Volume 3162 of *LNCS*, pp. 248–259. Springer.

Feige, Uriel (1998). A threshold of $\ln n$ for approximating set cover. *Journal of the ACM*, **45(4)**, 634–652.

Fellows, Michael R. (2002). Parameterized complexity: The main ideas and connections to practical computing. In *Experimental Algorithmics*, Volume 2547 of *LNCS*, pp. 51–77. Springer.

Fellows, Michael R. (2003*a*). Blow-ups, win/win's, and crown rules: Some new directions in FPT. In *Proc. 29th WG*, Volume 2880 of *LNCS*, pp. 1–12. Springer.

Fellows, Michael R. (2003*b*). New directions and new challenges in algorithm design and complexity, parameterized. In *Proc. 8th WADS*, Volume 2748 of *LNCS*, pp. 505–520. Springer.

Fellows, Michael R., Gramm, Jens, and Niedermeier, Rolf (2002). On the parameterized intractability of Closest Substring and related problems. In *Proc. 19th STACS*, Volume 2285 of *LNCS*, pp. 262–273. Springer. Long version to appear in *Combinatorica*.

Fellows, Michael R., Heggernes, Pinar, Rosamond, Frances A., Sloper, Christian, and Telle, Jan Arne (2004*a*). Finding k disjoint triangles in an arbitrary graph. In *Proc. 30th WG*, Volume 3353 of *LNCS*, pp. 235–244. Springer.

Fellows, Michael R., Knauer, Christian, Nishimura, Naomi, Ragde, Prabhakar, Rosamond, Frances A., Stege, Ulrike, Thilikos, Dimitrios M., and Whitesides, Sue (2004*b*). Faster fixed-parameter tractable algorithms for matching and packing problems. In *Proc. 12th ESA*, Volume 3221 of *LNCS*, pp. 311–322. Springer.

Fellows, Michael R., Szeider, Stefan, and Wrightson, Graham (2004*c*). On finding short resolution refutations and small unsatisfiable subsets. In *Proc. IWPEC*, Volume 3162 of *LNCS*, pp. 223–234. Springer. Long version to appear in *Theoretical Computer Science*.

Fernau, Henning (2004). A top-down approach to search-trees: Improved algorithmics for 3-Hitting Set. Technical Report ECCC-078, Electronic Colloquium on Computational Complexity.

Fernau, Henning and Niedermeier, Rolf (2001). An efficient, exact algorithm for constraint bipartite vertex cover. *Journal of Algorithms*, **38**(2), 374–410.

Flum, Jörg and Grohe, Martin (2004*a*). Parameterized complexity and subexponential time. *Bulletin of the European Association for Theoretical Computer Science*, **84**, 71–100.

Flum, Jörg and Grohe, Martin (2004*b*). The parameterized complexity of counting problems. *SIAM Journal on Computing*, **33**(4), 892–922.

Flum, Jörg and Grohe, Martin (2005). Model-checking problems as a basis for parameterized intractability. *Logical Methods in Computer Science*, **1**(1).

Flum, Jörg, Grohe, Martin, and Weyer, Mark (2004). Bounded fixed-parameter tractability and $\log^2 n$ nondeterministic bits. In *Proc. 31st ICALP*, Volume 3142 of *LNCS*, pp. 555–567. Springer.

Fomin, Fedor V., Grandoni, Fabrizio, and Kratsch, Dieter (2005). Measure and conquer: domination—a case study. In *Proc. 32d ICALP*, Volume 3580 of *LNCS*, pp. 191–203. Springer.

Fomin, Fedor V. and Thilikos, Dimitrios M. (2003). Dominating sets in planar graphs: branch-width and exponential speed-up. In *Proc. 14th SODA*, pp. 168–177. ACM/SIAM.

Fomin, Fedor V. and Thilikos, Dimitrios M. (2004*a*). Fast parameterized algorithms for graphs on surfaces: Linear kernel and exponential speed-up. In *Proc. 31st ICALP*, Volume 3142 of *LNCS*, pp. 581–592. Springer.

Fomin, Fedor V. and Thilikos, Dimitrios M. (2004*b*). A simple and fast approach for solving problems on planar graphs. In *Proc. 21st STACS*, Volume 2996 of *LNCS*, pp. 56–67. Springer.

Frances, M. and Litman, A. (1997). On covering problems of codes. *Theory of Computing Systems*, **30**(2), 113–119.

Frick, Markus and Grohe, Martin (2001). Deciding first-order properties of locally tree-decomposable structures. *Journal of the ACM*, **48**(6), 1184–1206.

Garey, Michael R. and Johnson, David S. (1979). *Computers and Intractability: A Guide to the Theory of NP-Completeness.* Freeman.

Garg, Naveen, Vazirani, Vijay V., and Yannakakis, Mihalis (1997). Primal-dual approximation algorithms for integral flow and multicut in trees. *Algorithmica*, **18**(1), 3–20.

Golumbic, Martin C. (1980, 2004). *Algorithmic graph theory and perfect graphs.* Academic Press, Elsevier, New York.

Gottlob, Georg, Scarcello, Francesco, and Sideri, Martha (2002). Fixed-parameter complexity in AI and nonmonotonic reasoning. *Artifical Intelligence*, **138**(1–2), 55–86.

Gramm, Jens (2003). *Fixed-Parameter Algorithms for the Consensus Analysis of Genomic Sequences.* Ph. D. thesis, WSI für Informatik, Universität Tübingen, Germany.

Gramm, Jens (2004). A polynomial-time algorithm for the matching of crossing contact-map patterns. *IEEE/ACM Transactions on Computational Biology and Bioinformatics*, **4**(1), 171–180.

Gramm, Jens, Guo, Jiong, Hüffner, Falk, and Niedermeier, Rolf (2004). Automated generation of search tree algorithms for hard graph modification problems. *Algorithmica*, **39**(4), 321–347.

Gramm, Jens, Guo, Jiong, Hüffner, Falk, and Niedermeier, Rolf (2005). Graph-modeled data clustering: Exact algorithms for clique generation. *Theory of Computing Systems*, **38**(4), 373–392.

Gramm, Jens, Guo, Jiong, and Niedermeier, Rolf (2002). Pattern-matching in arc-annotated sequences. In *Proc. 22nd FSTTCS*, Volume 2556 of *Lecture Notes in Computer Science*, pp. 182–193. Springer. Long version to appear in *ACM Transactions on Algorithms.*

Gramm, Jens, Guo, Jiong, and Niedermeier, Rolf (2003). On exact and approximation algorithms for Distinguishing Substring Selection. In *Proc. 14th FCT*, Volume 2751 of *LNCS*, pp. 261–272. Springer. Long version to appear in *Theory of Computing Systems.*

Gramm, Jens, Hirsch, Edward A., Niedermeier, Rolf, and Rossmanith, Peter (2003*a*). Worst-case upper bounds for MAX-2-SAT with an application to MAX-CUT. *Discrete Applied Mathematics*, **130**(2), 139–155.

Gramm, Jens and Niedermeier, Rolf (2003). A fixed-parameter algorithm for Minimum Quartet Inconsistency. *Journal of Computer and System Sci-*

ences, **67**(4), 723–741.

Gramm, Jens, Niedermeier, Rolf, and Rossmanith, Peter (2003*b*). Fixed-parameter algorithms for Closest String and related problems. *Algorithmica*, **37**(1), 25–42.

Gramm, Jens, Nierhoff, Till, Sharan, Roded, and Tantau, Till (2005). Haplotyping with missing data via perfect path phylogenies. To appear in *Discrete Applied Mathematics.*

Grohe, Martin (2002). Parameterized complexity for the database theorist. *SIGMOD Record*, **31**(4), 86–96.

Grohe, Martin (2004). Computing crossing numbers in quadratic time. *Journal of Computer and System Sciences*, **68**(2), 285–302.

Guha, Sudipto, Hassin, Refael, Khuller, Samir, and Or, Einat (2003). Capacitated vertex covering. *Journal of Algorithms*, **48**(1), 257–270.

Guo, Jiong (2005). *Algorithm Design Techniques for Parameterized Graph Modification Problems.* Ph. D. thesis, Institut für Informatik, Universität Jena, Germany.

Guo, Jiong, Gramm, Jens, Hüffner, Falk, Niedermeier, Rolf, and Wernicke, Sebastian (2005). Improved fixed-parameter algorithms for two feedback set problems. In *Proc. 9th WADS*, Volume 3608 of *LNCS*, pp. 158–168. Springer.

Guo, Jiong, Hüffner, Falk, and Niedermeier, Rolf (2004). A structural view on parameterizing problems: distance from triviality. In *Proc. IWPEC*, Volume 3162 of *LNCS*, pp. 162–173. Springer.

Guo, Jiong and Niedermeier, Rolf (2005*a*). Exact algorithms and applications for Tree-like Weighted Set Cover. *Journal of Discrete Algorithms*. To appear.

Guo, Jiong and Niedermeier, Rolf (2005*b*). Fixed-parameter tractability and data reduction for Multicut in Trees. *Networks*, **46**(3), 124–135.

Guo, Jiong and Niedermeier, Rolf (2006). A fixed-parameter tractability result for Multicommodity Demand Flow in Trees. *Information Processing Letters*, **97**, 109–114.

Guo, Jiong, Niedermeier, Rolf, and Raible, Daniel (2005*a*). Improved algorithms and complexity results for power domination in graphs. In *Proc. 15th FCT*, Volume 3623 of *LNCS*, pp. 172–184. Springer.

Guo, Jiong, Niedermeier, Rolf, and Wernicke, Sebastian (2005*b*). Parameterized complexity of generalized vertex cover problems. In *Proc. 9th WADS*, LNCS, pp. 36–48. Springer.

Hannenhalli, Sridhar and Pevzner, Pavel A. (1996). To cut... or not to cut (applications of comparative physical maps in molecular evolution). In *Proc. 7th SODA*, pp. 304–313. ACM/SIAM.

Haynes, Teresa W., Hedetniemi, Sandra M., Hedetniemi, Stephen T., and Henning, Michael A. (2002). Domination in graphs applied to electric power networks. *SIAM Journal on Discrete Mathematics*, **15**(4), 519–529.

Held, Michael and Karp, Richard M. (1962). A dynamic programming approach to sequencing problems. *Journal of SIAM*, **10**, 196–210.

Hirsch, Edward A. (2000). New worst-case upper bounds for SAT. *Journal of*

Automated Reasoning, **24**(4), 397–420.

Hochbaum, Dorit S. (ed.) (1997). *Approximation Algorithms for NP-hard Problems.* PWS Publishing Company.

Hoffmann, Michael and Okamoto, Yoshio (2004). The minimum weight triangulation problem with few inner points. In *Proc. IWPEC*, Volume 3162 of *LNCS*, pp. 200–212. Springer.

Hofri, Micha (1995). *Analysis of Algorithms: Computational Methods and Mathematical Tools.* Oxford University Press.

Hopcroft, John E., Motwani, R., and Ullman, Jeffrey D. (2001). *Introduction to Automata Theory, Languages, and Computation* (2nd edn). Addison–Wesley.

Horowitz, Ellis and Sahni, Sartaj (1974). Computing partitions with applications to the Knapsack problem. *Journal of the ACM*, **21**(2), 277–292.

Hromkovič, Juraj (2002). *Algorithmics for Hard Problems* (2nd edn). Springer.

Hüffner, Falk (2005). Algorithm engineering for optimal graph bipartization. In *Proc. 4th WEA*, Volume 3503 of *LNCS*, pp. 240–252. Springer.

Impagliazzo, Russell, Paturi, Ramamohan, and Zane, Francis (2001). Which problems have strongly exponential complexity? *Journal of Computer and System Sciences*, **63**(4), 512–530.

Iwama, Kazuo and Tamaki, Suguru (2003). Improved upper bounds for 3-SAT. Technical Report TR03-053, Electronic Colloquium on Computational Complexity. Also appears in *Proc. ACM/SIAM SODA 2004*.

Jia, Weijia, Zhang, Chuanlin, and Chen, Jianer (2004). An efficient parameterized algorithm for m-set packing. *Journal of Algorithms*, **50**(1), 106–117.

Jiang, Tao, Kearney, Paul E., and Li, Ming (2000). A polynomial time approximation scheme for inferring evolutionary trees from quartet topologies and its application. *SIAM Journal on Computing*, **30**(6), 1942–1961.

Jiang, Tao, Lin, Guohui, Ma, Bin, and Zhang, Kaizhong (2004). The longest common subsequence problem for arc-annotated sequences. *Journal of Discrete Algorithms*, **2**(2), 257–270.

Kannan, Ravi (1987). Minkowski's convex body theorem and integer programming. *Mathematics of Operations Research*, **12**, 415–440.

Kaplan, Haim, Shamir, Ron, and Tarjan, Robert E. (1999). Tractability of parameterized completion problems on chordal, strongly chordal, and proper interval graphs. *SIAM Journal on Computing*, **28**(5), 1906–1922.

Kellerer, Hans, Pferschy, Ulrich, and Pisinger, David (2004). *Knapsack Problems.* Springer.

Khot, Subhash and Regev, Oded (2003). Vertex Cover might be hard to approximate to within $2-\epsilon$. In *Proc. 18th IEEE Annual Conference on Computational Complexity.*

Khuller, Samir (2002). The Vertex Cover problem. *SIGACT News*, **33**(2), 31–33.

Kleinberg, Jon and Tardos, Éva (2005). *Algorithm Design.* Pearson, Addison Wesley.

Kloks, Ton (1994). *Treewidth: Computations and Approximations*, Volume 842 of *Lecture Notes in Computer Science*. Springer.

Kneis, Joachim, Mölle, Daniel, Richter, Stefan, and Rossmanith, Peter (2004, December). Parameterized power domination complexity. Technical Report AIB-2004-09, Deptartment of Computer Science, RWTH Aachen.

Küchlin, Wolfgang and Sinz, Carsten (2000). Proving consistency assertions for automotive product data management. *Journal of Automated Reasoning*, **24**(1–2), 145–163.

Kullmann, Oliver (1999). New methods for 3-SAT decision and worst-case analysis. *Theoretical Computer Science*, **223**(1–2), 1–72.

Kuo, Sy-Ken and Fuchs, W. Kent (1987). Efficient spare allocation for reconfigurable arrays. *IEEE Design and Test*, **4**, 24–31.

Křivánek, Mirko and Morávek, Jaroslav (1986). NP-hard problems in hierarchical-tree clustering. *Acta Informatica*, **23**(3), 311–323.

Lagarias, Jeffrey C. (1995). Point lattices. In *Handbook of Combinatorics* (ed. R. L. G. et al.), pp. 919–966. The MIT Press.

Lanctot, J. Kevin, Li, Ming, Ma, Bin, Wang, Shaojiu, and Zhang, Louxin (2003). Distinguishing string selection problems. *Information and Computation*, **185**(1), 41–55.

Langston, Michael A. and Suters, W. Henry (2005). Fast, effective VC kernelization: A tale of two algorithms. In *ACS/IEEE International Conference on Computer Systems and Applications*.

Lenstra, Hendrik W. (1983). Integer programming with a fixed number of variables. *Mathematics of Operations Research*, **8**, 538–548.

Li, Ming, Ma, Bin, and Wang, Lusheng (2002*a*). Finding similar regions in many sequences. *Journal of Computer and System Sciences*, **65**(1), 73–96.

Li, Ming, Ma, Bin, and Wang, Lusheng (2002*b*). On the closest string and substring problems. *Journal of the ACM*, **49**(2), 157–171.

Lin, Guo-Hui, Chen, Zhi-Zhong, Jiang, Tao, and Wen, Jianjun (2002). The longest common subsequence problem for sequences with nested arc annotations. *Journal of Computer and System Sciences*, **65**(3), 465–480.

Mahajan, Meena and Raman, Venkatesh (1999). Parameterizing above guaranteed values: MaxSat and MaxCut. *Journal of Algorithms*, **31**(2), 335–354.

Marx, Dániel (2004*a*, August). Chordal deletion is fixed-parameter tractable. Manuscript, Dept. Computer Science, Budapest University of Technology and Economics.

Marx, Dániel (2004*b*). Parameterized coloring problems on chordal graphs. In *Proc. IWPEC*, Volume 3162 of *LNCS*, pp. 83–95. Springer. Long version to appear in *Theoretical Computer Science*.

Marx, Dániel (2004*c*). Parameterized complexity of constraint satisfaction problems. In *Proc. 19th CCC*, pp. 139–149. IEEE Computer Society.

Marx, Dániel (2005). The Closest Substring problem with small distances. In *Proc. 46th FOCS*, pp. 63–72. IEEE Computer Society.

Matoušek, Jiří and Nešetřil, Jaroslav (1998). *Invitation to Discrete Mathematics*. Oxford University Press. Clarendon Press.

McCartin, Catherine (2002). Parameterized counting problems. In *Proc. 27th MFCS*, Volume 2420 of *LNCS*, pp. 556–567. Springer.

Mecke, Steffen and Wagner, Dorothea (2004). Solving geometric covering problems by data reduction. In *Proc. 12th ESA*, Volume 3221 of *LNCS*, pp. 760–771. Springer.

Mehlhorn, Kurt (1984). *Data Structures and Algorithms, Volume 2 : NP-Completeness and Graph Algorithms.* EATCS Monographs on Theoretical Computer Science. Springer.

Michalewicz, Z. and Fogel, B. F. (2004). *How to Solve it: Modern Heuristics* (2nd edn). Springer.

Mitzenmacher, Michael and Upfal, Eli (2005). *Probability and Computing: Randomized Algorithms and Probabilistic Analysis.* Cambridge University Press.

Mohar, Bojan (1999). A linear time algorithm for embedding graphs in an arbitrary surface. *SIAM Journal on Discrete Mathematics*, **12**(1), 6–26.

Mölle, Daniel, Richter, Stefan, and Rossmanith, Peter (2005, March). A faster algorithm for the Steiner Tree problem. Technical Report AIB-2005-04, Deptartment of Computer Science, RWTH Aachen.

Monien, Burkhard and Speckenmeyer, Ewald (1985). Ramsey numbers and an approximation algorithm for the vertex cover problem. *Acta Informatica*, **22**, 115–123.

Motwani, Rajeev and Raghavan, Prabhakar (1995). *Randomized Algorithms.* Cambridge University Press.

Nemhauser, George L. and Trotter, Leslie E. (1975). Vertex packing: structural properties and algorithms. *Mathematical Programming*, **8**, 232–248.

Nemhauser, George L. and Wolsey, Laurence A. (1988). *Integer and Combinatorial Optimization.* Wiley.

Nešetřil, J. and Poljak, S. (1985). On the complexity of the subgraph problem. *Commentationes Mathematicae Universitatis Carolinae*, **26**(2), 415–419.

Niedermeier, Rolf (2004). Ubiquitous parameterization—invitation to fixed-parameter algorithms. In *Proc. 29th MFCS*, Volume 3162 of *LNCS*, pp. 84–113. Springer.

Niedermeier, Rolf and Rossmanith, Peter (1999). Upper bounds for vertex cover further improved. In *Proc. 16th STACS*, Volume 1563 of *LNCS*, pp. 561–570. Springer-Verlag, Berlin.

Niedermeier, Rolf and Rossmanith, Peter (2000*a*). A general method to speed up fixed-parameter-tractable algorithms. *Information Processing Letters*, **73**, 125–129.

Niedermeier, Rolf and Rossmanith, Peter (2000*b*). New upper bounds for Maximum Satisfiability. *Journal of Algorithms*, **36**, 63–88.

Niedermeier, Rolf and Rossmanith, Peter (2003*a*). An efficient fixed-parameter algorithm for 3-Hitting Set. *Journal of Discrete Algorithms*, **1**, 89–102.

Niedermeier, Rolf and Rossmanith, Peter (2003*b*). On efficient fixed-parameter algorithms for Weighted Vertex Cover. *Journal of Algorithms*, **47**(2), 63–77.

Nikolenko, S. I. and Sirotkin, A. V. (2003). Worst-case upper bounds for SAT:

automated proof. In *15th European Summer School in Logic Language and Information (ESSLLI 2003)*.

Nishimura, Naomi, Ragde, Prabhakar, and Thilikos, Dimitrios M. (2001). Fast fixed-parameter tractable algorithms for nontrivial generalizations of Vertex Cover. In *Proc. 7th WADS*, Volume 2125 of *LNCS*, pp. 75–86. Springer.

Nishimura, Naomi, Ragde, Prabhakar, and Thilikos, Dimitrios M. (2002). On graph powers for leaf-labeled trees. *Journal of Algorithms*, **42**(1), 69–108.

Opatrny, Jaroslav (1979). Total ordering problem. *SIAM Journal on Computing*, **8**(1), 111–114.

Papadimitriou, Christos H. (1994). *Computational Complexity*. Addison-Wesley.

Papadimitriou, Christos H. and Yannakakis, Mihalis (1996). On limited nondeterminism and the complexity of the V-C dimension. *Journal of Computer and System Sciences*, **53**(2), 161–170.

Papadimitriou, Christos H. and Yannakakis, Mihalis (1999). On the complexity of database queries. *Journal of Computer and System Sciences*, **58**(3), 407–427.

Pearl, Judea (1984). *Heuristics*. Addison–Wesley, Reading, Massachusetts.

Pietrzak, Krzysztof (2003). On the parameterized complexity of the fixed alphabet Shortest Common Supersequence and Longest Common Subsequence problems. *Journal of Computer and System Sciences*, **67**(4), 757–771.

Prieto, Elena and Sloper, Christian (2003). Either/or: using vertex cover structure in designing FPT-algorithms—the case of k-Internal Spanning Tree. In *Proc. 8th WADS*, Volume 2748 of *LNCS*, pp. 474–483. Springer.

Prieto, Elena and Sloper, Christian (2004). Looking at the stars. In *Proc. IWPEC*, Volume 3162 of *LNCS*, pp. 138–148. Springer.

Prömel, Hans-Jürgen and Steger, Angelika (2002). *The Steiner Tree Problem*. Vieweg.

Raman, Venkatesh (1997). Parameterized complexity. In *7th National Seminar on Theoretical Computer Science (Chennai, India)*, pp. I–1–I–18.

Reed, Bruce (1993). Finding approximate separators and computing tree-width quickly. In *Proc. 24th STOC*, pp. 221–228. ACM Press.

Reed, Bruce, Smith, Kaleigh, and Vetta, Adrian (2004). Finding odd cycle transversals. *Operations Research Letters*, **32**(4), 299–301.

Robertson, Neil, Sanders, Daniel P., Seymour, Paul D., and Thomas, Robin (1996). Efficiently four-coloring planar graphs. In *Proc. 28th STOC*, pp. 571–575. ACM Press.

Robertson, Neil, Sanders, Daniel P., Seymour, Paul D., and Thomas, Robin (1997). The four-color theorem. *Journal of Combinatorial Theory, Series B*, **70**(1), 2–44.

Robertson, Neil and Seymour, Paul D. (1986). Graph minors. II: Algorithmic aspects of tree-width. *Journal of Algorithms*, **7**, 309–322.

Robson, John Michael (1986). Algorithms for maximum independent sets. *Journal of Algorithms*, **7**, 425–440.

Robson, John M. (2001). Finding a maximum independent set in time $o(2^{n/4})$. Technical Report 1251-01, Université Bordeaux, LaBRI.

Schaefer, T. J. (1978). The complexity of satisfiability problems. In *Proc. 10th STOC*, pp. 216–226. ACM Press.

Schroeppel, Richard and Shamir, Adi (1981). A $t = o(2^{n/2})$, $s = o(2^{n/4})$ algorithm for certain NP-complete problems. *SIAM Journal on Computing*, **10**(3), 456–464.

Scott, Jacob, Ideker, Trey, Karp, Richard M., and Sharan, Roded (2005). Efficient algorithms for detecting signaling pathways in protein interaction networks. In *Proc. 9th RECOMB*, Volume 3500 of *LNCS*, pp. 1–13. Springer.

Seymour, Paul D. and Thomas, Robin (1994). Call routing and the ratcatcher. *Combinatorica*, **14**(2), 217–241.

Shamir, R., Sharan, R., and Tsur, D. (2004). Cluster graph modification problems. *Discrete Applied Mathematics*, **144**, 173–182.

Sinz, Carsten, Kaiser, Andreas, and Küchlin, Wolfgang (2003). Formal methods for the validation of automotive product configuration data. *Artificial Intelligence for Engineering Design, Analysis and Manufacturing*, **17**(1), 75–97.

Skiena, Steven S. (1998). *The Algorithm Design Manual.* Springer-Verlag.

Spillner, Andreas (2005). A faster algorithm for the minimum weight triangulation problem with few inner points. In *1st ACiD*, pp. 135–146.

Steel, Mike A. (1992). The complexity of reconstructing trees from qualitative characters and subtrees. *Journal of Classification*, **9**, 91–116.

Szeider, Stefan (2004*a*). Minimal unsatisfiable formulas with bounded clause-variable difference are fixed-parameter tractable. *Journal of Computer and System Sciences*, **69**(4), 656–674.

Szeider, Stefan (2004*b*). On fixed-parameter tractable parameterizations of SAT. In *Proc. 6th SAT*, Volume 2919 of *LNCS*, pp. 188–202. Springer.

Vazirani, Vijay V. (2001). *Approximation Algorithms.* Springer.

Veinott, Arthur F. and Wagner, H. M. (1962). Optimal capacity scheduling. *Operations Research*, **10**, 518–532.

Weihe, Karsten (1998). Covering trains by stations or the power of data reduction. In *Proc. 1st ALEX'98*, pp. 1–8.

Weihe, Karsten (2000). On the differences between "practical" and "applied". In *Proc. WAE 2000*, Volume 1982 of *LNCS*, pp. 1–10. Springer.

West, D. B. (2001). *Introduction to Graph Theory (2nd Edition).* Prentice Hall.

Weston, Mark (2004). A fixed-parameter tractable algorithm for matrix domination. *Information Processing Letters*, **90**(5), 267–272.

Williams, Ryan (2004). A new algorithm for optimal constraint satisfaction and its implications. In *Proc. 31st ICALP*, Volume 3142 of *LNCS*, pp. 1227–1237. Springer.

Woeginger, Gerhard J. (2003). Exact algorithms for NP-hard problems: A survey. In *Proc. 5th International Workshop on Combinatorial Optimization – Eureka, You Shrink!*, Volume 2570 of *LNCS*, pp. 185–208. Springer.

INDEX

Lightning Source UK Ltd.
Milton Keynes UK
29 October 2009

145538UK00005B/7/P

9 780198 566076